RECENT ADVANCES IN EARTHQUAKE GEOTECHNICAL ENGINEERING AND MICROZONATION

GEOTECHNICAL, GEOLOGICAL AND EARTHQUAKE ENGINEERING

Volume 1

Series Editor

Atilla Ansal, *Kandilli Observatory and Earthquake Research Institute, Boğaziçi University, Istanbul, Turkey*

RECENT ADVANCES IN EARTHQUAKE GEOTECHNICAL ENGINEERING AND MICROZONATION

edited by

ATILLA ANSAL

*Kandilli Observatory and
Earthquake Research Institute,
Boğaziçi University,
Istanbul, Turkey*

SPRINGER SCIENCE+BUSINESS MEDIA, B.V.

A C.I.P. Catalogue record for this book is available from the Library of Congress.

ISBN 978-94-017-4038-8   ISBN 978-1-4020-2528-0 (eBook)
DOI 10.1007/978-1-4020-2528-0

Printed on acid-free paper

This book is dedicated to our dear friend and colleague Prof. Dr. Aykut Barka.

He took part in the first phase of this initiative
but regrettably passed away much too early to see the end.

PREFACE

Outstanding advances have been achieved on Earthquake Geotechnical Engineering and Microzonation in the last decade mostly due to the increase in the recorded instrumental in-situ data and large number of case studies conducted in analyzing the observed effects during the recent major earthquakes.

During the 15th International Conference on Soil Mechanics and Geotechnical Engineering held in Istanbul in August 2001, the Technical Committee of Earthquake Geotechnical Engineering, (TC4) of the International Society of Soil Mechanics and Geotechnical Engineering organised a regional seminar on Geotechnical Earthquake Engineering and Microzonation where an effort has been made to present the recent advances in the field by eminent scientists and researchers. The book idea was first suggested by the participants of this seminar.

The purpose of this book as well as of the seminar was to present the broad spectrum of earthquake geotechnical engineering and seismic microzonation including strong ground motion, site characterisation, site effects, liquefaction, seismic microzonation, solid waste landfills and foundation engineering. The subject matter requires multidisciplinary input from different fields of engineering seismology, soil dynamics, geotechnical and structural engineering.

The chapters in this book are prepared by some of the distinguished lecturers who took part in the seminar supplemented with contributions of few distinguished experts in the field of earthquake geotechnical engineering. The editor would like to express his gratitude to all authors for their interest and efforts in preparing their manuscripts. Without their enthusiasm and support, it would not have been possible to complete this book.

Atilla Ansal

TABLE OF CONTENTS

INTRODUCTION
ROLE OF GEOTECHNICS IN EARTHQUAKE ENGINEERING

Kenji Ishihara
Science University of Tokyo, Japan

The large earthquakes over the years have left many lessons to be learned which are essential in putting forward countermeasures or policy to mitigate similar calamities in future. The degree and nature of damage incurred by earthquakes depends largely upon states of social developments of the region in which an event occurs. The topography, ground water conditions and subsurface soil conditions are also important factors influencing features of the damage caused by great earthquakes. Needless to say, the most important would be the intensity of shaking of the ground at the time of the earthquake. There are so many factors as above to be considered that it is practically difficult to forecast the intensity of shaking and the level of the damage resulting form an earthquake at a given region.

Under the inherent circumstances as above, the earthquake engineering has been developed by reflecting on bitter experiences of calamity that occurred during past earthquakes. In this sense, the earthquake engineering could be cited as "experience engineering". It is thus mandatory for engineers to carefully investigate the damage feature, exercise deep insight into causes of the incident, come up with good ideas for mitigation and to implement them in the retrofit works that follows. The experiences should be reflected as well on implementation of countermeasures for existing facilities and structures and further on in renewing the design codes and regulations in future. It is without saying that the geotechnical engineers specializing earthquake engineering should recognize themselves to carry this responsibility and in this sense learning lessons from past earthquakes are the most important things assigned to our profession.

Since individual earthquake has its own characteristics, it would be necessary to learn new lessons as large earthquakes occur. In the development of earthquake geotechnology, for example, Niigata Earthquake in Japan 1964 could be cited as a milestone event in that it has first demonstrated the importance of liquefaction in sand deposits in bringing about various kinds of damage to the ground itself and structures thereupon. The subsequent earthquake in 1978 in Japan off Izu peninsula triggered the breach of a tailings dam located in the mountaintop, leading to widespread contamination of river beds downhill. The liquefaction of sand containing silt with low-plasticity fines was first identified to be of importance as well in generating a state of liquefaction in silty sand deposits. The Kobe Earthquake in Japan 1995 would be cited as the first event where man-made islands suffered catastrophic damage along their periphery where quay walls have grossly moved seaward involving large amount of soil deposits behind them. The lateral spreading of once liquefied soils was found to exert truly detrimental effects on the structures and facilities existing on such laterally moving soil ground. Since then, problems related with lateral spreading have become a subject of extensive studies and discussions in the international arena of the earthquake

A. Ansal (ed.), Recent Advances in Earthquake Geotechnical Engineering and Microzonation, 1–2.
© 2004 *Kluwer Academic Publishers.*

geotechnics. Performance of structures resting upon, or foundations embedded in liquefied deposit or those undergoing lateral spreading is now one of the major issues of consideration for which some solutions and consensus are in urgent need.

The damage by the earthquakes may be divided into two groups, structural injury due directly to inertia force during intense shaking and indirect damage due to liquefaction or lateral spreading of the ground. The features of these two kinds of damage have been found different between developing and developed countries. In the developed countries, seismic code or regulations for earthquake-resistant design has been put forward mainly for structures and implemented in the design of medium to large-scale buildings or facilities. Thus, the structural damage has become less and less pronounced and implementation of anti-seismic design is recognized to have contributed greatly for reduction of distress during earthquakes. In contrast, in developing countries codes or regulations have not yet been put into effect sufficiently and death tolls or property damage result mostly from the collapse of poorly constructed houses or buildings.

With respect to the geotechnics-associated damage, mitigation measures have not yet been implemented both in developed and developing countries to an extent to reduce the damage. Consequently, the damage due to geotechnical origin such as liquefaction and landslides forms a major part of the distress by earthquakes. From considerations as above, it may be mentioned that the ground damage due to liquefaction and landslides is still the cause of major damage not only in developing countries but also in developed region of the world, and there is a plenty of challenges emerging from one earthquake after another that is worthy of notice and requires further studies before relevant solutions become of use for mitigating the distress resulting from large earthquakes. In this context, geotechnical engineers should be encouraged to seek the problem areas and try to come up with some solutions in this unexplored area.

CHAPTER 1
MICROZONATION: DEVELOPMENTS AND APPLICATIONS

W. D. Liam Finn, *Kagawa University, Takamatsu, Japan*
Tuna Onur, *Pacific Geoscience Centre, Sidney, BC, Canada*
Carlos E. Ventura, *University of British Columbia, Vancouver BC, Canada*

1.1. Introduction

Building codes base seismic design forces on various seismic hazard parameters that describe the intensity of ground shaking during an earthquake. The design parameter is typically acceleration, velocity or spectral acceleration with a specified probability of exceedance. These parameters are mapped on a national scale for a standard ground condition, usually rock or stiff soil. Mapping to such a scale is called macrozonation.

Damage patterns in past earthquakes show that soil conditions at a site may have a major effect on the level of ground shaking. Mapping of seismic hazard at local scales to incorporate the effects of local soil conditions is called microzonation for seismic hazard. The analysis for calculating the probability of exceeding different levels of the mapped ground motion parameter is called seismic hazard analysis.

The basic structure of seismic hazard analysis is presented in this chapter and its evolution to the present state of the art will be described. The presentation is geared to the user, not the analyst. It attempts to give the user a useful level of understanding of how the seismic hazard parameter of the microzonation is determined, what it means, what uncertainties are associated with it and how they are handled in the analysis.

Microzonation for seismic hazard has many uses. It can provide input for seismic design, land use management, and estimation of the potential for liquefaction and landslides. It also provides the basis for estimating and mapping the potential damage to buildings. Mapping the losses expected from a particular level of seismic shaking is called microzonation for risk. The presentation of the procedures for microzonation for risk is also geared to the user. The procedures for estimating losses for a selected probability of exceedance of ground shaking level will be explained and the entire process illustrated by means of a case history of loss estimation conducted for the insurance industry in Canada.

Seismic hazard analysis, which is the major component of microzonation for seismic hazard and seismic risk, can be a very expensive and time consuming activity. Therefore the objectives of the microzonation and how the results are likely to be used should be clearly understood by analyst and user before the levels of effort and sophistication of the hazard analysis are decided. The potential range in useful effort is exemplified by the following two examples.

Hensolt and Brabb (1990) published a microzonation map of San Mateo County, California, showing the distribution of the site factors, S, in the Uniform Building Code. These site factors define the amplification of ground motions by four different soil profiles compared to the motions in rock or stiff soils. Therefore the map, in effect, shows the relative seismic hazards at different locations in terms of S. In addition, if

3

A. Ansal (ed.), Recent Advances in Earthquake Geotechnical Engineering and Microzonation, 3–26.

this map is overlaid on the basic hazard map for stiff ground, a revised map can be drawn that reflects in a significant way the effects of local soil conditions. Such a map is feasible in most metropolitan areas as the basic soil data is available from construction records. This represents a very basic, elementary, and affordable way of microzoning a metropolitan area for hazard, while taking into account local soil conditions. The other extreme is represented by the probabilistic seismic hazard analysis for ground motions at Yucca Flat, Nevada. This site is a potential geologic repository for spent nuclear fuel. The seismic hazard study for this site has been described as the largest and most comprehensive analysis ever conducted for assessing the hazard from ground shaking (Stepp et al., 2001). The huge effort was driven by the critical need to provide a stable basis for assessing the impact of ground motions on the long term performance of the containment facility for the spent fuel.

1.2. The Structure of Probabilistic Seismic Hazard Analysis

The methodology for conducting probabilistic seismic hazard analysis was developed by Cornell (1968) and an updated summary of the state of the art was prepared by the EERI Committee on Seismic Risk (EERI, 1989). The method has 4 steps as shown in Figure 1.1. The first step is to identify the active faults and areal seismic sources that may affect the site (Figure 1.1a). The second step is to characterize the recurrence rates of earthquakes of different magnitudes in each source. This involves specifying an earthquake occurrence relation for each source and a maximum magnitude. The distribution of occurrences is often given by the Gutenberg and Richter (1954) relation in Equation (1.1),

$$Log_{10}N(M) = a - bM \qquad (1.1)$$

Here $N(M)$ is the number of earthquakes per year with a magnitude equal to or greater than M and a and b are constants for the seismic zone. N is associated with a given area and time period. The constant 'a' is the logarithm of the number of earthquakes with magnitudes equal to or greater than zero. The constant 'b' is the slope of the distribution and controls the relative proportion of large to small earthquakes (Figure 1.1b).

An alternative formulation for $N(M)$ is given in Equation (1.2),

$$N(M) = e^{\alpha - \beta M} = N_o e^{-\beta M} \qquad (1.2)$$

The third step is to select an appropriate attenuation relationship that relates the median value of the seismic motion parameter to be mapped to the magnitude of the earthquake and distance from the source. An attenuation relationship for median acceleration is shown in Figure 1.1c for magnitude M. Finally the fourth step is to compute the hazard curve shown in Figure 1.1d taking into account all the data provided by the first three steps.

The hazard curve gives the probability that a given level of acceleration will be exceeded in a given time period.

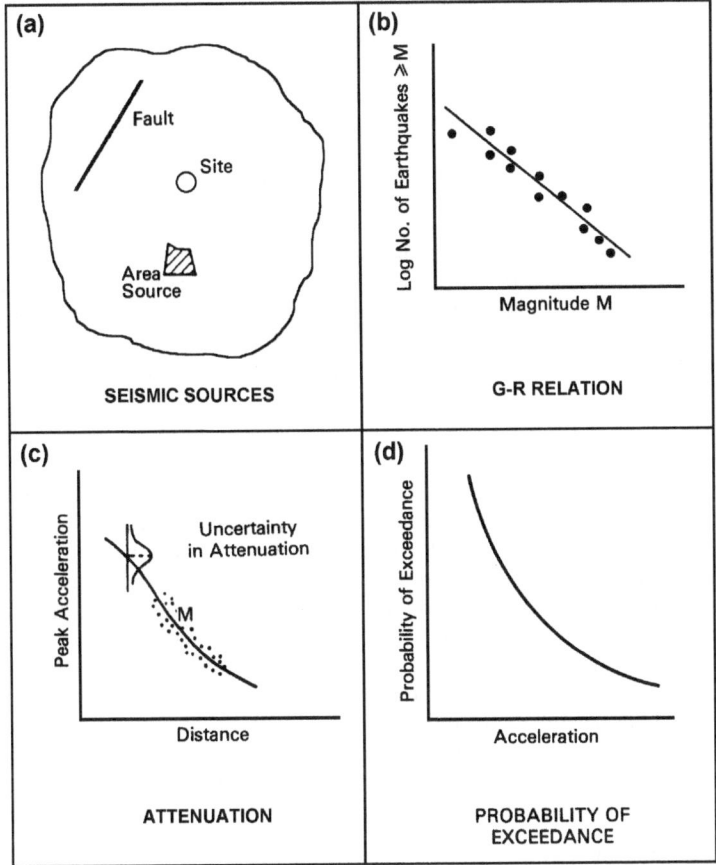

Fig. 1.1. The four steps in seismic hazard analysis

1.3. Developments in Seismic Hazard Analysis

Two major factors have shaped developments in seismic hazard analysis, since Cornell introduced the methodology in 1968. The first is the huge increase in the number of strong motion records available, since the 1971 San Fernando earthquake in California. This has led to improved forms of attenuation relations that take into account different types of faulting, different tectonic environments and different soil conditions. The second is the need of the nuclear power industry for stable estimates of ground motion at low probabilities of exceedance. This need has shaped the standards for implementing every element of the Cornell structure for seismic hazard analysis at the highest level of practice.

The distinguishing characteristic of the best modern practice is the formal treatment of the uncertainty associated with almost every aspect of seismic hazard analysis. Uncertainty was not considered formally in the original Cornell paper. There are two kinds of uncertainty, aleatory and epistemic. Aleatory uncertainty is due to the random nature of seismic events. No matter how much data are accumulated on ground motions for example, the standard deviation about the median in attenuation relations remains significant. Epistemic uncertainty is considered to be due to a lack of scientific knowledge and is established by processing opinions from a number of experts. The manner in which such opinions are elicited and interpreted is considered so crucial to the final assessment of hazard that the nuclear industry has established standards for it (SSHAC, 1997).

The process of getting the data for each step in the seismic hazard analysis, including considerations of both aleatory and epistemic uncertainty is described in a general way here. The objective is to show that though the analytical techniques for conducting seismic hazard analysis are well established and accepted, getting the appropriate input is often a process of chasing a very elusive target. This is something the user needs to understand.

1.3.1. SEISMIC SOURCES

There are two kinds of seismic sources, areal sources and faults. Generally areal sources are used to represent distributed seismicity that cannot be associated with known faults. These sources are delineated by experts who take into account the distribution of historical earthquakes and the seismotectonic regime and geology. There can be wide differences in the expert definitions of the areal sources, reflecting the lack of scientific information to constrain the delineation of sources. Source definition is one major source of epistemic uncertainty. Expert opinions are used to deal with epistemic uncertainty in each step of the hazard analysis. The differences in opinions at each stage lead ultimately to wide differences in hazard estimation. To aggregate these opinions to arrive at a stable hazard estimate, the different hazard estimates are combined by a weighting process. In the Yucca Mountain study, for example, the decision to give equal weight to all experts was taken at the beginning of the study.

All faults are potential sources. The challenge is to decide whether faults with no history of earthquake occurrence should be considered active to meet the objectives of the study. In this case it is necessary to rely on paleoseismicity. Trenching across a fault will reveal the dislocation of strata by past earthquakes that can be dated. When this information is considered in combination with geological or geodetic data on slip rates, estimates of magnitude and frequency of past events outside the historical record can be made.

Another area of uncertainty is how a fault will break; along the total length of the fault or just along a segment. If potential breaking segments are identified, how many may break at once. What about adjacent faults? As stresses readjust in the fault under consideration as a result of an earthquake, will the changes in the stress regime of nearby faults lead to sympathetic rupture? This again calls for the reconciliation of different expert opinions.

1.3.2. RECURRENCE RELATIONS

Recurrence relations are a crucial component of seismic hazard analysis. They are the means of defining the relative distribution of large and small earthquakes and incorporating the seismic history into the hazard analysis. On the basis of worldwide seismicity data, Gutenberg and Richter established the loglinear relation (G-R line) given by Equation (1.1). This relation has been assumed to apply to individual areal and fault sources also. One of the steps in characterizing seismic sources is the assignment of a maximum magnitude to each source. This requires the G-R line to taper into the maximum value as shown in Figure 1.2. This distribution is called the truncated exponential and is given in exponential form in Equation (1.3),

$$N(M) = [\ \beta exp(-\beta(M-M_{min}))]/[1- exp(-\beta(M_{max}-M_{min}))] \qquad (1.3)$$

Here M_{max} is the assigned maximum magnitude, M_{min} is the smallest earthquake that needs to be considered, $\beta = b\ ln(10)$ and b is the slope of the G-R line in Figure 1.1b. Source specific values of b are used in this equation. The divisor renormalizes the distribution so that integration between M_{min} and M_{max} gives unity.

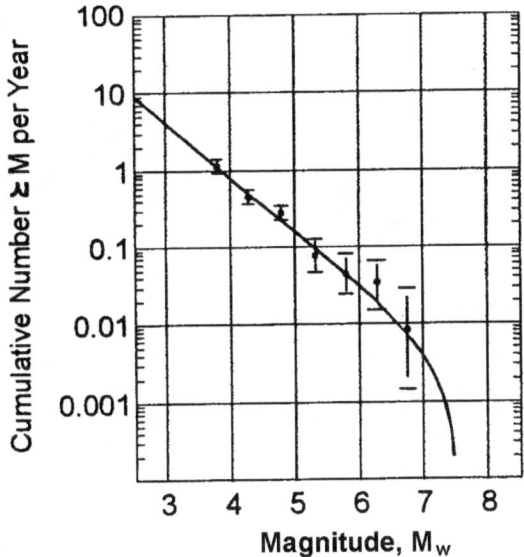

Fig. 1.2. Truncated exponential distribution of recurrence rates

An important problem in defining a recurrence relation is ensuring completeness. For the time period under consideration, it is crucial to ensure that all earthquakes have been recorded for each magnitude range of interest. For example, there is a threshold magnitude below which earthquake occurrences have not been recorded completely in a given time frame for a given layout of seismographs and a given distribution of population. Fitting the Gutenberg –Richter relation to incomplete data at the lower end will flatten the slope of the line which leads to an inflated estimate of the occurrence of large earthquakes. Because of the long recurrence time, the historical record may be too short to ensure completeness in the large earthquake range. It is imperative that any data used to establish occurrence rates of earthquakes should be complete for the time period under consideration for each magnitude range.

The G-R line does not always apply. Some fault segments tend to have occurrences of earthquakes of similar size or within a narrow range of magnitudes. These earthquakes are called characteristic earthquakes. Typically smaller earthquakes on the fault follow the G-R line and the characteristic earthquakes occur at higher rates. A typical combined distribution is shown in Figure 1.3. A commonly used characteristic model is that of Youngs and Coppersmith (1985) shown in Figure 1.4. Note the higher occurrence rates of the characteristic earthquakes. More seismic energy is released by the larger earthquakes according to this model than is released in the truncated exponential model (Abrahamson, 2000). An alternative recurrence model can be formulated by assuming that all the strain energy is released in characteristic earthquakes (CDMG, 1996).

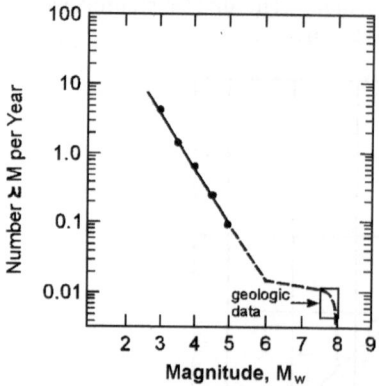

Fig. 1.3. Geological documentation of characteristic earthquakes (after Schwartz and Coppersmith, 1984; Courtesy of the American Geophysical Union)

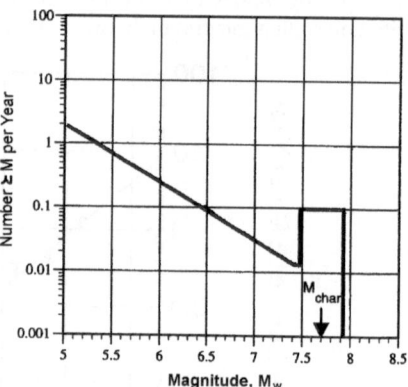

Fig. 1.4. Characteristic earthquake occurrence model (after Youngs and Coppersmith, 1985; Courtesy of the Seismological Society of America)

1.3.3. ATTENUATION RELATIONS

A state-of-the-art assessment of the main attenuation relations in use in North America may be found in a special issue of Seismological Research Letters (SSA, 1997). Modern attenuation relations typically give the natural logarithm of a ground shaking parameter such as acceleration or spectral acceleration as a function of magnitude and distance. The dispersion about the median is characterized by a lognormal distribution and the value of the standard deviation, as shown in Figure 1.1c. The lognormal distribution is symmetrical. Therefore the distribution of the underlying ground motion parameter itself is not symmetrical. Therefore the mean is greater than the median. As pointed out by Abrahamson (2000),

$$\text{Mean} = \text{Median} * \exp(- \sigma^2/2) \tag{1.4}$$

Here σ is the standard deviation. The standard deviation is a measure of the aleatory uncertainty associated with the ground motion parameter. In addition, there is also epistemic uncertainty regarding the coefficients in the attenuation relation itself.

It is very important for the end user to understand what acceleration is being estimated by the attenuation relation. In the United States, the attenuation relations give the geometric mean of the two horizontal components of motion. Engineers are often under the impression that it is the largest component. Practice varies in other countries where the largest component is often used. According to Abrahamson (2000), the largest component is on the average 15% greater than the mean. The difference becomes greater at periods longer than 5s, especially for near fault ground motions.

Attenuation relations tend to be regionally specific. For example, in North America, very different attenuation relations are used in the west and the east because of the radically different geological and tectonic structures in these regions. Relations also tend to be specific with respect to the type of faulting. Attenuation from subduction sources is different than from strike-slip sources. There are also differences between attenuation from strike-slip sources and reverse or thrust faults. Attenuation relations may be site-specific in the sense that the relation may be established for a particular soil condition such as rock, soft soil, deep stiff soil, shallow stiff soil, etc. Some attenuation relations define these conditions by means of descriptive adjectives. Boore et al. (1993) characterized the site conditions by means of the time averaged shear wave velocity, V_s, in the top 30m of a site. The characteristics of attenuation relations will be illustrated using the attenuation relation for crustal sources by Boore et al. (1997),

$$Ln\, Y = b_1 + b_2 \left(M - 6\right) + b_3 \left(M - 6\right)^2 + b_5 Ln\, r + b_v Ln\frac{V_s}{V_A} \tag{1.5}$$

Here:

$$r = \sqrt{r_{jb}^2 + h^2} \tag{1.6}$$

$$b_1 = \begin{cases} b_{1SS} & \text{for strike slip earthquakes} \\ b_{1RS} & \text{for reverse slip earthquakes} \\ b_{1ALL} & \text{if mechanism is not specified} \end{cases} \tag{1.7}$$

In this equation, Y is the ground-motion parameter (peak horizontal acceleration or pseudo-spectral acceleration response at a particular period in g); the predictor variables are moment magnitude (M_w), distance (r_{jb} in km), and average shear-wave velocity to 30 m (V_s in m/s). Note that h is a regression parameter. Values of the coefficients in Equation (1.5) are given in Table 1.1 for the attenuation of spectral accelerations at periods from 0 to 2s to illustrate the structure of such tables. The entries for zero period are the coefficients for peak horizontal acceleration. In Table 1.1, σ_{lnY}, is the square root of the overall variances of the regression. The complete table of coefficients is given in Boore et al. (1997). Values of the parameter V_s will be presented later, when the effects of local soil conditions on ground motions are considered.

The many different definitions of distance, r, from the earthquake source used in attenuation relations are illustrated in Abrahamson and Shedlock (1997). The Boore et al. (1997) distance, r_{jb}, is the closest horizontal distance to the vertical projection of the rupture surface. In the case of a vertical fault, this is the same as the distance to the fault break. For dipping faults, the distance can be as low as zero when the source is within the vertical projection of the rupture surface. The attenuation relation (Equation 1.5) is valid for earthquake magnitudes ranging from $M = 5.5$ to $M = 7.5$ and for distances $D \leq 80$ km.

Table 1.1. Smoothed values of the coefficients in the Boore et al. (1997) attenuation relation

Period	b_{1SS}	b_{1RV}	b_{1ALL}	b_2	b_3	b_5	b_V	V_A	h	σ_{LnY}
0.000	-0.313	-0.117	-0.242	0.527	0.000	-0.778	-0.371	1396.	5.57	0.520
0.200	0.999	1.170	1.089	0.711	-0.207	-0.924	-0.292	2118.	7.02	0.502
1.000	-1.133	-1.009	-1.080	1.036	-0.032	-0.798	-0.698	1406.	2.90	0.613

Youngs et al. (1997) provide attenuation relations for horizontal response spectral acceleration (5% damping) for subduction earthquakes applicable to rock and soil sites. These attenuation relationships are considered appropriate for earthquakes with $M = 5$ and greater, and for distances to the rupture surface from 10km to 500km. Most attenuation models for subduction zone events are based on recordings from Japan and South America. Most of these events were recorded at large distances. However, recordings were made during the 1985 Michoacan earthquake at distances as small as 13km, although recordings within 30km were sparse. Youngs et al. (1997) show that peak ground motions from subduction zone earthquakes attenuate more slowly than those from shallow crustal earthquakes and that intraslab earthquakes produce larger peak ground motions than interface earthquakes for the same magnitude and distance. However, the database contains a very limited number of intraslab recordings.

1.3.4. EFFECTS OF LOCAL SOIL CONDITIONS

Site conditions play a major role in establishing the damage potential of incoming seismic waves from major earthquakes. Damage patterns in Mexico City after the 1985 Michoacan earthquake demonstrated conclusively the significant effects of local site conditions on seismic response of the ground. Peak accelerations of incoming motions in rock were generally less than 0.04g and had predominant periods of around 2s. Many clay sites in the dried lakebed on which the original city was founded had site periods also around 2s and were excited into resonant response by the incoming motions. As a result the bedrock outcrop motions were amplified about 5 times. The amplified motions had devastating effects on structures with periods close to site periods. In the 1989 Loma Prieta earthquake, major damage occurred on soft soil sites in the San Francisco-Oakland region where the spectral accelerations were amplified 2 to 4 times over adjacent rock sites (Housner, 1989). Clearly any assessment of seismic hazard or seismic risk should incorporate the amplification effects of local soil conditions. The crucial question is how this can be done effectively without unduly complicating the hazard assessment process or increasing the cost significantly. There are three effective ways to include the effects of local soil conditions in hazard and risk studies: use an attenuation relationship that incorporates a variety of soil classes, use well-documented empirical amplification factors or conduct site response analyses. The latter procedure is time consuming and expensive and it is difficult to get reliable results for large areas. Site response analysis is more appropriately used for estimating hazard to individual structures.

Attenuation laws that incorporate site parameters for different types of soil conditions offer a convenient way to include site effects. The BJF (1997) attenuation relation described above includes a site parameter V_s that is related to site classes proposed by NEHRP (BSSC, 1994). The site classes are given in Table 1.2. Values of V_s for the NEHRP site classes of major interest, B, C and D and for rock and soil categories are given in Table 1.3.

Table 1.2. NEHRP (BSSC, 1994) site classes

Soil Class	Description	Properties
A	Hard rock	$V_s > 1500$ m/sec
B	Rock	760 m/sec $< V_s \leq 1500$ m/sec
C	Very dense soil and soft rock	360 m/sec $< V_s \leq 760$ m/sec $N > 50$, $s_u \geq 100$ kPa
D	Stiff soil	180 m/sec $< V_s \leq 360$ m/sec $15 \leq N \leq 50$, 50 kPa $\leq s_u \leq 100$kPa
E	Soil	$V_s < 180$ m/sec >3m of soft clay (PI > 20, w $\geq 40\%$ and $s_u < 25$ kPa)

Table 1.3. Values of average shear velocity, V_s, for use in the BJF attenuation relation

NEHRP site class B	1070 m/sec
NEHRP site class C	520
NEHRP site class D	250
Rock	620
Soil	310

The nonlinear behaviour of soils causes amplification factors to be dependent on the intensity of shaking. This was demonstrated very clearly by Jarpe et al. (1989) by comparing the amplification factors for a site on Treasure Island in San Francisco Bay relative to the rock motions at adjacent Yerba Buena Island, using data from the main shock of the 1989 Loma Prieta earthquake and 7 aftershocks. The amplification factors for surface motions recorded at the Treasure Island site during the 1989 Loma Prieta earthquake are shown in Figure 1.5. The solid line shows the variation in the NS spectral ratio for the first 5 seconds of the shear wave in the main shock before any liquefaction took place at the site.

W. D. L. Finn, T. Onur and C. E. Ventura

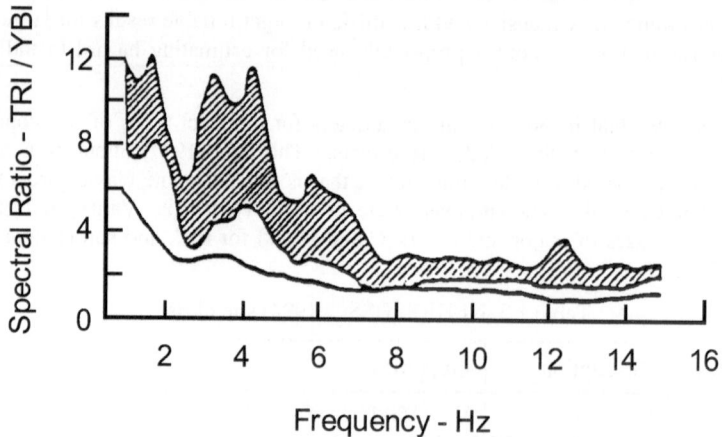

Fig. 1.5. Amplification of ground motions at Treasure Island site (after Jarpe et al., 1989; Courtesy of the Seismological Society of America)

The shaded area in Figure 1.5 shows the 95% confidence region for the spectral ratios of 7 aftershocks. The amplification factors are drastically reduced in the strong motion phase, although still 2 or greater over a wide frequency band of engineering interest. The reduction in amplification with increased intensity of shaking is due to the nonlinear stress-strain response of the soil, resulting from reduced effective shear moduli and increased damping. The peak acceleration at the surface is only 0.16g, so the amplification factors are associated with fairly low levels of earthquake shaking.

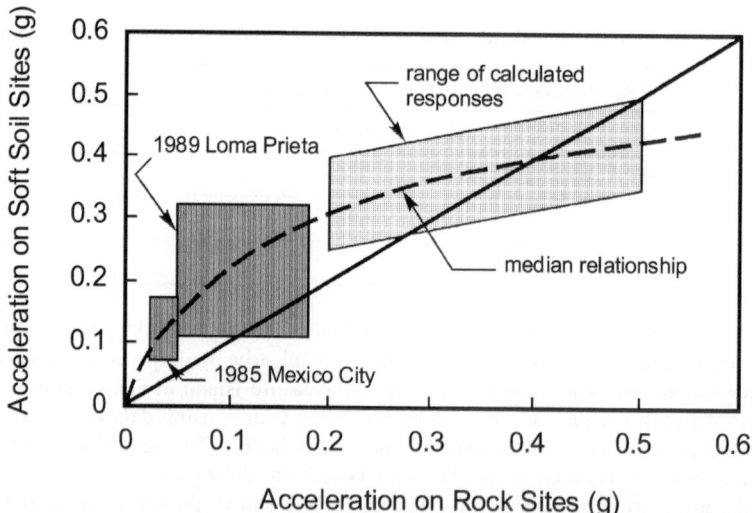

Fig. 1.6. Accelerations on soft soil and associated rock sites (after Idriss, 1991; Reproduced with permission from University of Missouri and Ed. S.Prakash)

Idriss (1990) has summarized the relationship between peak accelerations on soft soil sites and on associated bedrock sites in Figure 1.6. The median curve is based on data recorded in Mexico City during the 1985 Michoacan earthquake, strong motion data from the 1989 Loma Prieta earthquake and data from equivalent linear site response analyses. The curve suggests that, on the average, the bedrock accelerations are amplified in soft soils until the peak rock accelerations reach about 0.4g. The higher amplification ratios between rock and soil sites, in the range of 1.5 – 4, are associated with levels of rock acceleration less than 0.10g, when the response is more nearly elastic. The increased nonlinearity of soft soil response at the higher accelerations reduces the amplification ratios.

1.3.5. NEHRP AMPLIFICATION FACTORS

After major studies of site amplification following the Loma Prieta earthquake, NEHRP (BSSC, 1994) published the two sets of amplification factors given in Tables 1.4 and 1.5 that are dependent on the NEHRP site classes (Table 1.2) and shaking intensity. One set is for short period motions and one for long period motions, centered on periods of 0.2s and 1.0s respectively.

Table 1.4. Short period amplification factors; $T = 0.2s$

Site Class	Shaking Intensity				
	$A_a = 0.1$	$A_a = 0.2$	$A_a = 0.3$	$A_a = 0.4$	$A_a = 0.5$
A	0.8	0.8	0.8	0.8	0.8
B	1.0	1.0	1.0	1.0	1.0
C	1.2	1.2	1.1	1.0	1.0
D	1.6	1.4	1.2	1.1	1.0
E	2.5	1.7	1.2	0.9	-

Table 1.5. Long period amplification factors; $T = 1.0s$

Site Class	Shaking Intensity				
	$A_v = 0.1$	$A_v = 0.2$	$A_v = 0.3$	$A_v = 0.4$	$A_v = 0.5$
A	0.8	0.8	0.8	0.8	0.8
B	1.0	1.0	1.0	1.0	1.0
C	1.7	1.6	1.5	1.4	1.3
D	2.4	2.0	1.8	1.6	1.5
E	3.5	3.2	2.8	2.4	-

1.3.6. EVALUATING THE HAZARD

Seismic hazard curves are first obtained for each seismic source and then combined to obtain the total hazard. If the probability that a given level of seismic shaking k is exceeded by a particular earthquake of magnitude m at distance r in source 'i' is given by $P(K>k /m,r)$, then the expected number of times $K>k$, termed λ_k, is

$$\lambda_k = \sum_{i=1}^{N} \lambda_i \int_{m_e}^{m_u} \int_{r=0}^{r=\infty} P(K>k|m,r) \, f_i(m) \, f_i(r) \, dm dr \qquad (1.8)$$

Here λ_k is the expected rate of exceedance of k in a given time, taking all sources into account. For source i, λ_i is the mean rate of occurrence of earthquakes between the upper and lower bound magnitudes, m_l and m_u respectively and $f_i(m)$ and $f_i(r)$ are the probability density functions for magnitude and distance. The summation is carried out over all sources. The analysis is conducted using one of the commercial computer programs such as EZ-FRISK[TM] (Risk Engineering, 1997) or SEISRISK III (Bender and Perkins, 1987).

The occurrence of earthquakes is assumed to follow the Poisson probability density function. Therefore the probability that the ground motion k will be exceeded at least once (called the probability of exceedance) in a time period T, given the annual rate of exceedance λ_k, is

$$P(K>k) = 1 - e^{-\lambda_k T} \qquad (1.9)$$

Here $e^{-\lambda kT}$ is the probability that the ground motion level will not be exceeded. If the probability, $P(K>k)$ is set for a time period, T, then the associated exceedance rate given by Equation (1.8) is

$$\lambda_k = \ln(1 - P(K>k)/ T \qquad (1.10)$$

If the desired probability of exceedance is a 10% chance in 50 years, then the annual exceedance rate is $\lambda_k = 0.00211$.

1.4. Microzonation for Risk

Microzonation for seismic risk is a mapping of the distribution of potential monetary losses associated with the occurrence of a mapped distribution of seismic hazard. In effect, microzonation for risk adds another layer of information to microzonation for seismic hazard.

The crucial element in any loss estimation study is a correlation between an index of seismic hazard and damage. This requires a definition of different damage states or levels of damage. Experience in past earthquakes has demonstrated that different classes of buildings sustain different levels of damage for the same intensity of shaking. Therefore building class is also an independent variable.

Modified Mercalli Intensity (MMI) has been used most frequently as the index of ground motion intensity in damage estimation methodologies (ATC-13, 1985; King and Kiremidjian, 1994; Rojahn et al., 1997; Dowrick and Rhoades, 1997; Blanquera, 1999). The MMI scale for intensities VI and up is presented in Table 1.6. MMI VI is the level of shaking intensity at which structural damage begins to occur.

Table 1.6. Modified Mercalli Intensity Scale (MMI) from Level VI to XII

MMI	DESCRIPTION OF EFFECTS
VI	Felt by all, many frightened. Some heavy furniture moved. A few instances of fallen plaster. Damage slight.
VII	Damage negligible in buildings of good design and construction; slight to moderate in well-built ordinary structures; considerable in poorly-built structures. Some chimneys broken.
VIII	Damage slight in specially-designed structures; considerable in ordinary substantial buildings with partial collapse; great in poorly-built structures. Fall of chimneys, factory stacks, columns, walls. Heavy furniture overturned.
IX	Damage considerable in specially designed structures; well-designed frame structures thrown out of plumb. Damage great in substantial buildings, with partial collapse. Buildings shifted off foundations.
X	Some well-built wooden structures destroyed; most masonry and frame structures with foundations destroyed. Rails bent.
XI	Few, if any masonry structures remain standing. Bridges destroyed. Rails bent greatly.
XII	Damage total. Lines of sight and level are distorted. Objects thrown into air.

The damage estimation methodology developed by the Applied Technology Council (ATC-13, 1985) is widely used. In ATC-13, the intensity of ground shaking is specified by MMI and the potential damage to different classes of buildings is described by Damage Probability Matrices (DPMs). The DPMs describe for each building class, the probability that a building is in a specified damage state given the level of ground shaking intensity (MMI). The seven distinct damage states recognized in ATC-13 are given in Table 1.7. Each of these damage states is associated with a range of damage factors (DF), which are defined as the ratio of dollar loss to the replacement value. These ranges and the corresponding central damage factors (CDF) are given in Table 1.8.

W. D. L. Finn, T. Onur and C. E. Ventura

Table 1.7. Definition of damage states in ATC-13

DAMAGE STATE	DAMAGE DESCRIPTION
1. None	No damage.
2. Slight	Limited localized minor damage not requiring repair.
3. Light	Significant localized damage of some components generally not requiring repair.
4. Moderate	Significant localized damage of many components warranting repair.
5. Heavy	Extensive damage requiring major repairs.
6. Major	Major widespread damage that may result in facility being demolished or repaired.
7. Destroyed	Total destruction of the majority of the facility.

Table 1.8. Damage factors for different damage states

DAMAGE STATE	DF RANGE (%)	CENTRAL DF (%)
1. None	0	0
2. Slight	0 - 1	0.5
3. Light	1 - 10	5
4. Moderate	10 - 30	20
5. Heavy	30 - 60	45
6. Major	60 - 100	80
7. Destroyed	100	100

The Damage Probability Matrix (DPM) for one of the building classes (WFLR = Wood Frame - Low Rise) in ATC-13 (1985) is shown in Table 1.9 to illustrate the structure of such matrices. The total level of damage in a building is described by mean damage factors (MDF), in terms of the ratio of dollar loss to replacement cost. The MDFs for each MMI level are the product of the CDFs and their corresponding probabilities of occurrence for all distinct damage states. The total damage for a give prototype building is given by the sum of the MDFs for all seven damage states,

$$MDF^{MMI}_{prototype} = \frac{1}{100} \cdot \sum_{j=1}^{7} CDF_j \cdot P\left(ds_j\right) \tag{1.11}$$

Here CDF_j is the central damage factor for damage state j and $P(ds_j)$ is the probability of the prototype being in damage state j.

As an example, for WLFR at an MMI level of VIII, the MDF is calculated as follows:

$$MDF^{VIII}_{URMLR} = \frac{0.0 \cdot 0.0 + 0.5 \cdot 1.6 + 5.0 \cdot 94.9 + 20.0 \cdot 3.5 + 45.0 \cdot 0.0 + 80.0 \cdot 0.0 + 100.0 \cdot 0.0}{100} = 5.35\% \tag{1.12}$$

Here the first number of each pair in the numerator is the CDF (%) and the second number is the probability (%) of occurrence of that CDF. The MDFs are calculated in a similar way for all prototype building classes in the study area at the MMI levels of interest.

The application of the ATC-13 methodology will be illustrated by an example from engineering practice.

Table 1.9. Damage probability matrices for Wood Frame – Low Rise class

Central DF	Modified Mercalli Intensity						
	VI	VII	VIII	IX	X	XI	XII
Wood Frame - Low Rise							
0	3.7	***	***	***	***	***	***
0.5	68.5	26.8	1.6	***	***	***	***
5	27.8	73.2	94.9	62.4	11.5	1.8	***
20	***	***	3.5	37.6	76.0	75.1	24.8
45	***	***	***	***	12.5	23.1	73.5
80	***	***	***	***	***	***	1.7
100	***	***	***	***	***	***	***

1.5. Case History

1.5.1. BACKGROUND

After the 1989 Loma Prieta, 1994 Northridge and 1995 Kobe earthquakes, the insurance industry in Canada became concerned about the potential for catastrophic loss due to major earthquake impacting South-Western British Columbia where most of the population is concentrated in the major cities. The industry began discussions with the federal and provincial governments for cooperation in dealing with their concerns. To assist in making their case, the insurance industry needed assessments of their potential losses from insured buildings in key cities and commissioned a risk study at the University of British Columbia which was funded by the federal government, the insurance industry and the City of New Westminster. The study covered three cities; Vancouver, Victoria and New Westminister. The city of Victoria was selected as a case history here because its compact size and variety of soil conditions make it a very clear example of how the ATC-13 methodology is used in practice and how the results may be presented. The insurance risk analysis for this study demonstrates the close connection between the objectives of a risk study and the way the risk study is carried out.

An essential requirement of the study was that neither the elements of the seismic hazard study and nor the risk assessment methodology should give grounds for controversy over the findings. It was essential that the predicted losses should be acceptable to all parties in the discussions. The objectives of the insurance industry were met in the following way. The seismic source zones and the associated recurrence rates and maximum magnitudes adopted for the hazard study were those developed by the federal government scientists for the National Building Code of Canada (NBCC, 1995; Adams et al., 1996). The attenuation relation for peak ground acceleration used in the building code (Boore et al., 1993) was also used in the study. The aleatory uncertainty given by the lognormal distribution about the expected value used in the hazard calculations but no allowance was made for epistemic uncertainty. The widely used ATC-13 loss estimation methodology was employed for the risk estimation. This required a distribution of MMI in the study area, and a set of building classes and their associated damage probability matrices. The task of classifying the buildings and modifying the ATC-13 damage probability matrices was subcontracted to a local consultant with wide experience in the seismic design and evaluation of structures. Buildings in British Columbia were classified into 31 different types and associated damage probability matrices were developed (Bell, 1998). Damage probabilities for three prototype classes, unreinforced masonry-low rise (URMLR), wood light frame-residential (WLFR) and concrete frame-low rise (CFLR) are given in Table 1.10 as an example. The risk study for Victoria was carried out by Onur and is described in detail in her PhD dissertation (Onur, 2001).

Table 1.10. MMI based damage probabilities

Building Prototype	MMI	Discrete Probabilities (%) for Different Damage States						
		None	Slight	Light	Moderate	Heavy	Major	Destroyed
URMLR	VIII	0.0	0.0	21.0	60.0	15.0	2.0	2.0
WLFR	VIII	1.0	6.0	86.0	5.0	2.0	0.0	0.0
CFHR	VIII	0.0	2.0	57.0	40.0	1.0	0.0	0.0

1.5.2. VICTORIA RISK STUDY

Risk analysis

The city of Victoria is located on the Southern tip of Vancouver Island on the West Coast of Canada. It was incorporated as a City in 1862 and was proclaimed the capital of British Columbia in 1871. It has a population of over 77,000 people as of 1999 and has an area of roughly 23.5 square kilometres. The building inventory of the City has of over 13,000 structures. The risk study area was confined to the downtown area and adjacent districts and contains about 2,500 structures. A probabilistic seismic hazard analysis for Victoria (48.5° N, 123.3° W), carried out for firm ground conditions and a 10 % chance of exceedance in 50 years, gave a PGA of 0.31g. This corresponds to MMI VIII based on the Neumann (1954) correlation between PGA and MMI.

$$MMI = [\log(PGA) + 0.041]/0.308 \qquad\qquad (1.13)$$

An alternative hazard analysis based on attenuation relations for MMI (Atkinson, 1997) also gave an MMI VIII.

The Planning Department of the City of Victoria gave access to their building database, which included location (street address), zoning and land use information on every structure in the municipality. However, the database did not contain any information on the structural properties of the buildings, such as building material, construction date, load bearing system, height (or number of stories), and footprint area. These data were collected by sidewalk surveys covering about 2,500 buildings in and around the downtown core of the city. The most prevalent material type and building prototype classes were determined for each block and mapped independently for the study area. The distribution of prevalent building prototype classes is presented on a block-by-block basis in Figure 1.7.

About 65% of the buildings surveyed in Victoria are wood and it is the prevalent material in about 51% of the blocks studied. In downtown Victoria, where most of the historical buildings are located, the number of masonry buildings is quite high. About 28% of the buildings are masonry, and it is the prevalent material type in about 42% of the blocks studied. Concrete buildings constitute about 7% of the buildings in the study area. There are a few steel buildings in the study area but steel is not the prevalent material type in any of the blocks.

The most common prototype is WLFR, which constitute about 45% of all the buildings in the study area and about 67% of the wood buildings. The second most common building prototype class is URMLR which represents about 20% of all buildings and 57% of the masonry buildings in the study area. Among concrete buildings, CFMR is the most common prototype. It makes up about 3% of all buildings and 41% of concrete buildings in the study area.

Damage estimation was carried out, using the MMI-based damage matrices developed by Bell (1998). Total damage levels were estimated as a percentage of replacement cost using mean damage factors, MDFs, for different MMI levels for each building prototype. The mean MDF for each block was calculated by averaging the MDFs of the buildings within that block, taking either the average or a weighted average obtained by weighting the MDFs by the footprint areas of the buildings in the block.

The estimated structural damage distribution on a block-by-block basis, based on weighted average MDFs, is shown in Figure 1.8. The majority the buildings in downtown Victoria, about 35% of all the blocks in the study area, have MDFs between 10% and 30%. The surrounding neighbourhoods have lower MDFs, in the range of 5% to 10%.

Fig. 1.7. Prevalent buildings prototypes by block

Non-structural damage distribution was calculated in a similar manner using appropriate MDFs for each of displacement-sensitive components, acceleration-sensitive components and building contents (Cook, 1999). About half the blocks are expected to have MDFs in the 20%-30% range and the other half in the 15%-20% range for displacement-sensitive components. About half the blocks are estimated to have MDFs between 5% and 10%, and the rest between 0% and 5% for acceleration-sensitive components. Damage to building contents remain below 5% in all the blocks.

Fig. 1.8. Structural damage distribution by average MDF weighted by footprint area

Effects of soil conditions

In Victoria, the effect of geology was considered important because half the city rests on softer material than the "firm soil" assumed in the hazard calculations. Therefore, the effects of possible soil amplification were investigated. The geological units that appear in the study area (Monahan et al., 2000) are presented in Table 1.11 and are shown on a block by block basis in Figure 1.9.

The amplification depends on the geological unit, level of ground shaking and the period of the ground motion. In Victoria, the expected PGA is equal to 0.31g. For this level of shaking, short-period ground motions are not amplified considerably, however long-period ground motions are amplified roughly by 1.5 for C1 and F, 2.0 for C2 and 2.5 for O1 (Monahan et al., 2000). The geological unit R2 is not expected to amplify the ground motion.

Table 1.11. 6 Main types of geological units in Victoria

Geologic Unit	Description	NEHRP Site Class Range	F_a	F_v
R2	Thin soil over bedrock with scattered outcrops; generally <5m of Victoria clay over < 10m of older Pleistocene	A to C	1.0	1.0
C2	>3m of the grey clay facies of the Victoria clay, under the brown clay facies and over thin (<10m) older Pleistocene deposits	D to E	1.0	2.0
C1	Areas where units R2 & C2 cannot be differentiated; also areas with >5m of the Victoria clay but <3m of grey clay facies	C to E	1.0	1.5
F	Anthropogenic fill with variable amplification	C to E	1.0	1.5
O1	Holocene peat over the grey clay facies of the Victoria clay	E to F	1.0	2.5

The PGA's were multiplied by the amplification factors corresponding to each geological unit and the resulting PGA values were converted into MMI for use with the damage probability matrices. The resulting damage distribution map was calculated using these MMI levels. Total monetary losses resulting from the estimated structural and non-structural damages were calculated for Victoria for MMI VIII, taking into account the effects of soil amplification and summed over each block to display the total loss in each block. The results are shown in Figure 1.10.

Fig. 1.9. Geological units in Victoria

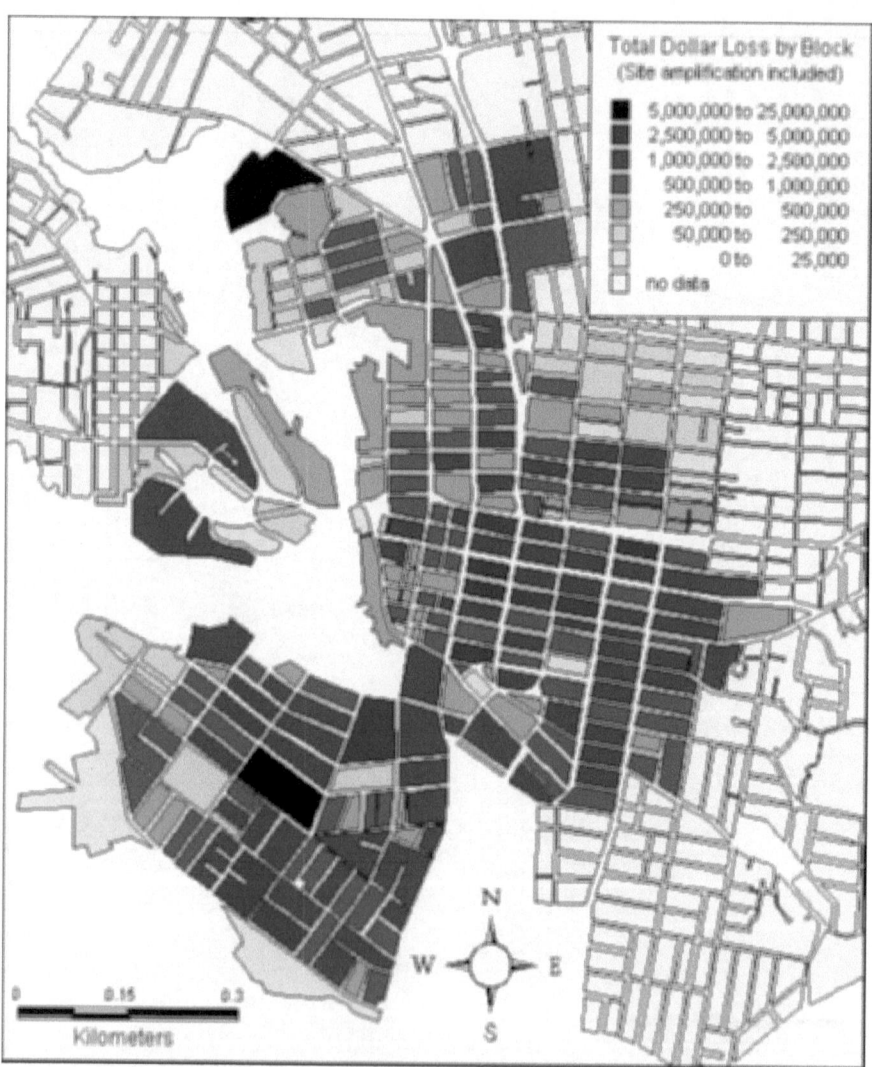

Fig. 1.10. Total monetary losses in Victoria, both structural and nonstructural, taking soil amplification into account

1.6. Final Remarks

Seismic hazard analysis is the crucial element in a microzonation study. To plan and use a microzonation study effectively requires an understanding of how the input to the hazard analysis is developed, the ways in which the analysis may be carried out and the uncertainties associated with almost every component of the analysis. The formal treatment of uncertainty is one of the major conceptual changes since Cornell introduced the method in 1968.

Today the aleatory uncertainty in attenuation relations is nearly always incorporated in the analysis through the assumed lognormal distribution about the mean but the formal inclusion of epistemic uncertainty is still rare except for the high level of practice associated with critical structures. Epistemic uncertainty in seismic source definition is a major source of uncertainty in final results. Here and in the other components of the analysis, expert opinion is used to handle epistemic uncertainty. The different opinions are weighted to arrive at a final judgment. Several opinions are required to make the process viable, and it is a complex and expensive undertaking.

The elements of seismic risk have also been presented and illustrated with a case history from practice. A key lesson from the case history is that the objectives of the microzonation for risk and how the results are to be used should be clearly understood before planning how the study will be conducted and at what level of sophistication.

Acknowledgement

The seismic risk study for South-Western British Columbia, referred to above, was funded by grants to the first author from the National Science and Engineering Council, the insurance industry and the City of New Westminster. The study was a joint collaborative effort with Professor C. E. Ventura, Director of the Earthquake Engineering Facility at the University of British Columbia and Professor Gail Atkinson, Carleton University, Ottawa. A. Blanquera, S. Cook and T. Onur worked on different aspects of the study for their theses. Their outstanding contributions were responsible for the success of the project.

The assistance of Noboru Fujita, Kagawa University, in the preparation of this paper was invaluable.

Appendix 1

BC Building Classification

No.	Material	Building Type	Code
1	Wood	Wood Light Frame Residential	WLFR
2		Wood Light Frame Low Rise Commercial/Institutional	WLFCI
3		Wood Light Frame Low Rise Residential	WLFLR
4		Wood Post and Beam	WPB
5	Steel	Light Metal Frame	LMF
6		Steel Moment Frame Low Rise	SMFLR
7		Steel Moment Frame Medium Rise	SMFMR
8		Steel Moment Frame High Rise	SMFHR
9		Steel Braced Frame Low Rise	SBFLR
10		Steel Braced Frame Medium Rise	SBFMR
11		Steel Braced Frame High Rise	SBFHR
12		Steel Frame with Concrete Walls Low Rise	SFCWLR
13		Steel Frame with Concrete Walls Medium Rise	SFCWMR
14		Steel Frame with Concrete Walls High Rise	SFCWHR
15		Steel Frame with Concrete Infill Walls	SFCI
16		Steel Frame with Masonry Infill Walls	SFMI
17	Concrete	Concrete Frame with Concrete Walls Low Rise	CFLR
18		Concrete Frame with Concrete Walls Medium Rise	CFMR
19		Concrete Frame with Concrete Walls High Rise	CFHR
20		Reinforced Concrete Moment Frame Low Rise	RCMFLR
21		Reinforced Concrete Moment Frame Medium Rise	RCMFM
22		Reinforced Concrete Moment Frame High Rise	RCMFHR
23		Reinforced Concrete Frame with Infill Walls	RCFIW
24	Masonry	Reinforced Masonry Shear Wall Low Rise	RMLR
25		Reinforced Masonry Shear Wall Medium Rise	RMMR
26		Unreinforced Masonry Bearing Wall Low Rise	URMLR
27		Unreinforced Masonry Bearing Wall Medium Rise	URMMR
28	Tilt Up	Tilt Up	TU
29	Precast	Precast Concrete Low Rise	PCLR
30		Precast Concrete Medium Rise	PCMR
31	Mobile	Mobile Homes	MH

CHAPTER 2

THE INFLUENCE OF SCALE ON MICROZONATION AND IMPACT STUDIES

Carlos Sousa Oliveira
DECivil/ICIST, Instituto Superior Técnico, Lisbon, Portugal

Contents

A general overview of the methods for estimating earthquake impact in large urban areas is presented. This overview includes the most important aspects, from geophysical insight to engineering seismology, and the vulnerability of the existing stock of buildings and other structures, as well as infrastructures. The role of each factor is analysed, but emphasis is given to soil properties in the context of definition of strong motion acting on the structure foundation and on possibilities for potential surface rupture, liquefaction, land-sliding and subsidence.

As an introductory part, a brief analysis on the effects of earthquakes on the built environment during the XXth century is presented and policies for mitigation of earthquake risk are summarised, constituting Part I – "Earthquakes and the impact on societies". It calls the attention to the communities that seismic risk has been increasing along the times, in spite of all the great advancements achieved in scientific and technical grounds. Problems of bad use of "good engineering knowledge" and lack of quality control are behind these observations.

Part II – "Definition of problems and techniques", develops the main concepts of this presentation. The first topic to be dealt with is the scale of analysis. Depending on the level of detail (national, regional, local, site), hypotheses are different and so are data quality, methods, uncertainties and conclusions. A second topic concerns seismic scenarios which can be obtained through different processes, such as taking into consideration historical seismicity, de-aggregation techniques, hazard analysis or having in mind the minimization of any other objective function as total losses inflicted within a given time period.

The way to incorporate soil influence into impact studies depends on the amount of detailed work performed prior to the study. Sometimes only general geological information exists; in other cases geological maps at a good scale or information on borehole data are available. Methods for soil analysis are though very different and require different analytical tools. A general analysis will act as a first filter to indicate the areas that are more prone to amplification, attenuation of seismic waves, etc. On the other hand, a more detailed method will clarify the zones of doubts. A large part of this chapter is dedicated to the soil problem.

Building and infrastructure (lifeline) stocks are analyzed into their main topical issues namely, classification of typologies, inventories and vulnerabilities.

Finally, in Part III – "Examples for illustration", examples of impact studies illustrate

A. Ansal (ed.), Recent Advances in Earthquake Geotechnical Engineering and Microzonation, 27–65.
© 2004 *Kluwer Academic Publishers.*

the effect of scale, scenario type, etc., as referred in Parts I and II. Applications to four different cases emphasize each one with a different detail of analysis. Comparison and discussion on the results and on the level of accuracy obtained are presented.

This chapter is essentially devoted to methods and does not develop the mathematical algorithms to obtain the results presented. These can be found in the listed references.

2.1. Part I – Earthquakes and the Impact on Societies

2.1.1. EARTHQUAKES IN THE WORLD AND IN EUROPE IN THE XXTH CENTURY

A simplified analysis of the evolution of human casualties and economic losses all around the world caused by the seismic activity during the XXth century (Pinto, 1998, Oliveira and Sánchez-Cabañero, 2002) (Figures 2.1 and 2.2), clearly indicates a steady increase of economic losses (Figure 2.2), especially in the last decade, in contrast with a slight decrease in human casualties (Yong et al., 1997, Figure 2.1). In fact, while casualty figures oscillate around the 150 000 per decade (in a total of 1.5 million) and are marked by the occurrence of very large events (Japan, Kwant, 1923, China, 1920, etc., in the decade 1920-30 and Tangshan, China, 1976, decade 1970-80), the economic losses, corrected to the year 1997, show an exponential increase. Such increase can be attributed to earthquakes striking regions of high urban concentration, for which no seismic protection has been implemented due to difficulty in transferring of technology to the construction industry.

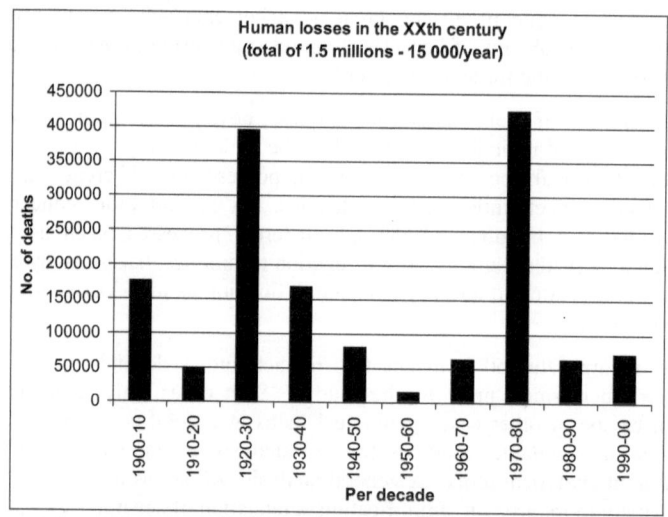

Fig. 2.1. Human losses in the world during the XXth century

Even though great advances in seismology and earthquake engineering have been acquired in the last 20 years, a great deal of implementation is still missing. Many international organizations have spoken out for this problem, but results from these campaigns are still difficult to judge. It is worth noting that the same pattern of damage (human and economical) has been observed in the first years of the XXIth century.

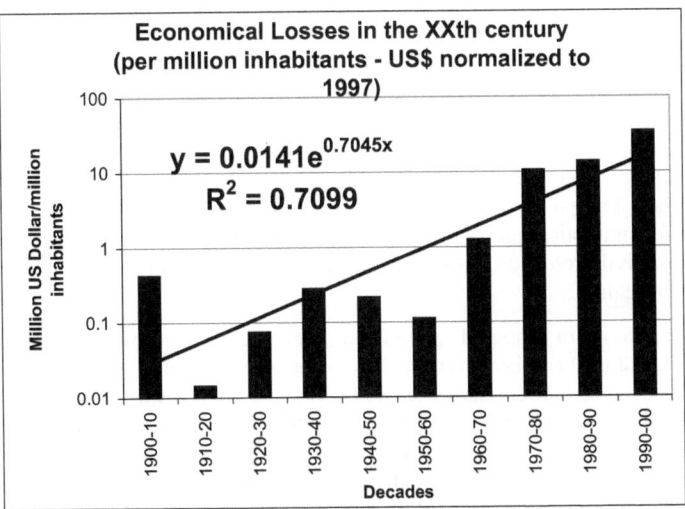

Fig. 2.2. Economical losses in the world during the XXth century

In relation to earthquake risk, it is interesting to analyse the tendency to divide the political world into two large geographic areas: the world of poor countries and the world of rich countries. For the former ones, human casualties are increasing throughout the century and are one order of magnitude above the rich countries, whereas, for these ones, the opposite tendency is observed when dealing with economic losses. On the other hand, it should be important to emphasize that in the last decade of the XXth century very large events took place with catastrophic consequences, both in human and in economic terms. Considering only the six earthquakes with magnitudes in the range 6.7 to 7.5, human casualties attained more than 70 000 and losses were above 150.000 x 10[6] Euros.

Looking in a more detailed way to the economic losses, the Kobe 1995 earthquake alone was responsible for more than 100 x 10[9] Euros, which is equivalent to almost 2% of the Japanese Gross National Product (GNP). This earthquake affected 3.6 million inhabitants from a total of 4.5 million residents in the region, causing 5500 casualties and 41 000 injures, important damage to 3500 buildings in reinforced concrete and steel, and destroying around 80 000 dwellings. These incredible numbers when transferred to other regions such as the metropolitan area of Lisbon, with a slight smaller population, would cause losses of the order of the Portuguese GNP.

The Northridge 1994 earthquake caused about one third of the Kobe losses, but economic loss estimation corresponding to the repetition of the Kwant (Tokyo) 1923 earthquake would surpass, in an order of magnitude, those numbers.

The XXth century has finished with two very large events with magnitude greater than 7.5, separated in time by just less than a month. The Kocaeli earthquake in Turkey occurred in August 1999 and caused tremendous impact in the area east of Istanbul. Over 15 000 people were killed, about 24 000 were injured and 600 000 people become homeless. About 120 000 buildings and houses were considered beyond repair, among which about 5000 were seriously damaged or completely collapsed. The total economic impact of this earthquake is still difficult to establish. The second event, the Chi-Chi earthquake of September 1999, affected a large area of Taiwan causing over 2000 deaths and an economical impact of the order of 3.2×10^9 Euros.

Whatever location around the world, the impact of earthquake activity is so large that, in recent years, a great concern has led to the development of impact studies in large metropolitan areas such as Mexico City, Tokyo, San Francisco, Istanbul, Bogotá, etc. These studies, known as scenario evaluations for Megacities, are essential tools for several different applications, which run from simple evaluations of earthquake risks, to exercises for civil protection, indications for insurance companies, and re-evaluations of mitigation measures.

Earthquake activity in time and space for a given region is not well known and cycles with certain stability can be interrupted by other type of cycles. In recent times, large earthquakes occurred in regions where no historical evidence was present. The cases of Kobe and Athens earthquakes in 1995 and 1999, respectively, are among these events occurring in regions of low seismicity. A great discussion was initiated on the reliability of hazard methods, which cannot present good results if long period observations are not taken into consideration. To avoid such difficulties, studies including paleo-seismology and arqueo-seismology information should be made for regions where long return period activity is suspected.

The lack of regularity in the pattern of earthquake activity has been a characteristic of the seismic activity in continental Portugal (Oliveira and Sánchez-Cabañero, 2002). In fact, periods of long quiescence alternate with periods of great activity. From the end of the XIXth century to the second decade of the XXth century a high rate of activity was observed throughout the country, with the occurrence the Benavento M=6.3 earthquake in 1909, followed by an enormous number of aftershocks. Since then, approximately 80 years of very low seismic activity have passed, only interrupted by two isolated episodes, one large M=8.0 event in 1941 with epicentre mid distance between the Azores and the Continent, and a M=7.2 event in 1969, in the Gorringe Bank, near the most seismic active interplate region SW of the Continent (see Example 1, Part III). But a similar pattern can be visualized for the large area running from the Azores to the Greek islands, in the neighbouring of the Euro-Asiatic and African plates. With the exclusion of Greece, no event with magnitude larger than 5.8 was recorded in that area from 1920 till 1980. In the year of 1980, three larger events occurred: in the Azores (Jan, 1st), in El Asnam Algeria (Oct 10th), and in Irpinia, south Italy (Nov 23th). Since then, the most important events occurred in Algeria (1994) and in Italy (1997), the latter causing great impact over the historical heritage in the centre of Italy (http://emidius.itim.mi.cnr.it/, 1997, 1998)[1]. In summary, one can say that, in the period 1980 to 1995, about 5000 people lost their lives in the territory of the European Union due to earthquake activity leading to a total economic loss above 430×10^6 Euros (Ghazi and Yeroyanni, 1997).

In the topic of natural catastrophes, earthquakes play a very important role, world-wide. As a matter of fact, statistics taken from the period 1973-1997 (http://www.cred.be),

---

[1] This last period was interrupted by a M_w=6.7 earthquake in Algiers in 21 May, 2003 causing over 2200 deaths and more than 10 000 injures.

organized by 5-year bins, show that earthquakes are among the disasters with larger death impact (Figure 2.3), even though the total number of flood events is twice per year.

This analysis on the effects of earthquakes on the built environment can be traced for the entire XXth century. This calls the attention to the communities that seismic risk has been increasing along the times in spite of all the great advancements achieved in scientific and technical grounds. Problems of bad use of "good engineering knowledge" and lack of quality control are behind these poor results.

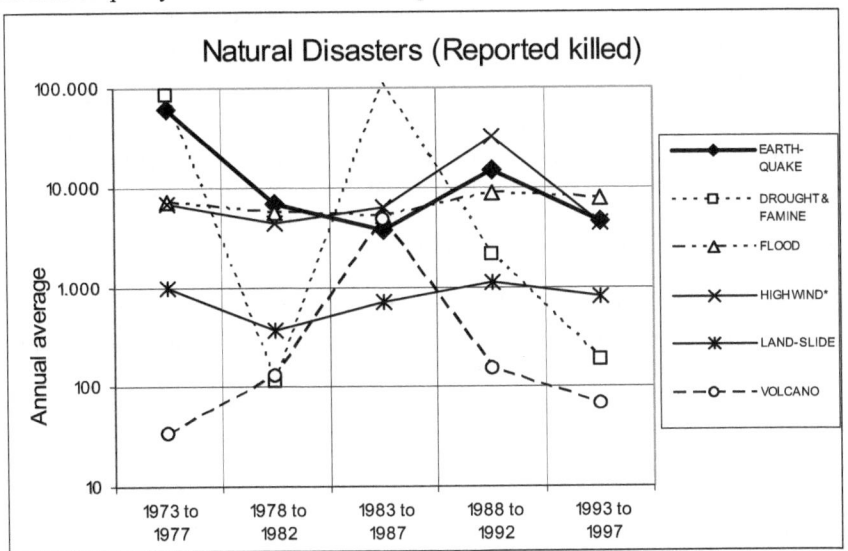

Fig. 2.3. Comparison among different types of natural catastrophes

2.1.2. THE SOIL EFFECT ON THE CATASTROPHIC EVENTS

The effect of soil conditions on the general impact of large earthquakes has been observed in several occasions in the XXth century, attributing to it the responsibility for the great damage inflicted in several areas of the territory. Wide spread destruction, such as in the Guerrero earthquake (1985) in Mexico City, the Spitak earthquake (1988) in Leninakan, the Loma Prieta (1989) earthquake in the San Francisco Bay Area, the Kobe earthquake (1995) in the coastal areas of the city, and the Kocaeli earthquake (1999) in Adapazari are important examples of the influence of soil conditions on ground motion acting in the foundations of structures located in areas away from the epicentre.

But other cases not so well publicised in the literature are also good examples of this important parameter. For example the Erzincan 1992 event, one of many in the Anatolian fault, caused heavy damage, part of which was attributed to site conditions and resonance; other reasons are poor workmanship quality of construction (Ansal et al., 1993). Part of the large destruction caused by the Tangshan earthquake of July 28, 1976, near Beijing, can be attributed to the existence of sedimentary deposits from the Quaternary period (Chen, 1988). Not only the large magnitude (M=8.2) has contributed to the wide spread damage (the 1 million industrial city was reduced to rubble), but the soil influence was clearly marked in the geography of damage. Finally, the recent 2001 Gujarat-Bhuj earthquake (Narula et al., 2002) also caused great damage at a distance of 250 km from the epicentre, denoting a marked influence of soil conditions. This circumstance, together with the resonance effects acting on several structures, caused important damage.

The influence of soil conditions on the ground motion can be translated into the modification of amplitude, of spectral content and of duration of incident motion due to the presence of soft surface layers. But other problems related to the soil can also be considered, such as potential surface rupture, liquefaction, lateral spreading, land-sliding, etc. This is the reason all these features should always be considered in any seismic study. They affect in a definite way the territory and contribute to the seismic risk of the populations, of the building stock and of the lifelines. The incorporation of soil influence in these studies can be seen directly in codes for new or existing construction, in urban development layout and, in more generic sense, in scenario studies. If they refer a particular area and invoke a great deal of specific studies, they are usually referred in the literature as microzonation.

2.1.3. MITIGATION OF EARTHQUAKE RISK AND PREPAREDNESS

In order to mitigate the earthquake risk as seen in the previous chapters it is necessary to act at several levels of the society, in a pure scientific/technical point of view, involving the social, fiscal and political issues (SPES, 2001; "*A Contribution to the reduction of seismic vulnerability of the building stock*", http://www.spes-sismica.org/).

What can we do to reduce the impact of future earthquakes in the building stock and in the monumental structures? The following general topics are of most importance: (i) perception of the origin of earthquakes and of propagation of seismic waves; (ii) understanding of the behaviour of all kind of structures under seismic action; (iii) rehabilitation and retrofit of existing structures; (iv) development of appropriated code of practice; and (v) development of quality control to insure a correct application of all legislation.

In terms of earthquake preparedness, one can act at two different levels:
- Institutional
  - Different Ministries (risk mitigation)
  - Civil Protection:
    - Risk study;
    - Information and education;
    - Response preparedness (EMERGENCY PLANNING).
- Individual
  - Home preparation;
  - Family emergency planning;
  - Self-protection measures.

Because it is not possible to predict earthquakes, it is necessary to minimize the risk, preparing a *Preventive Planning* and to minimize the effects of the event, developing an *Operational Planning*.

In order to minimize seismic risk, one should: (i) develop and enforce preventive measures; (ii) improve building regulations for construction and reinforcement; (iii) develop appropriate land use plans; and (iv) carry out civil protection awareness and educational programs for the population, civil protection entities and decision-makers.

The measures to minimize the effects after the occurrence of the event should be prepared: (i) plan civil protection actions to activate when an earthquake occurs; (ii) organise civil protection entities involved in aid operations, concerning its mission and operational procedures; (iii) plan emergency means and resources and their allocation, and plan management.

These last issues require *Emergency Master Plans* and *Detailed Response Plans* for specific risks - *i.e.* the *Seismic Risk Emergency Plan*.

2.2. Part II – Definition of Problems and Techniques

2.2.1. SCENARIO STUDIES – GEOGRAPHIC SCALE OF INTERVENTION

Seismic scenarios for impact studies are essential exercises for a number of reasons: (i) Help public authorities preparing emergency planning; (ii) define *means* and *resources* to cope with potential earthquakes; (iii) accelerate "on-line" damage assessment caused by an earthquake; and (iv) quantify the extent of any programme for repair and retrofit of structures.

In the following, the several parts intervening in the process will be presented, with greater emphasis on the soil component. Part III, by presenting several cases at different scales, will develop in more detail the various subjects referred.

Elements of society in risk and overall impact estimation

1. Identification and characterization of elements of society in risk (vulnerability identification)

The elements of society in risk considered for impact studies due to a serious seismic action are the ones which can be seriously affected, causing human life loss and society disruption, some times for a long period of time. The vulnerable elements to be considered are:
- Building stock and industrial plants (dwelling with partial or total collapse);
- Lifelines: roadways; railways, including the subway; energy (electricity, gas, and other combustibles); water; sewerage and telecommunications (disturbing seriously the social tissue due to operation collapse, temporarily malfunction, etc.);
- Population (producing death, injures of various kinds, and homeless);
- Vital or important structures (those that are relevant for emergency management, either due to its operational or political role).

2. Vulnerability evaluation

The vulnerability of each element in risk is analysed using functions that relate the expected damages of each type of element with its characteristics and subjected to the seismic action acting at the foundation level.

The building stock is statistically analysed having as reference the unit of the smallest administrative territorial division, depending on the scale of work, and a diversity of parameters of the buildings such as age, constructive typology, number of floors, etc. Several techniques have been developed to derive these vulnerability functions, from simple curves taken from observed behaviour during past earthquakes, to complex engineered structured response analyses, or directly obtained from definitions of intensity scales. Several typologies have been developed to consider the seismic behaviour. Here, again, depending on the scale of work and on the knowledge on individual units, building typologies range from 5 or 6 main categories (essentially by epoch of construction and attending to the type of material) to a detailed analysis with a few dozen cases (detailed construction types within, for instance, reinforced concrete), as referred in the *Hazus* 99 methodology.

A similar methodology has been adopted when dealing with lifelines, by classifying them according to types, materials, geometries, epoch of construction, etc., and defining the corresponding vulnerability functions. For emergency planning purposes, the results of each lifeline damage estimation, for each scenario, aims at a quick preliminary restoration, in order to assure basic indispensable services.

To estimate the impact of earthquakes on the human life, it is indispensable to characterise the demographic distribution for different periods of the day and around the year. Not only the population living within the study area, but also the population commuting to/from other adjacent areas, are of utmost importance to define their geographical location at the time of the seismic event.

Vital or important structures have to be identified, and their vulnerability functions need to be attributed. This is a very difficult task because only an individual detailed study can more effectively produce reliable results, requiring the participation of the entities responsible for them. The use of vulnerability functions adapted from the general above mentioned ones has been commonly practised in impact studies, increasing the errors associated with their evaluations.

3.   Establishment of seismic occurrence scenarios

The establishment of feasible scenarios requires the study of past seismic occurrences, the definition of different seismogenic zones affecting the area under study, and the characterisation of their most important parameters (maximum expected magnitude and frequency or probability). It may also involve the knowledge of attenuation functions, if scenarios have to do anything with ground motion parameters, or the minimization of any other objective function, as total losses inflicted within a given time period.

In many applications not only one single criterion prevails. Hypotheses to be analysed are: the consideration of a largest historical event, as a measure of an extreme type event; the 50, 100 or 1000-year mean return period event, requiring an hazard and de-aggregation analysis; the most probable measure of impact over the entire stock in the study area; or the eventual rupture of a possible fault structure.

4.   Propagation of seismic energy

The release of seismic energy from their different possible sources (geotectonic structures) and its propagation through the crustal Earth layers to the bedrock underneath the site under analysis is generally considered as the attenuation of seismic waves. Local effects representing the transference of energy from bed-rock through the upper soils, with the potential for ground motion amplification or reduction, liquefaction and land-slide potential, are consequences of soil properties and characteristics of input ground motion (level and energy content).

Several techniques can be used in the context of propagation of seismic waves, ranging from peak ground values (acceleration - PGA, velocity - PGV or displacement - PGD), spectral values, or seismic intensities (Mercalli Modified intensities - MMI, European Macroseismic Scale - EMS-98, etc.). Primarily they relate these values with magnitude and epicentral distance and, in the case of better tectonic definition, also with other source parameters.

This task is of great importance to reduce uncertainties in the entire process of impact estimation. Existing data from past earthquakes, even of small magnitude values, should exercise the calibration of attenuation relationships.

5. Damage evaluation for the established scenarios

The evaluation of damages of the selected elements in risk, for each scenario, is obtained using a simulation model that integrates all the above mentioned aspects, by summing up all possible contributions. The damages of each element in risk are classified in several different limit states, usually defined at five levels: no damage, slight damage, moderate damage, heavy damage, and collapse. Depending on the element under study, this damage classification may be lumped into coarser categories or adapted to operational/non-operational terms, as in the case of sections of lifelines, etc.

For a rapid visualisation and treatment of all information, Geographical Information Systems (GIS) techniques are generally used. This requires a great deal of effort due to the need of digital vectorization of many layers, but is of great benefit for future applications, updating, corrections, alteration of algorithms, etc.

The GIS allows the visualisation of all types of variables organized by "layers". Examples of these are: intermediate results such as the geographical distribution of characteristic motion parameters (maximum acceleration, response spectra, etc.), either for bedrock and soils (local effects); the intensity distribution (Mercalli, EMS, etc.); any damage category for any specific layer, etc. It can be made at the work-unit level or can aggregate various units, depending on the objective pursued. Smoothing can also be applied for geographical interpretation. At the same time statistical analyses, cross-references, etc., are easily obtained with any standard statistical package.

Scenario studies are viewed in different perspectives according to the objectives to be targeted. These define the scale of intervention which is very critical in the way to obtain the elements necessary for the analysis. In this overview, special attention is given to the problem of scale, with the presentation of impact studies at four different scales: (i) scale of a country or large region, on the order 1:1 000 000; (ii) scale of a region, 1:25 000; (iii) scale of a city or of a block, 1:5 000; and (iv) scale of a large building, 1:1 000.

GIS techniques can combine different scales, but the accuracy of final results is determined by the one with poorer geographical detail.

2.2.2. SOIL INFORMATION

The soil information in each scale-case is analysed in a very different form: for the case of a country scale the elements are only descriptions of the large geological units, whereas for the regional scale a mixed large description with some more detailed analysis is necessary; for the case of a city or block a more detailed description of geotechnical units should be available, and for a building a detailed description including borehole information might be important. The detail of this information should be settled at a degree similar to the detail of the other types of information for the impact studies, such as the building stock, the infrastructure network and the population geographical location.

Geological published maps are used as fundamental tools for any of the analysis, but the degree of detail has to be found in more specific studies. In this case information obtained in construction sites where boreholes were made is of most interest, together with data on the geological setting, and some other collateral information. Regionalization is also made using other techniques for identification of soil properties, among which are geophysical prospecting. Historical information on the zones of systematic higher seismic intensities should be regarded as very important pieces of evidence which deserve the most detailed analysis.

Several techniques for soil analyses, analytical as well as experimental, have been used in connection with impact studies.

Among the analytical techniques, a first one considers soils as a one-dimensional (1-D) representation defined by horizontal layers characterized by thickness and shear wave velocity. It is only done for specific studies such as for important or critical structures and includes *in-situ* and laboratory testing for determining the main geotechnical properties and their associate mechanical properties such as the shear modulus of elasticity G_0 and density ρ, and a measure of their non-linear behaviour. This technique, implying very good information, is only available for a reduced number of sites.

In most places where surface geology indicates stronger formations not deeper than 30m, knowledge of shear velocity down to 30m is considered to be sufficient.

An interpolation of the characteristics found in these sites to cover wider areas can only be done by expert analysis based on lithographic description of the geological horizon and on knowledge from other similar situations. If we are working in a more general framework (e.g. in a larger scale), these descriptions are necessarily more vague, smoothly considered and, consequently, reflecting average values.

The experimental technique which by far became very popular in the last 20 years, considers the possibility of knowing the physical properties underlying the soil substratum by obtaining, using ambient noise, the lower natural frequency of vibration of the soil layer. This technique, known as Nakamura spectral ratio of horizontal to vertical components (Nakamura, 1989 and 2000), can be used to identify frequencies and possible amplification of ground motion in geotechnical zones of soft/hard "impedance contrast" for moderate to large scale projects at low cost.

In all cases, the motion at the surface requires the definition of ground motion at the bed-rock which depends essentially on the magnitude, source properties and properties of the path medium. The convolution of the input motion at the bed-rock with the response of the upper soil layers will give the final surface result. More sophisticated models, with 2-D and 3-D geometry, with linear and non-linear constitutive relations, considering topographic implications, etc., are not used in studies of moderate to large scale nature, even though great influence on the results may occur, such as in cases of alluvial basins where gravity waves possibly dominate.

In impact studies several of the above mentioned techniques have been applied, depending on the degree of knowledge and the scale of work. A convergence of results from different techniques should be achieved when the information is of greater quality. Naturally, working in wide regions, uncertainties may be quite large due not only to the fact of the difficulty in generalising to the whole area the information from singular points, but also due to the intrinsic uncertainty in the scrutiny methods used. This is the main reason why results should always be used with caution and, if possible, margins of uncertainty should be given.

Using the stratigraphy/lithology of the upper layers, Medvedev (1962) proposed a simplified method that produces the macroseismic intensity at the soil surface by increasing the intensity at the bed-rock horizon by an increment, function of the class of soil. As an example, for a "rigid granite" there is no increase; "sandy soils" may increase 1.2 to 1.8 (for MSK scale); "uncontrolled fill" 2.3 to 3. These numbers were taken from empirical observations and reflect the propagation on a two-layer soil system.

In all cases one has to deal with the process of knowing the seismic action that can be acting at the bed-rock level, essentially as a function of magnitude and epicentral distance. Of course other parameters may intervene in quite drastic way, such as the source mechanism including the rupture process and the radiation pattern, but these parameters are only considered at special situations when the geodynamic information is good enough to be included. This issue involves the consideration of attenuation of peak ground motion parameters, spectral ordinates or more simply intensities (MM, MSK, EMS-98, etc.). Oliveira and Sanchez-Cabañero (2002) discuss these issues in great detail.

An alternative way to this procedure is to use the concepts developed in the recent codes for definition of seismic action, such as the approved version of Eurocode-8, where soils are classified in great detail covering most common situations observed in many different regions. Ground motion in terms of response spectra for each soil class at the surface is also given, consequently avoiding the difficult part of analysing the wave propagation within the soil layers. In this respect, the information leading to the proposed spectra was supported in a large collection of strong motion data recorded at the top of the various types of soil profiles. US policies through recent UBC97 legislation (FEMA, 1997, SEAOC, 1998) also consider a variety of soil classes, which can be used in connection to the type of studies under analysis.

A classification of soil properties made within Eurocode-8 (2002) is as follows, Table 2.1 (Sabetta and Bommer, 2002):

Table 2.1. Classification of subsoil classes (EC8, 2002)

Subsoil class	Description of stratigraphic profile	Parameters		
		$V_{S,30}$ (m/s)	N_{SPT} (bl/30cm)	c_u (kPa)
A	Rock or other rock-like geological formation, including at most 5 m of weaker material at the surface	> 800	–	–
B	Deposits of very dense sand, gravel, or very stiff clay, at least several tens of m in thickness, characterised by a gradual increase of mechanical properties with depth	360 – 800	> 50	> 250
C	Deep deposits of dense or medium-dense sand, gravel or stiff clay with thickness from several tens to many hundreds of m	180 – 360	15 - 50	70 - 250
D	Deposits of loose-to-medium cohesionless soil (with or without some soft cohesive layers) or of predominantly soft-to-firm cohesive soil	< 180	< 15	< 70
E	A soil profile consisting of a surface alluvium layer with V_s values of class C or D and thickness varying between about 5 m and 20 m, underlain by stiffer material with V_s > 800 m/s	–	–	–
S_1	Deposits consisting – or containing a layer at least 10 m thick – of soft clays/silts with high plasticity index (PI > 40) and high water content	< 100	–	10 - 20
S_2	Deposits of liquefiable soils, of sensitive clays, or any other soil profile not included in classes A–E or S_1	–	–	–

Table 2.1, besides presenting a classification based on the lithology/stratigraphy description of the layers and on the average shear wave velocity, also considers the N_{SPT} values (Standard Penetration Test) and the "undrained shear strength", c_u. Consistency between the velocity bounds, N_{SPT} and c_u has been made for the different soil classes (Ohta and Goto, 1976 and Clayton, 1995).

The consideration of five subsoil classes for the purpose of defining the elastic response spectrum is the result of the latest developments in data collecting of recent events, the Kobe, 1995, the Kocaeli, 1999, and the Chi-Chi, 1999. The classes are defined by the average shear wave velocity of the upper 30 metres of soil ($V_{S,30}$), by improved descriptions of the stratigraphy, and by ranges of values of geotechnical parameters.

For special classes S_1 and S_2, which describe other geotechnical situations, special studies for the definition of the seismic action, are required. For these classes, and particularly for S_2, the possibility of soil failure under the seismic action must be considered.

2.2.3. SPECTRAL SHAPES

Within the scope of Eurocode 8, the earthquake motion is represented by an elastic ground acceleration response spectrum, dependent of sub-soil class defined in Table 2.1 and on the magnitude value.

For application purposes only, two different response spectra, Type 1 and Type 2, have been introduced, Figures 2.4 and 2.5, to be adopted respectively in high (M_w>5.5) and low seismicity regions (M_w<5.5). The epicentral distance is not considered because it controls the amplitude of the spectra but not their shape, and therefore has no effect on normalized spectra. For other applications and tectonic environments, it may be recommended the adoption of more spectral shapes, especially for the very large magnitude values (M>7.5), for which Tc and Td may be higher.

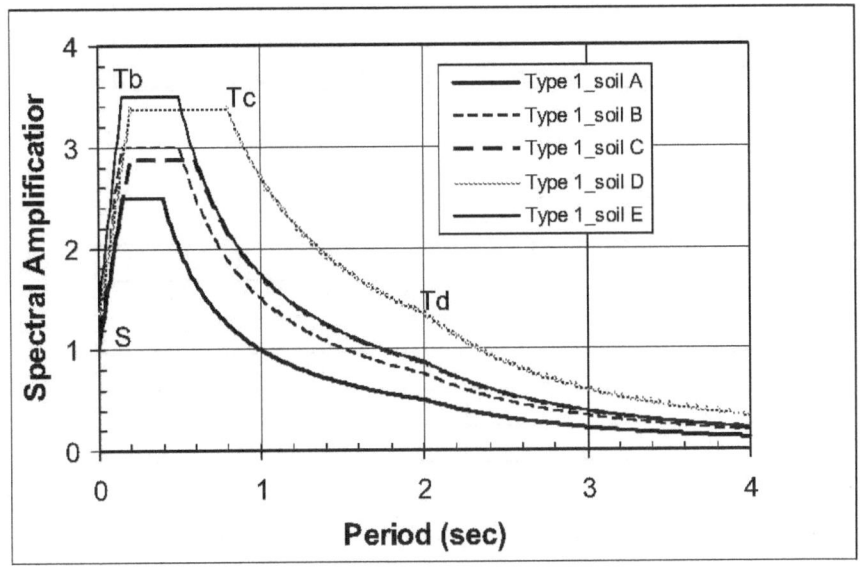

EC8-00 TYPE 1	S	Tb	Tc	Td
soil A Vs > 800 m/s	1,00	0,15	0,4	2,0
soil B 360<Vs<800 m/s	1,10	0,15	0,5	2,0
soil C 180<Vs<360 m/s	1,35	0,20	0,6	2,0
soil D Vs < 180 m/s	1,35	0,20	0,8	2,0
soil E (h < 20 m)	1,40	0,15	0,4	2,0

Fig. 2.4. Type 1 elastic response spectra for the 5 subsoil classes and corresponding soil parameter (S) and control periods (Tb, Tc, Td)[2]

---

[2] S *- is the spectral value for period zero*
   Tb, Tc - are period limits (sec) for constant spectral acceleration branch,
   Td   - is the period value defining the beginning of the constant displacement response range
   of the spectrum.

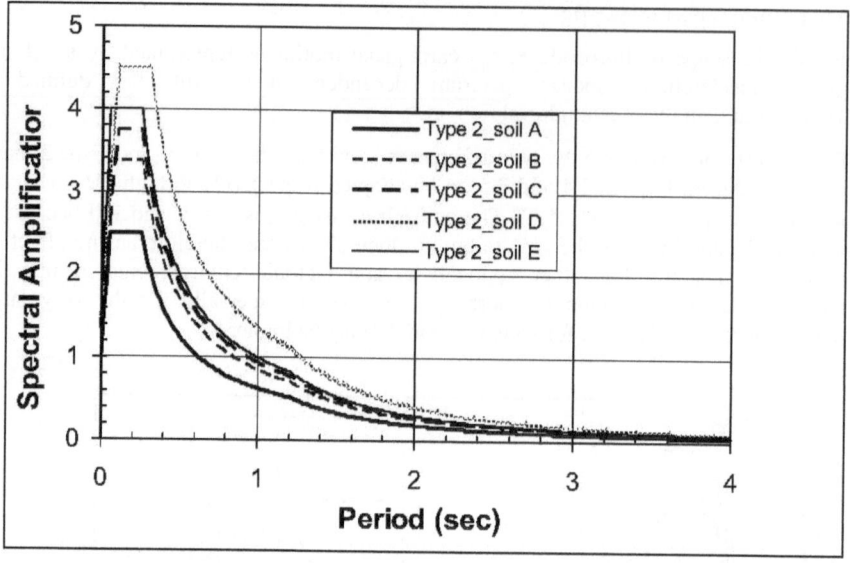

EC8-00 TYPE 2	S	Tb	Tc	Td
soil A Vs > 800 m/s	1,0	0,05	0,25	1,2
soil B 360<Vs<800 m/s	1,2	0,05	0,25	1,2
soil C 180<Vs<360 m/s	1,5	0,10	0,25	1,2
soil D Vs < 180 m/s	1,8	0,10	0,30	1,2
soil E (h < 20 m)	1,6	0,05	0,25	1,2

Fig. 2.5. Type 2 elastic response spectra for the 5 subsoil classes and corresponding soil parameter (S) and control periods (Tb, Tc, Td)

It is important to point out that the value of *S* is larger for the Type 2 spectrum than for the Type 1, for all classes other than A. This reflects the non-linear response of soil layers and the fact that weak motion is amplified more than strong motion (Rey et al., 2002).

In situations where the 30 m depth are not enough to reach a bed-rock type soil formation, as it happens in certain geological environments in the centre of Europe such as in deep sedimentary basins, it might be necessary to introduce more classes, which are essentially the same as above but with specific deep ground geological constitution.

In conclusion, one can say that there are several possibilities to deal with the soil problem in impact studies, depending on the working scale, the knowledge of the geotechnical situation and the software availability. But it also depends on the detail of knowledge of the other components of the entire process for estimating the seismic impact. There is no point in studying the soil with great detail and treat the stock of buildings by large geographical units.

Liquefaction, land-slides and subsidence are other topics connected to soil performance with great importance essentially for stabilisation of foundations and performance of lifelines. The level of water content in the soil stratum is of major importance and, consequently, the epoch of the year when the event takes place does have a great influence on the potential for liquefaction, land-slides, etc. We will brief refer this topic when analysing lifelines.

Tsunami and fires may aggravate the entire situation contributing to a chaotic environment. They are not looked up in this review.

2.3. Part III – Examples for Illustration

Four examples will be analysed to illustrate the methodologies and problems encountered in cases where scales of intervention are very different. The first case shows an application to a large area which covers the entire Continental Portugal; the second is the case of a regional-scale, the Metropolitan Area of Lisbon (AML); the third case looks up to the Lisbon County, with the inclusion of a detailed geotechnical description and the treatment of blocks of buildings; the last case discusses the situation of blocks of buildings on the basis of individual structures.

In each example attention is given to a special topic, different from others to avoid repetition of situations and to permit an analysis of broader subjects. Results obtained using analyses at different scales are also discussed and compared.

2.3.1. EXAMPLE 1. STUDIES AT THE COUNTRY LEVEL: PORTUGAL

The first example is taken from a study done in late nineties for the entire Continental Portugal (Campos-Costa et al., 1998). The seismicity of the country is moderate to strong, alternating periods of large events with long periods of quiescence (Oliveira and Sánchez-Cabañero, 2002). Figures 2.6a and 2.6b present a general view of the more active zones of the surrounded area, showing the geodynamic and seismological environments.

The attenuation models were evaluated at the country scale (Sousa and Olievira, 1997) refined with the inclusion of three classes of subsoil conditions taken from Eurocode 8, Part 1.1 (Eurocode 8, 1994). These subsoil classes are labelled A, B and C, and, roughly speaking, corresponds to hard, intermediate and soft subsoil classes.

Portugal was divided into sub-regions corresponding to 275 administrative counties. The class of soil assigned to each county was the most representative in terms of its spatial distribution. Figure 2.7a presents the distribution of local ground conditions in Portugal according to the three classes referred to above.

Figure 2.7b presents the hazard map for 1000 years return period. The seismic hazard evaluation already comprehends the average ground condition of each county, as it is included in the attenuation model.

Among the current distributions, it was verified that the one that globally fits the evaluated seismic hazard annual distribution is the Beta distribution with, adjusted parameters from county to county.

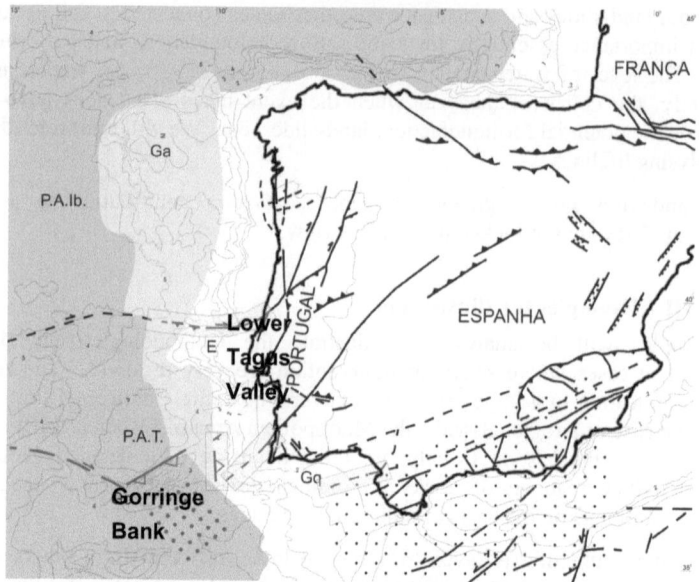

Fig. 2.6a. Geodynamic environment for Continental Portugal (after Cabral, 1996;
Courtesy of "Colóquio/Ciências", Lisbon)

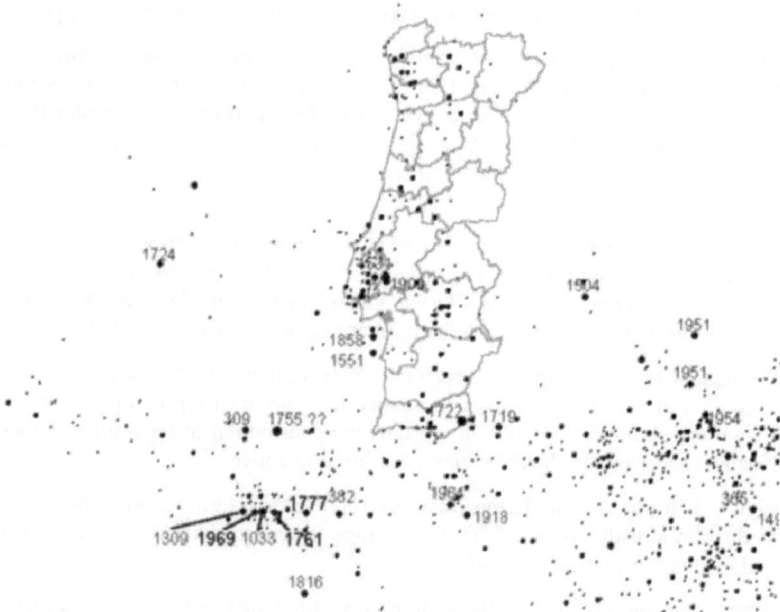

Fig. 2.6b. Seismological environment for Continental Portugal (adapted from Sousa et
al., 1992)

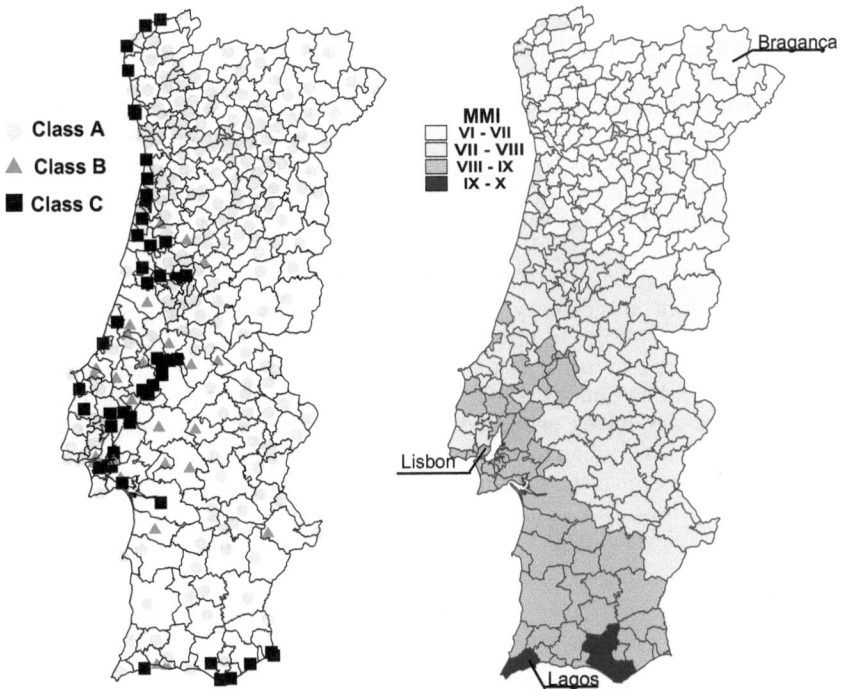

Fig. 2.7 a. Average ground conditions in Portugal according to Eurocode 8 classes
(Eurocode 8, 1994); b. Hazard map for 1000 years return period; macroseismic intensity
in MM scale

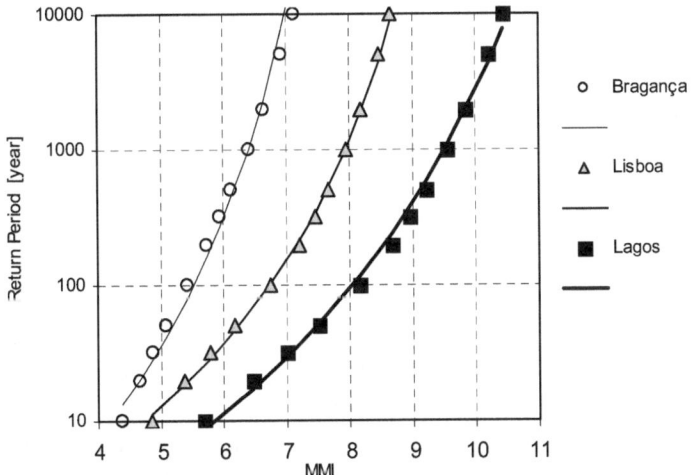

Fig. 2.8. Hazard curves for Portugal: Lisbon County, higher (Lagos) and lower
(Bragança) seismic hazard conditions; best fit of Beta distributions (continuous lines)

Figure 2.8 exhibits the hazard curve for Lisbon County and for the counties with higher and lower seismic hazard in Portugal for the return period of 1000 years, respectively Lagos in the Southern cost and Bragança in the Northwest. Figure 2.8 also illustrates the Beta distributions that best fit the hazard distributions. The ground conditions at Lisbon, Lagos and Bragança are averaged as classes B, C and A, respectively. As it can been seen the Beta distribution fits adequately the hazard evaluation. The mean, maximum and minimum standard errors of the fitting, among the 275 Portuguese counties, are 0.021, 0.043 and 0.012, respectively.

Housing stock and probabilistic vulnerability analysis

The above-described methodology was preliminarily applied to the Lisbon County (Sousa and Olievira, 1997). In that work a detailed analysis of the existing stock of buildings was made, due to the fact that a considerable amount of information was available for the Lisbon County. Unfortunately, the same can not be said on behalf of the other 274 counties in the Continent due to: (i) lack of knowledge of the distribution of the housing stock according to a set of typologies and (ii) problems in adapting the world-wide damage earthquake statistics to the vulnerable typologies that were feasible to identify. The consequence of the above encountered difficulties is the increase of the dispersion of the vulnerability distribution, as it will be mentioned later on.

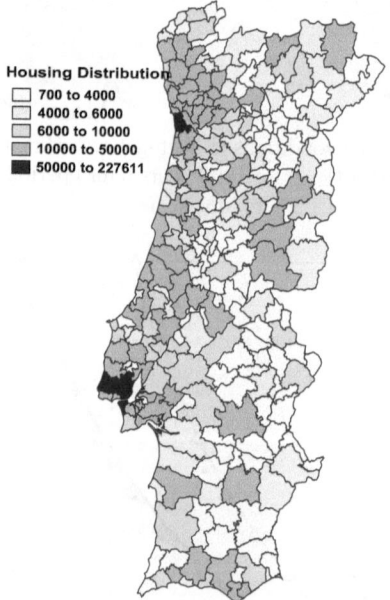

Fig. 2.9. Distribution of number of housing units per county (data from Census 1991; INE, 1994)

In the present study, the data available in the Portuguese 1991 Census (INE, 1994) were adopted and the typological categories were grouped taking into account just one vulnerability factor, the building age. The vulnerability curves were adapted from world-wide earthquake statistics compiled by Tiedemann (1992), bearing in mind the characteristics of the Portuguese housing stock.

Figure 2.9 shows the distribution of total number of housing units per county (the total number in the country is approximately 3 millions). Notice the high concentration of housing units at Oporto and Lisbon metropolitan areas and in most littoral counties.

Beyond these elements, the Census also provides information at the county level to classify the housing stock into five different typologies, according to the vulnerability factor V_1 = *building age*:

- T1 - Buildings constructed prior to 1919.
- T2 - Buildings constructed during the period 1919-1945.
- T3 - Buildings constructed during the period 1946-1960.
- T4 - Buildings constructed during the period 1961-1985.
- T5 - Buildings constructed during the period 1985-1991.

Figure 2.10 illustrates the distribution of housing units per typology in the main five Portuguese geographic regions.

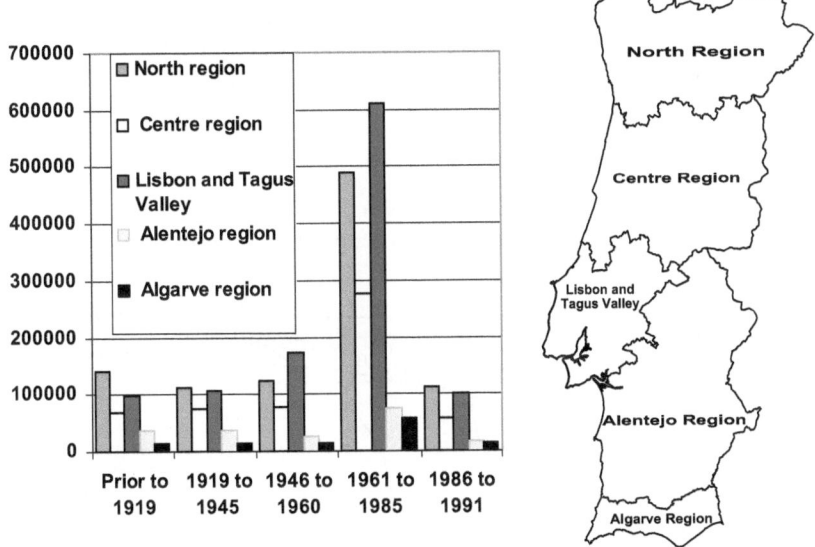

Fig. 2.10. Distribution of housing units per typology in the main five Portuguese geographic regions (data from Census 1991; INE, 1994)

Another vulnerability factor, related to the main materials used in the building construction, could be identified in the Census in order to reflect the main structural properties of each typology (V_2 = *building materials*). However, as building materials can only be related to the number of buildings per county, and not with the housing units, the vulnerability factor *building materials* was not considered in the present analysis.

Figure 2.11 shows the vulnerability curves per typology. Notice that typology T4 corresponds to the first Portuguese seismic resistant code (RSCCS, 1958) and the typology T5 corresponds to the actual code (RSA, 1983). The RSCCS code comprises three different seismic regions and imposes an increase of the building resistance in accordance with the region seismicity. The RSA code includes four different seismic regions and three subsoil classes, compelling the building vulnerability to decrease with the subsoil softness and/or with the increase of the region seismicity. For those reasons, the actual ranges of the vulnerability curves for typologies T4 and T5 are grey shadowed in Figure 2.11, as they correspond to different seismic designs from North to South of the country, conjugated with different subsoil conditions, in case of typology T5.

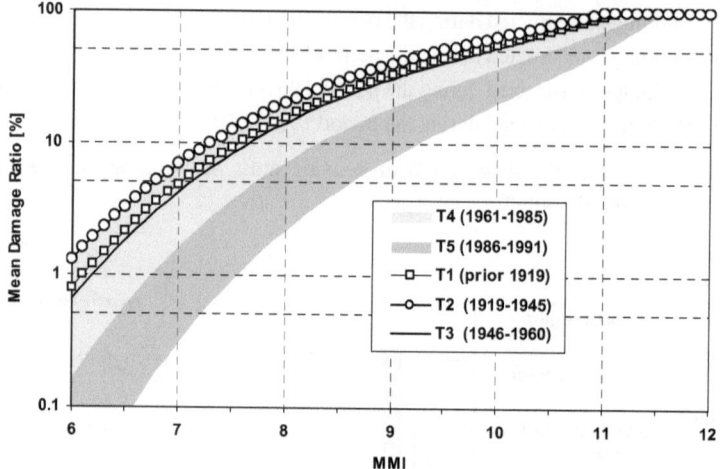

Fig. 2.11. Vulnerability curves per typology (Census 91)

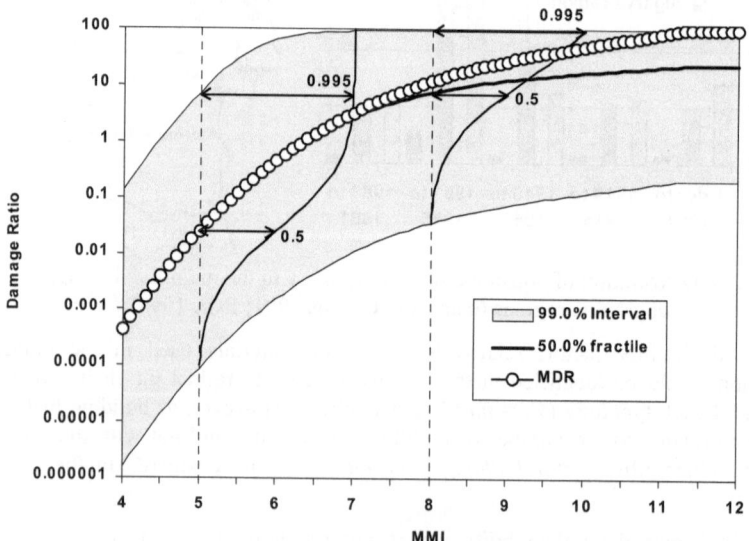

Fig. 2.12. Averaged Damage Ratio (DR) distribution for Lisbon County

Figure 2.12 shows the Mean Damage Ratio (*MDR*) weighted average for Lisbon County. It also shows the 99% variation domain and the median of the Damage Ratio distribution (Lognormal distribution, $\sigma = 2.2\%$) as a function of macroseismic intensity. For macroseismic intensities V and VIII the figure also presents the distribution function of *DR|MMI* and the corresponding fractiles of 50% and 99.5% (For detailing the computational algorithms, see Campos-Costa et al., 1998).

Probable Losses for various reference time intervals

Figure 2.13 shows the probability distribution of the Losses for the entire country for the time reference intervals of 10 and 50 years. The shadowed grey areas, in the figure, contain the domain of the distribution of Losses, county per county, denoting the highest and lowest trend lines in the Losses distributions.

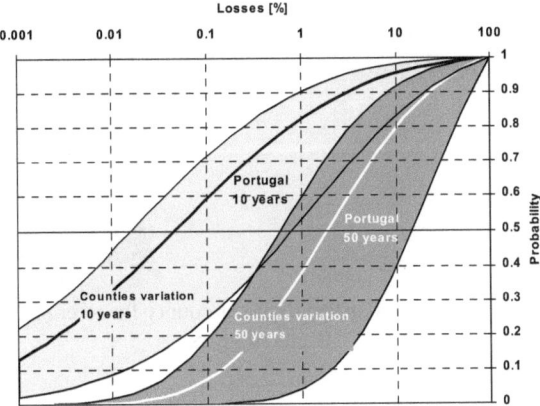

Fig. 2.13. Distribution of the overall Losses in Portugal and counties envelopes; 10 and 50 years reference time intervals

From this figure one can conclude that there is a ratio of 40 between the 50% fractile of the 50 years distribution and the 10 years distribution for Portugal. Referring to the 10 years distribution, the median of the overall Losses in Portugal is 0.05% and there is a 90% probability that Losses are between $2 \cdot 10^{-4}\%$ and 9%. If the 50 years distribution is considered, those values are 2%, 0.08% and 41%, respectively.

These ranges reflect the high dispersion that one deals with in this work, and reveals the importance of considering uncertainty in the vulnerability functions.

To visualise the relative importance of each county on the total value, Figure 2.14 shows the median Losses distribution for reference time intervals of 10 and 50 years, and Figure 2.15 the geographic distribution of the Losses (median) normalized (*Losses incidence*).

As expected, Figure 2.15, showing higher Losses incidence in the South and in zones around Lisbon, reflects both the higher hazard region in the South and the higher concentration of housing units in the metropolitan area of Lisbon. The highest value was observed in the Lisbon County which is responsible for 11% of the total Losses of the entire country.

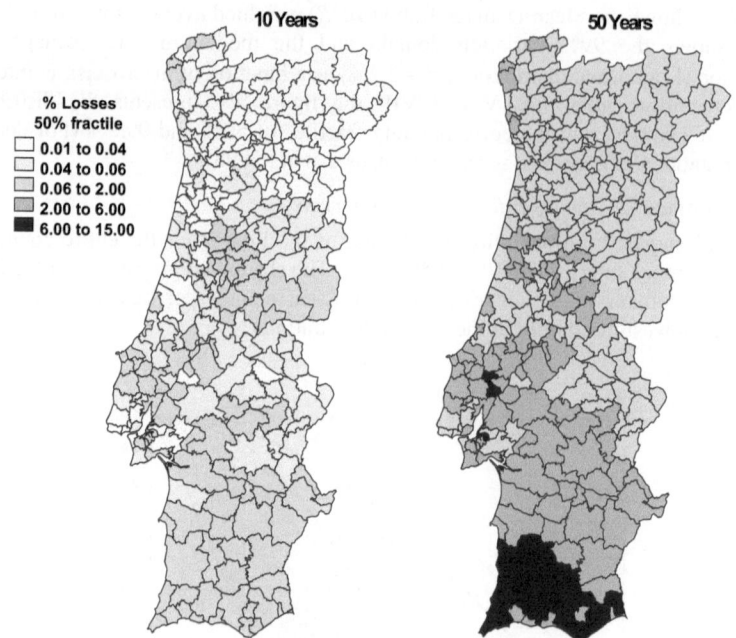

Fig. 2.14. Median of the Losses distribution per county for reference time intervals of
10 and 50 years

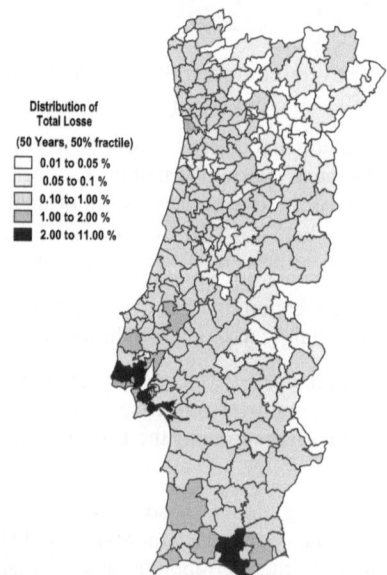

Fig. 2.15. Geographic distribution of incidence of the Losses; normalisation through the
median of total Losses in Portugal for the reference time interval of 50 years

Comments

This methodology presents a wide range of possible applications and results. However, caution should be exercised in relation to the results shown, because data used in the application must be calibrated against local situations, e.g. historical seismic scenarios. In fact, the confrontation of these results with the values that can be obtained directly from the historical seismicity will be a further step to bring an additional degree of validation.

From the application of this methodology to Portugal one can conclude that, for a time reference interval of 50 years, the median of the total housing Loss in Portugal is 2% and there is a 90% probability that housing Losses are between 0.08% and 41% for the same time window. This high range of uncertainty is mostly attributed to the uncertainties in vulnerability assessment of the building stock, emphasising the importance of considering this randomness whenever seismic risk studies are carried out.

In order to obtain more precise estimates of losses, the present application requires further refinements at the data level. A better knowledge of the building inventory in Portugal, namely the distribution of buildings considering other vulnerability factors beyond building age, such as building materials and number of storeys, is essential. Data from 2001 Census is available at a more detailed geographical unit, allowing a better accuracy in geographical terms. It also permits an updating of the presented results, as significant changes in the stock of buildings have taken place during the 1990-2000 decade (M. L. Sousa, work in progress, 2003).

Sensitivity studies should be performed in order to identify the most important parameters controlling losses, particularly: (i) in the hazard model the maximum probable magnitudes and the dispersion of attenuation laws and (ii) in the vulnerability analysis the adequate choice of the distribution of damage and its variance. In fact, these last two parameters require a more profound study in order to reduce the high dispersion of the results.

The present model can easily be extended to obtain the number of casualties and injuries per county or for any other group of counties, the number of homeless, and the number of destroyed or damaged housing, etc.

Other applications can be made in different fields such as insurance, urban planning, disaster mitigation, earthquake preparedness, emergency planning, etc. The study of extended structures such as lifelines is also of interest. To perform similar studies at a more refined scale such as for a region around Lisbon, a better definition of the parameters modelled at a more refined geographic scale are required, as will be shown in Example 2.

2.3.2. EXAMPLE 2. STUDIES AT THE REGIONAL LEVEL: THE METROPOLITAN AREA OF LISBON (AML)

This study, with phases similar to the ones of Example 1, developed recently, culminated with the development of a *simulator* to be used for the elaboration of an emergency plan to face the seismic risk in the region, Figure 2.16. The *simulator* allows the damage evaluation (to humans, to the building stock, to various lifelines, to vital structures, etc.) for given scenarios defined by a magnitude and an epicentral location.

The geographical unit was of three types according to the layers under consideration: (i) for soil characterization a detailed analysis of seismic profiles was made on the basis of a 1:50 000 geological mapping and the information on boreholes was collected from construction sites; (ii) the building stock was obtained from the 1991 Census at parish level (INE, 1994), digitized from 1:25 000 scale, with a total of 277 units; the same can be said about the areas that suffered land-slides in the past; (iii) the tracing of a few lifelines, obtained by the entities in care, was based on a vectorization, sometimes at a 1:1000 scale.

The Metropolitan Area of Lisbon (AML) is a large area of approximately 100 km diameter, with about 430 000 buildings, 1.1 Million dwellings and a resident population of the order of 2.7 Million inhabitants (INE, 1994), representing 27% of the population of the country.

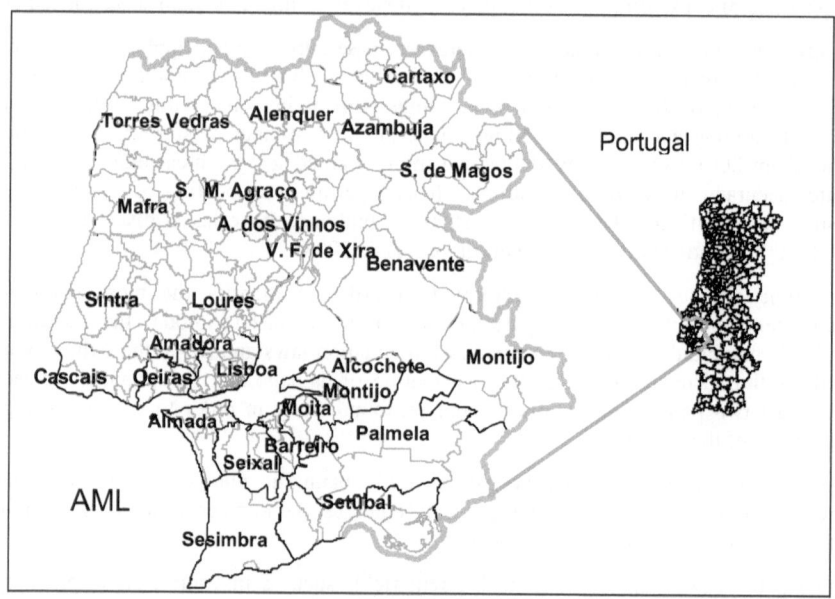

Fig. 2.16. Localization of the Metropolitan Area of Lisbon (AML)

All data, collected and produced, were implemented in a GIS environment. The detail of the work was impressive in all respects (Rocha et al., 2002, and 2003 for a summary). In this chapter, attention is essentially given to the soil analyses and briefly to the vulnerability methodology. A few results are shown, together with discussion on uncertainties and how to deal with them.

The soil classes were obtained by assigning a geotechnical profile to a geographical unit of homogeneous characteristics (Campos-Costa et al., 2002), Figure 2.17, and treating this area as a 1-D non-linear model subjected to a ground motion defined by a response spectrum acting at the bed-rock level. Spectral attenuation relationships (Bommer et al., 1998) were used to obtain the spectral contents at bed-rock as a function of magnitude and epicentral distance.

Soil profiles, in general with information down to the bed-rock, were identified, in a total of 36 profiles, Figure 2.17 (A, AA, AB,...AK, B, C,...Z). These profiles, which can be associated to the geomorphology of the area, as seen in the figure, vary from "rigid" (*A*), to "sandy" (*M*) (V_s=250 m/s; h=15 m), to "very soft alluvial" (*V*) (V_s=150 m/s, h= 40 m). Figure 2.18 presents the profiles more common in the area, for which the average geotechnical properties of the upper layer are given in Table 2.2.

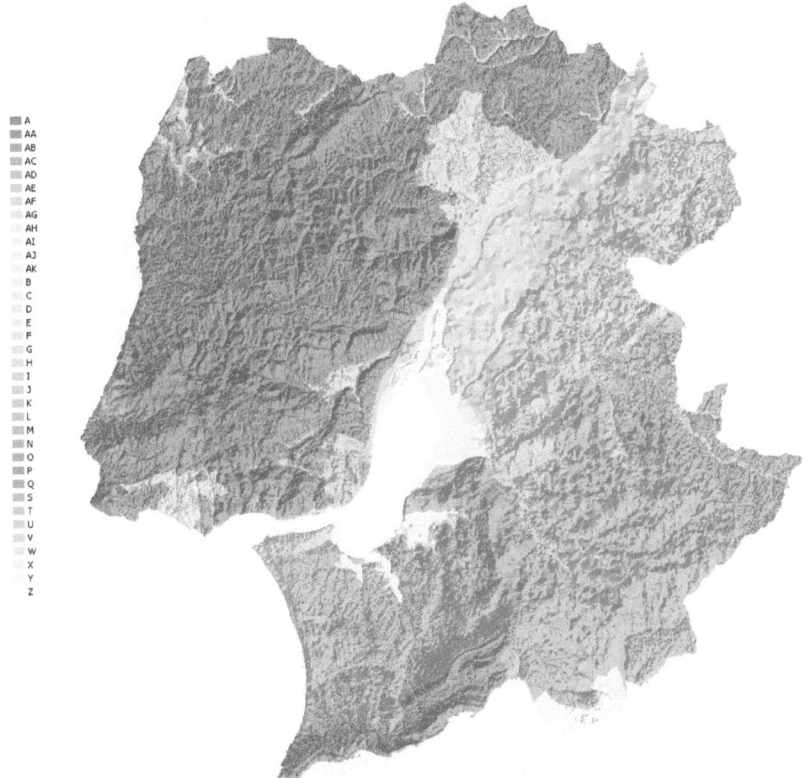

Fig. 2.17. Soil profiles and topography for the "AML model" (adopted from Campos-Costa et al., 2002; Reproduced with permission from Elsevier)

An example for illustration of the results obtained at the soil surface by means of the *simulator* (called the "AML model") is presented in Figure 2.19 (Rocha et al., 2003). It corresponds to a seismic scenario of a strong historical event which took place in 1531 (magnitude 7.2 Richter). The seismic motion was transformed into MMI, for easier comparison with the isosseismals of the historical event, Figure 2.20 (Justo and Salwa, 1998). Comparing the simulated with the historical, it is clear that the simulation captures several patterns of the historical, even though the remarquable NNE-SSW predominant propagation observed in the historical may suggest an important fault rupture mechanism, not possible to simulate with a point source. On the other hand, the softer soils to the East support the idea of higher intensities in these areas.

Carlos Sousa Oliveira

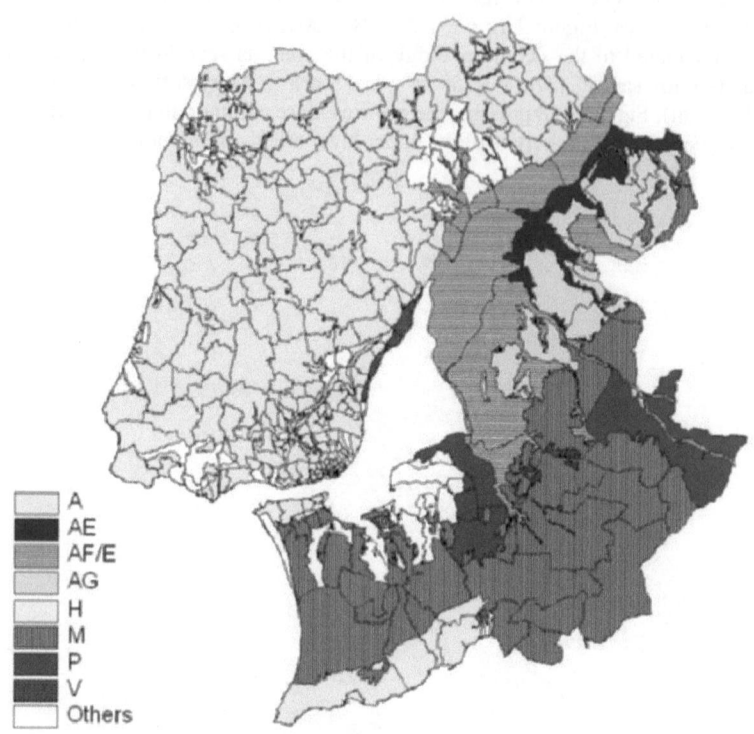

Fig. 2.18. Detailed soil profiles for the AML (adopted from Campos-Costa et al., 2002; Reproduced with permission from Elsevier)

Table 2.2. Characteristics of soil profiles presented in Figure 2.17 (approximate values of upper layers - "AML model")

Soil profile type(AML)	A	AE	AF/E	AG	H	M	P	V
V_S (km/s)	rock	190	220	200	300	270	250	150
Depth, h (m)	-	60	26	19	19	17	37	40
EC-8 classif. (Table 2.1)	A	D	C	E	D/E	E	C	D

This simulation also permits to visualize the locations more prone to soil amplification, by computing the ratio of top to bottom PGA and PGV at each location, Figure 2.21. In this figure only the *amplification/attenuation* of PGA is presented, indicated the zones with more influence on low rise buildings, which show high frequency content. The pattern of Figure 2.21 follows closely the soil distribution of Figure 2.18. *Amplification/attenuation* of PGV (not presented) will affect less rigid buildings.

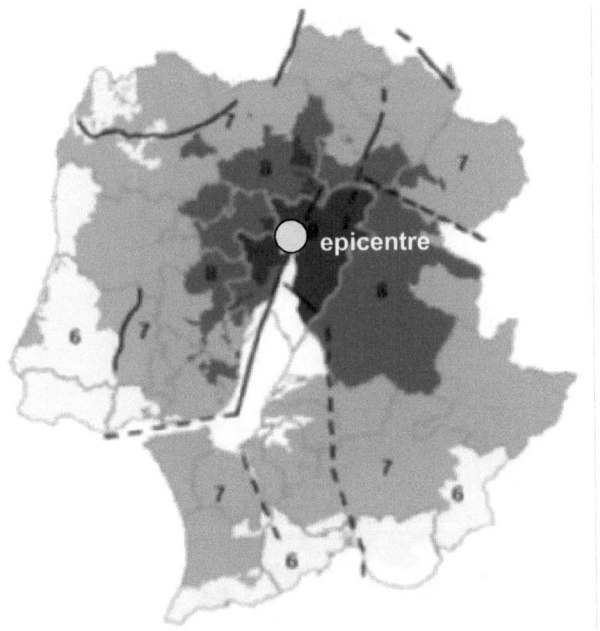

Fig. 2.19. 1531 earthquake scenario – IMM ("AML model")

Fig. 2.20. Isosseismals of the 1531 earthquake – IMM (Justo and Salwa, 1998; Courtesy of the Seismological Society of America)

Behind all the above referred topics and results there are several processes not discussed here, which deal with earthquake source, spectral attenuation, non-linear soil modelling, etc., and can be found in Campos-Costa et al. (2002), Carvalho et al. (2002) and M.L. Sousa (work in progress, 2003).

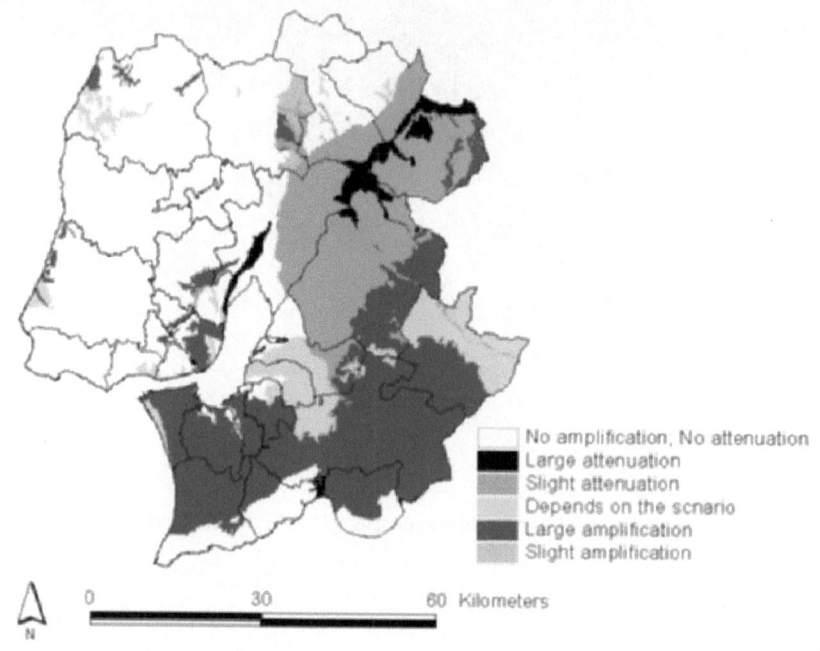

Fig. 2.21. 1531 earthquake scenario – predominant amplifications (PGA's) "AML model"

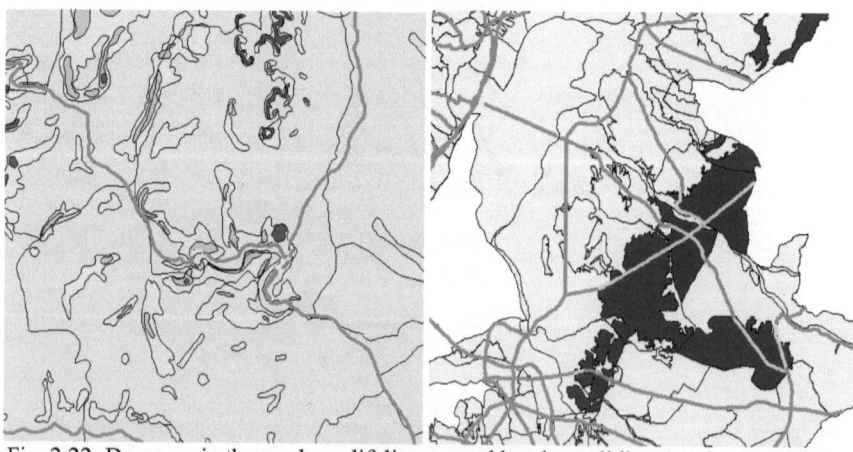

Fig. 2.22. Damages in the roadway lifeline caused by slope sliding (on the left side) and due to liquefaction (on the right side) (light gray: no problem; darker: higher potential for sliding/liquefaction)

Land-sliding and liquefaction were analysed under different perspectives. The first one was based on the existence of past scars detected from aerial photography, together with simple Hazus 99 techniques based on magnitude and distance. The second one was based on the soil profiles defined earlier, for which the potential for liquefaction derives directly. Water content in the soils determined by the epoch of the year is also considered in the analysis. Figure 2.22 shows, in a zoom, a detail of damage inflicted to a roadway due to land-side and liquefaction.

The classification of elements at risk was made according to their seismic vulnerability, with building classes based on age, and structural type, as in Example 1, but also on the number of storeys (Carvalho et al., 2002). The definition of fragility curves to adopt in each class was based on Hazus 99 methodology, permitting the computation of probabilities associated to each damage state, for a given seismic scenario. A performance-based assessment for each building type was adapted from Hazus 99 proposals to define individual capacity curves. Calibration of these curves based on more realistic models and on empirical observations is a matter of urgency.

Figure 2.23 illustrates the geographic distribution of total collapsed buildings for the 1531 scenario, and Figure 2.24 shows a comparison in numbers between the distribution of damage among the different limit states for the entire AML and for the Lisbon County. From Figure 2.23 one can observe the concentration of damage around the epicentral region due to the proximity to the release of energy, and in the southern ring, consequence of the soil profile type. (The characteristics of the building stock in these two large areas are essentially of the same type).

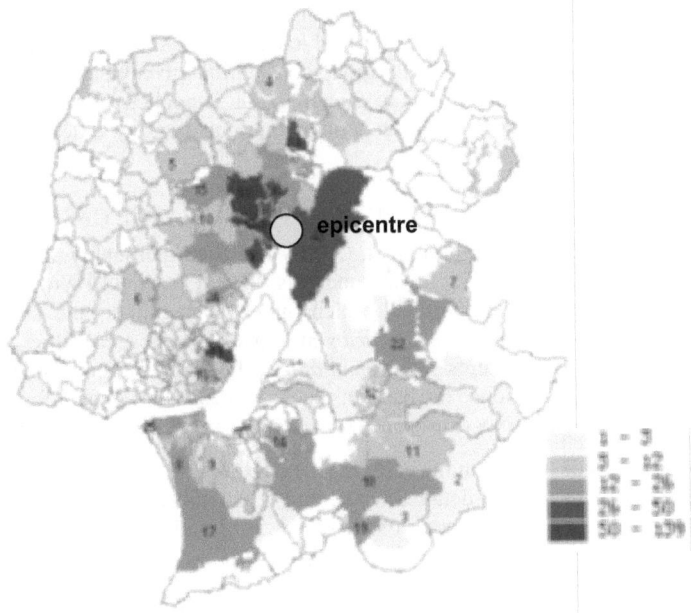

Fig. 2.23. 1531 earthquake scenario – total number of collapsed buildings, "AML model"

This *simulator* is now under probation in many of its functions in order to reproduce not only historical earthquake experience, but also to be adjusted to data from recent small magnitude events. In special for those events for which there exists strong motion monitoring at free-field and in structural elements as well as direct field information. Running several different seismic scenarios will help creating a body of knowledge capable of providing the necessary confidence for pursuing with the setting of *Emergency Planning*.

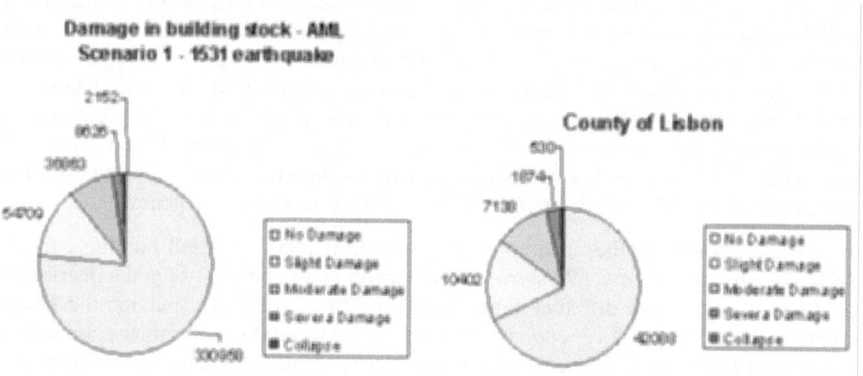

Fig. 2.24. 1531 Earthquake scenario – Comparison between AML and Lisbon County

The *simulator* has also helped defining the scenarios to be used in conjunction with the pair (M-magnitude; E-epicentral location) which produces the most probable global Losses in the entire or part of the region, considering given mean return periods (Campos-Costa et al., 2002). To do so, deaggregation is applied having as object that particular global characteristic. This scenario is one out of many others dealing with what may happen in the region.

2.3.3. EXAMPLE 3. STUDIES AT THE COUNTY LEVEL: THE CASE OF LISBON

The topic of impact studies in the Lisbon County has been thoroughly analysed in the past (Mendes-Victor et al., 1993, Oliveira and Pais, 1993, Pais et al., 1996). It gave rise to the first GIS *simulator* in Portugal which, given a magnitude and an epicentral distance, develops a set of *damage scenarios* in terms of victims, casualties, destroyed facilities and any other structures. This *simulator* (called the "Lisbon Council model") has been widely used to produce the basis of the current *Emergency Plan* for the Lisbon County.

Even though it was developed in the early nineties, Example 3 is referred in this chapter because it deals with a more detailed scale, and still retains a lot of good and solid scientific background. New updating of the model, now under study, will be discussed at a later stage.

The model, as it is nowadays, considers: (i) an attenuation of MMI with one single parameter (affecting the hypocentral distance) which can be changed according to the earthquake source area; (ii) soil characterization considering several classes reflecting the "impedance contrast" of the upper layers.

Soil was analysed in great detail on the base of the information compiled from hundreds of boreholes across the county. Figure 2.25a presents the geotechnical map of Lisbon showing a great detail on the geological units. The influence of soil on ground motion transmission from bed-rock to the surface was made taking into account the amplification of energy in a 2-layer system defined by impedance $[(V_{S1} \times \rho_1)/(V_{S2} \times \rho_2)]^2$, ($V_{Si}$ and ρ_i are the S-wave velocity and density of layer i, respectively) as in Medvedev (1962). The transformation into MMI was then made through a logarithm operation. For the particular case of Lisbon County, 9 different situations with combination of the 2 layers were considered. A comparison with the classification made in Example 2 is presented in Figure 2.25b. The main differences are: (i) in the tracing of zones, the first linked to the geology and the second more related to the administrative units; and (ii) the thickness of the upper layers, which the first does not consider and should be introduced in updated versions of the model.

Figure 2.26a shows the distribution of MMI computed with the above mentioned method ("impedance contrast") for the earthquake similar to the 1531 already referred in Example 2 (Figure 2.19). A zoom of Figure 2.19 for the Lisbon County is presented in Figure 2.26b. It can be observed that the two methodologies lead to important differences of MMI within the region, even though the overall values are similar. This shows the significance of the scale as well as the method of analysis.

Buildings were classified in 5 categories, **A** to **F** (**A** being the oldest masonry construction prior to the strong Lisbon earthquake of 1755, **B** the post-1755 to 1870 corresponding to the reconstruction, **C** the 1870-1930 with poor masonry and larger number of storeys, **D** the 1930-1960, with introduction of reinforced concrete, **E** the 1960-1980, with the first codes applying lateral forces, and **F** the recent reinforced concrete structures built in the last decade according to the updated seismic code of actions, RSA, 1983), aggregated into a geographical area corresponding to the parish.

The Lisbon County is divided into 52 parishes, with a resident population of around 0.6 Million. Data on building and population were obtained from the Census 91 at the parish level, together with other partial inventories to older buildings allowing corrections of local indexes. It should be referred that when working at towns with a particular important old stock of buildings, the existing published Building Census does not cover well this portion of data because all buildings constructed before 1919 are all in the same class. The building classification used in this Example reflects this preoccupation and is much more detailed than the Census information.

Vulnerability and fragility functions used to compute damage inflicted were taken from Coburn and Spence (1992), based on limit states D3 for severe damage and D5 for collapse. The population present in 5 different periods of the day was obtained from a study on its mobility. The *simulator* computes the percentage of damage per typology in each parish, the number of buildings in class D3 and D5, and the costs of repair based on average costs for reconstruction per m^2, number of storeys and area in plant. It also estimates the damage to population (deaths, injuries, homeless). Table 2.3 presents the damage estimation to buildings (average with large dispersion) for four typical earthquake scenarios affecting Lisbon (Oliveira et al., 2000).

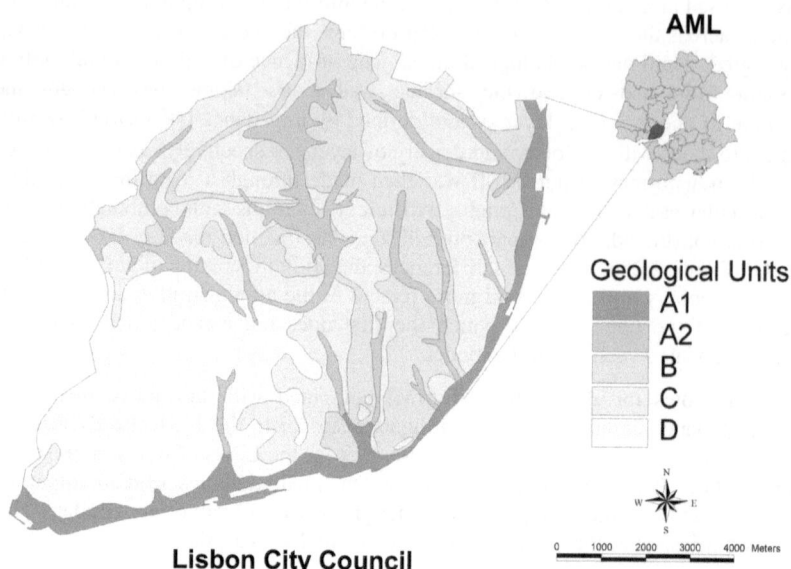

Lisbon City Council

Fig. 2.25a. Detail surface geological units in Lisbon County: A1 – Alluvium, Reclaimed Land, mainly mud and sand ($90<V_s<150$ m/s); A2 – Alluvium, Reclaimed Land, mainly clay and sand ($150<V_s<200$ m/s); B – Non-cohesive sandy soils and weakly cemented sandstones ($400<V_s<600$ m/s); C – Dense non-cohesive soils stiff clays and weak rock ($1000<V_s<1500$ m/s); D – Cretaceous limestones and marnly limestones and volcanic rocks ($V_s>1500$ m/s) - "Lisbon Council model"

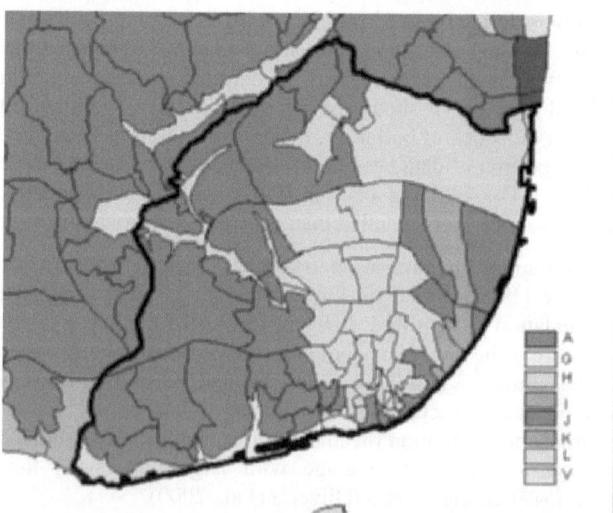

Fig.2.25b. Detail surface geological units in Lisbon County taken from a zoom of Figure 2.18 of Example 2, "AML model"

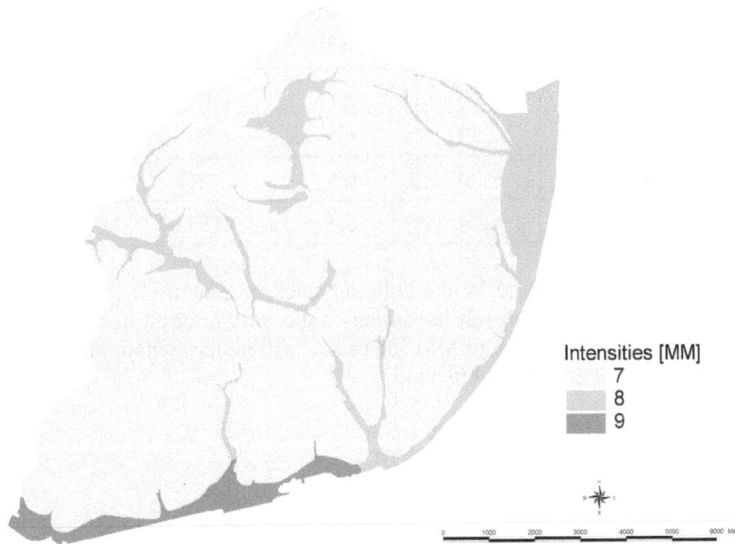

Fig. 2.26a. 1531 earthquake scenario – MM Intensities: Detailed soil description-simplified soil analysis ("impedance contrast"), "Lisbon Council model"

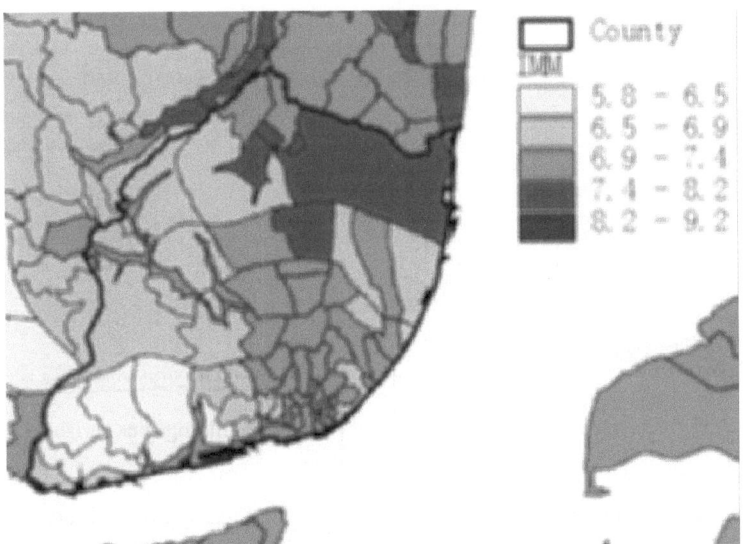

Fig. 2.26b. 1531 earthquake scenario – MM Intensities: Zoom of Figure 2.19 for Lisbon Council; simplified soil description-detailed soil analysis (non-linear 1-D column), "AML model"

Table 2.3. Total number of damage occurrences in the area of the Lisbon County

Scenario	Epicentral distance (km)	Magnitude (local)	Buildings with D3 (Coburn and Spence)	Buildings with D5 (Coburn and Spence)
Gorringe – 1969+	150	7.5	1673	61
Tagus Valley (1531)	30	6.5	346	11
Setúbal	20	6.5	808	27
Gorringe – 1755	150	8.5	10214	1236

Damage of type D5 suffered by the building stock in Lisbon for the 1531 scenario is showed in Figure 2.27, with results similar to the ones referred in Example 2 (Figure 2.23). As referred in relation to MM Intensities, differences within the county are also important here, due to scale and method.

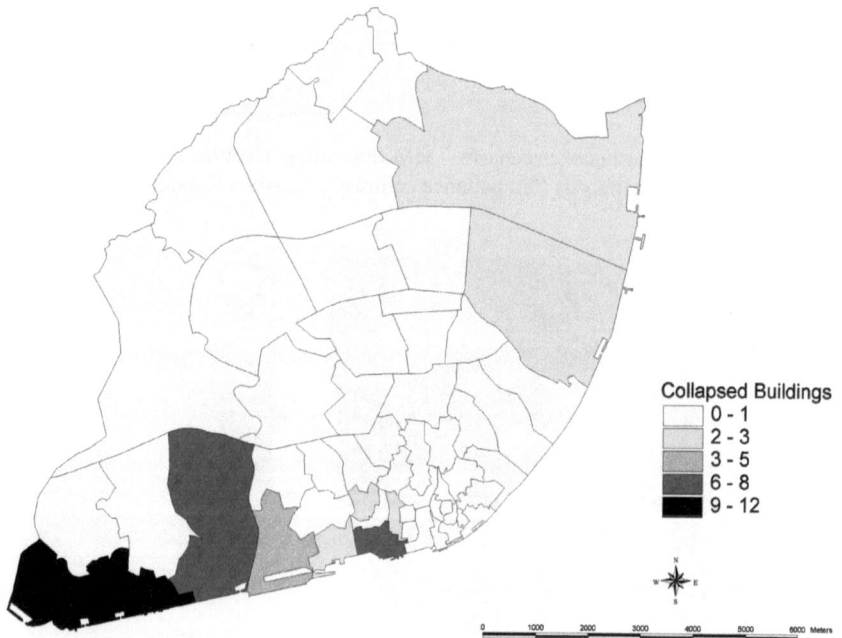

Fig. 2.27. 1531 earthquake scenario – Total collapsed buildings per parish, "Lisbon Council model"

Figures 2.28 presents the MM Intensities for the 1755 earthquake, observed (Figure 2.28a) and computed (Figure 2.28b) with the "impedance-contrast" method. The comparison between the two Figures shows important differences, but the general pattern can be attributed to the soil influence.

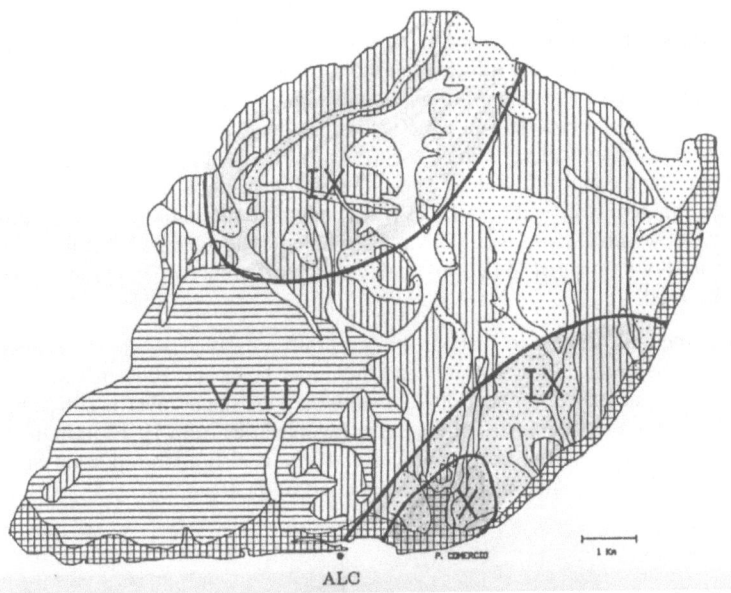

Fig. 2.28a. Observed MM Intensities for the 1755 earthquake (from Pereira de Sousa, 1932)

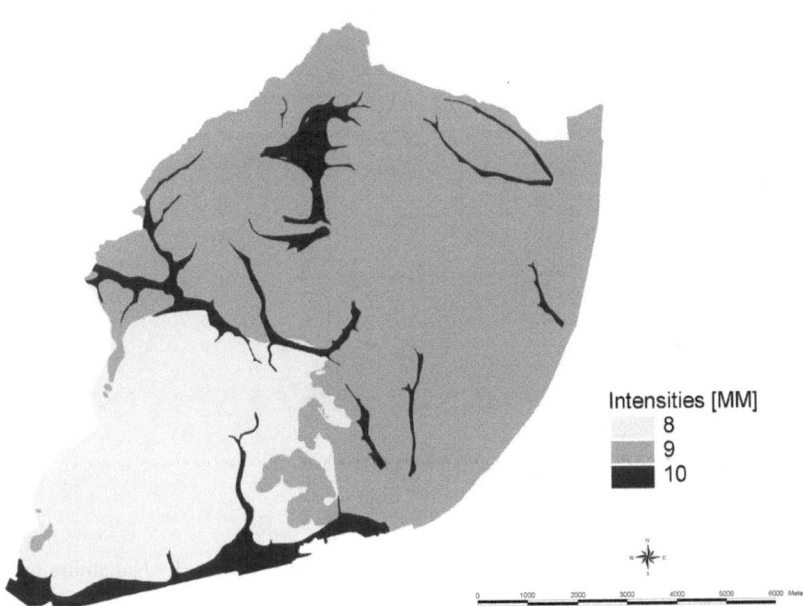

Fig.2.28b. Computed MM Intensities for the 1755 earthquake (Impedance-contrast), "Lisbon Council model"

Let's now compare the two large earthquakes of 1531 and 1755, as far as soil influence can be considered. For 1531, the "impedance-contrast" model gives MMI variations VII to IX, while the historical information indicates a general figure of IX, Figure 2.20, dropping rapidly to VI westbound. In relation to 1755, Figure 2.28a presents the areas in downtown Lisbon (Pereira de Sousa, 1932) with higher intensities attaining degree X. The observed geographical variations, according to Pereira de Sousa (1932), are similar to the "impedance contrast" model. The 1-D non-linear model for the 1531 is slightly more discriminative, with values VI to VIII. If now one refers to the recent earthquake of Jan 24, 1983, with a $M_L=5.8$ at 380 km from Lisbon, the average MMI in Lisbon was III, with variations from I-II to IV. So, in all cases, a difference of 3 degrees in the MMI scale is observed.

Looking to the soil profiles in Lisbon, it is generally observed a 2-layer situation with velocities on the order 250 to 300 m/s and the upper layer and 1000 m/s in the deeper layer. There is one exception near downtown, where the upper layer is softer, reaching values of 150 m/s. The thickness of the upper layers varies from 10-12 m in the thinner cases to 50 m in the referred downtown case. There is also a large portion of the city directly founded in "hard material", to the west part of the town.

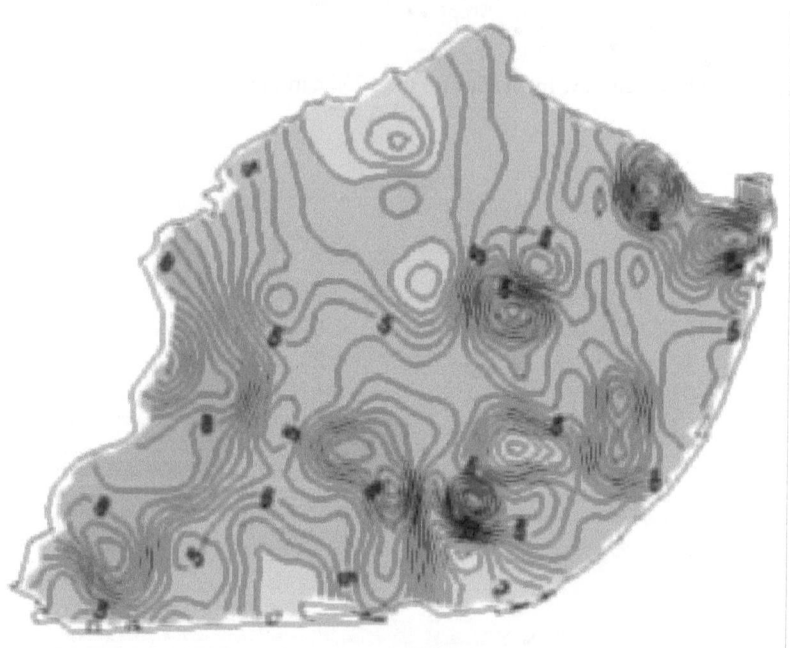

Fig. 2.29. Predominant frequencies (Hz) in the Lisbon County by Nakamura method
(Teves-Costa et al., 1995)

Frequencies of the first mode, apart the "hard" region, are essentially on the order of 4 to 5 Hz in the eastern part. On the soils bordering the river and along the main small river outlets downtown, frequencies drop to around 3 Hz to the west of centre downtown and to 1 Hz in the centre where soft material go to depths of 50 m. This analysis is generally confirmed by field measurements using the Nakamura technique (Teves-Costa et al., 1995), Figure 2.29. In fact, discounting the high values of 8 Hz in three points to the northeast part, the general pattern of frequencies agree with the above description, with the exception of the top centre north where Nakamura shows a lower frequency. But on the other hand this area coincides with the zone which showed an increase of intensities to a value of IX for the 1755 event.

Lifelines were studied in detail (Pais et al., 1999) in the context of this project. Ground motion was transformed into PGV in order to apply the methodology of Isoyama et al., 1998. Stratigraphy and morphology, to take into account soil and the presence of valleys, hills, etc., were used to adapt ground motion for lifeline vulnerability analyses (For more details, see Pais et al., 1999).

2.3.4. EXAMPLE 4. STUDIES AT THE BUILDING BLOCK LEVEL

The information available now at the Lisbon City Council (Câmara Municipal de Lisboa) contains data from the Census 91 at the level of the *statistic sub-section* (block of buildings), organized in a GIS, with digital cartography at 1:1 000 basis. Table 2.4 shows the main numbers related to buildings by statistical sub-sections. Census 91 contains information on the epoch of construction according to 6 typologies, on the height according to 5 classes, and on the resident population.

Table 2.4. Data on statistical sub-sections, Lisbon County (Census 91)

Variables	Lisbon County
Number of Buildings	61575
Number of statistical sub-sections	3686
Average area of statistical sub-sections	22918 m^2
Total area	84,5 km^2
Average number of buildings per statistical sub-section	16,7 ± 19,8

Using the methodology summarized earlier, the computation of damages to the building stock from any single earthquake scenario is easily obtained. Figure 2.30 presents the distribution of *severe damage* (D3) per sub-section (block) for one earthquake, 270 km southwest of Lisbon, M=8.5, showing the detail of the obtained results. The system will allow the estimation of human casualties and injuries, but this topic, exhibiting a great deal of uncertainty, is not referred here. It also can be used to estimate other consequential problems related to emergency, such as the volume of debris from collapsed facades and buildings, or the determination of possible location of obstructions.

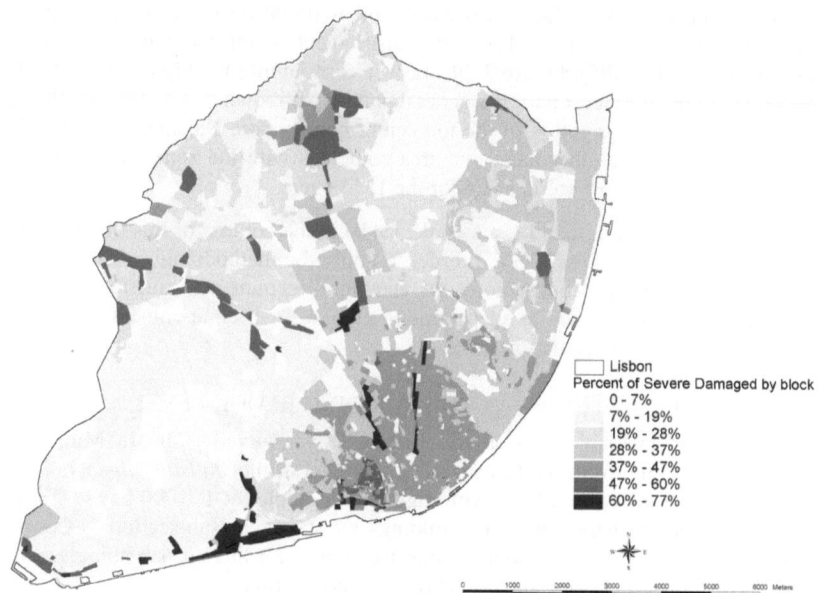

Fig. 2.30. Severe damage (D3 – in % of buildings block by block) inflicted by an event
similar to 1755 in Lisbon County, "Lisbon Council model"

It is interesting to notice that, in a recent work, Giovinazzi and Lagomarsino (2003),
developed a method for obtaining vulnerability functions based on the concepts
supporting the EMS-98 scale (Grunthal, 1998), by using fuzzy theory to interpret
incomplete information in the scale. An application of this method to the Lisbon
County led to approximately the same results as with the 1993 vulnerability curves,
slightly aggravating the older construction (in 10 to 20%, depending on the MMI) and
disaggravating the new construction (on approximately the same amount). This
indicates that both procedures lead to similar results.

In the following, several new refinements to be introduced in the model, at the level of
characterization of the building stock, are briefly referred.

- Using a digital model of the terrain and the level of the roof of each building, one can
  estimate the height of the buildings and the corresponding number of storeys.
- The computation of individual areas and lateral discontinuities between adjacent
  buildings is of great importance for a more detailed algorithm.
- The information at individual buildings has to be cross-correlated with information
  already available at the sub-sections in order to validate the models.

An average vulnerability index can be assigned to a block of buildings, which would
make it possible to propose the definition of a *typical block*. This concept would be of
most importance for rapid assessment of damage after the occurrence of an event.
Monitoring the performance of various *typical blocks* could accelerate the process of
estimation of damage, by feeding back this information into the *simulator*, correcting
the overall picture of damage.

2.4. Final Considerations and Future Developments

Further studies in progress are improving the concept of spectra at the source, attenuation to the firm stratum beneath each site and propagating the waves to the surface using more sophisticated models especially when the scale of interest is of great detail. Calibration with simpler models as described in Example 3 should be done. The selected pair *earthquake source – magnitude* is viewed as an event with a certain probability of occurrence which is given to the operator (Sousa et al., 1997). Other developments in progress refer to the calibration of *capacity curves* for the different typologies. These curves, function of the initial frequency of vibration (given by the height of the building) and of the non-linear behaviour of the structural system, can be obtained in a first step by performing "pushover analysis".

Also the collection of recent earthquakes that struck several areas around the world has produced a large amount of data on vulnerability of different construction types as well as of human casualties. This information will for sure increase quite significantly the knowledge on damage, and will help to calibrate the entire damage estimation process. The relation with the macroseismic scale should be pursued.

All these refinements are part of this new approach to the problem using the new tools in the area of GIS and the new findings from recent earthquakes and research. Working at several scales enriches the knowledge and should be well articulated at all levels. This coordination of efforts is the only way to maximize the resources in case of emergency.

A final word to call the attention to the importance of understanding the level of quality control practiced in a region. This is an essential matter because, if quality control is not considered in the analysis, the damage estimates may be completely erroneous. The example of the Turkey earthquake of 17 August, 1999 is extremely clear, as in many zones damage inflicted to the built stock was essentially attributed to poor quality control in code enforcement. This is why a correction factor in vulnerability curves has to be included in countries or regions where there is suspicion that codes are not fully enforced.

Acknowledgement

This overview chapter corresponds to studies developed by the author for more than one decade in the area of seismic scenarios in collaboration with two institutions, the Serviço Nacional de Protecção Civil (SNPC) and Serviço Municipal de Protecção Civil da Câmara Municipal de Lisboa. Many people have participated in the developments summarized here, to whom I want to express my sincere acknowledgement:

Dr. A. Campos-Costa, M. L. Sousa and Anabela Martins from LNEC, Lisbon; Isabel Pais from CNPCE, Lisbon; Fernanda Rocha, Sandra Serrano and Maria Anderson from SNPC, Lisbon; F. Mota de Sá from "Fuzzy, Ltd", Lisbon; Prof. Jorge Proença, from IST, Lisbon; Prof. Paula Teves-Costa from FCUL, Lisbon; "Chiron, Ldt", Lisbon, collaborated in software development; and Gonçalo Caiado, Gonçalo Pais, Mónica Ferreira, Mónica Oliveira and Paula Pestana, former students at IST, brought great enthusiasm to these matters. This work was partially supported by Fundação para a Ciência e a Tecnologia, Lisbon, "Programa Pluri-Anual". Prof. Isabel Viseu helped revising the final text.

CHAPTER 3
STRONG GROUND MOTION

Mustafa Erdik and Eser Durukal
Boğaziçi University, Kandilli Observatory and Earthquake Research Institute
Department of Earthquake Engineering, Istanbul, Turkey

3.1. Introduction

From engineering point of view strong ground motion study is concerned with the understanding of the characteristics and effects of potentially damaging earthquake ground motions. For earth sciences strong ground motion investigations provide information on the source failure process and the near field wave propagation. Today there exist about 20,000 strong motion instruments operating worldwide, 10% of which are located in Europe.

With the introduction of performance based earthquake resistant design for buildings and other civil engineering structures the capability of simulating realistic ground motions has been indicated. Especially with the recent developments in software tools and structural modelling techniques for time domain transient non-linear dynamic analysis, the use of simulated time histories of ground motion gained utmost importance. Although the use of recorded ground motion under similar conditions with the design earthquake is appealing, there may never be an adequate suite of such data in terms of tectonic structure, earthquake size, local geology and near-fault conditions.

This paper will first review the elements of the earthquake source physics important to the characteristics and modelling of the strong ground motion. The time and frequency domain characteristics and the attenuation of the strong ground motion will be covered. An approach for the simulation of the strong ground motion is elaborated with an example.

3.2. Attenuation

Attenuation relationships are empirical descriptions providing the median and standard deviation of various intensity measures of the strong ground motion, assumed to be log-normally distributed, in terms of earthquake size, distance, source mechanism and site conditions.

The moment magnitude is currently the preferred scale for the size of the earthquake. For the distance parameter distance to fault has gained importance for correlation with ground motion characteristics. The deviation of the observed from the predicted strong ground motion (residual) generally fits to a lognormal distribution for up to two standard deviations. The standard deviation of the predictions is in the order of 0.5 natural logarithm units, corresponding to a multiplicative factor of 1.6 (times the mean value) to obtain the value, which exceeds 84 % of the data. Large degree of uncertainty because of other source (near-fault rupture directivity), propagation path (crustal wave guide), basin response and site effects are not treated as parameters.

A. Ansal (ed.), Recent Advances in Earthquake Geotechnical Engineering and Microzonation, 67–100.

Shallow earthquakes in active tectonic regions have provided the largest amount of ground motion data and hence the largest number of ground motion attenuation relationships. Most of the strong motion data used in attenuation relationships are obtained from reverse and strike-slip earthquakes. However, due to regional differences in some of the factors affecting earthquake ground motions, different ground motion attenuation relationships have been developed for different regions, such as Europe, Western United States (shallow crustal earthquakes), Eastern United States, and subduction zones.

Most attenuation relationships are characterized by:

"Distance saturation" where the function slope decreases at close distances, reflecting the fact that the earthquake is a distributed source, "magnitude saturation" where PGA increase more gradually with magnitude for large magnitudes, reflecting the fact that magnitude is not well correlated with PGA. Although the current attenuation relationships use moment magnitude, various magnitude definitions used in earlier attenuation relationships should be carefully studied (Figure 3.1). The distance definitions used in the attenuation relationships also differ especially on near fault conditions (Figure 3.2). It should also be noted that, ground motions in the near-source region of earthquakes have certain characteristics not found in ground motions at more distant sites, especially directivity, as evidenced by a high-energy intermediate-to-long-period pulse that occurs when fault rupture propagates toward a site. Directivity effect is currently only indirectly incorporated in the attenuation relationships.

The general for of the attenuation relationships used by the researchers has been of the following form:

$$Y = b_1 \ f_1(m) \ f_2(r) \ f_3(M, r) \ f_4(P) \ E \qquad\qquad (3.1)$$

where: Y is the strong ground motion parameter to be predicted,

$f_1(m)$ is a function of the earthquake size M, usually given by the form
$f_1(m){=}\exp(b_2 \ m)$,
$f_2(r)$ is a function of the distance r, the most common form being
$f_2(r) = \exp(b_4 \ r) \ (r{+}b_5)^{b3}$,
where b_3 and b_4 represent respectively the geometric and anelastic attenuation rates,
$f_3(M,r)$ accounts for the possible variation of earthquake size measure with distance,
$f_4(P)$ is the function accounting for the propagation path and site parameters,
E is a random variable representing the uncertainty in Y.

Following is short description of the currently used attenuation relationships.

Boore et al. (1997) PGA and Spectral Acceleration attenuation relationship is based on the selected strong motion data from western North America. The equations predict the random horizontal component peak acceleration and 5% damped pseudo acceleration response spectra in terms of moment magnitude, distance and site conditions for strike-slip, reverse slip or unspecified faulting mechanism. Site conditions are represented by the shear wave velocity averaged over 30 m. The smoothed coefficients in the equations for predicting ground motion were determined using a weighted, two-stage regression procedure. In the first stage, the distance and site condition dependence were determined along with a set of amplitude factors, one for each earthquake. In the second stage, the amplitude factors were regressed against magnitude to determine the magnitude dependence. The general form of the ground motion estimation equation used in the study is:

$$\ln(Y) = b_1 + b_2 (M-6) + b_3 (M-6)^2 + b_5 \ln r + b_V \ln (V_S / V_A) \qquad (3.2)$$

where:
$$r = (r_{jb}^2 + h^2)^{1/2} \qquad (3.3)$$

In this equation:

    Y = peak ground motion measure,

    M = moment magnitude M ≥5.0,

    r = closest distance from rupture to the station in km r ≥ 20 km,

    r_{jb} = closest horizontal distance from the station to a point in km,

    V_S = average shear-wave velocity (m/s) to a depth of 30 m,

    b_1 = parameter related to fault mechanism,

    b_{1SS}, b_{1RS}, b_{1ALL}, b_2, b_3, b_5, b_V, V_A and h are regression coefficients provided in tabular form.

Campell (1997) study develops empirical attenuation relationships for horizontal and vertical PGA, PGV, and SA using accelerograms generated by western USA and other worldwide earthquakes of moment magnitude greater than 5 and sites with distances to seismogenic rupture within 60 km.

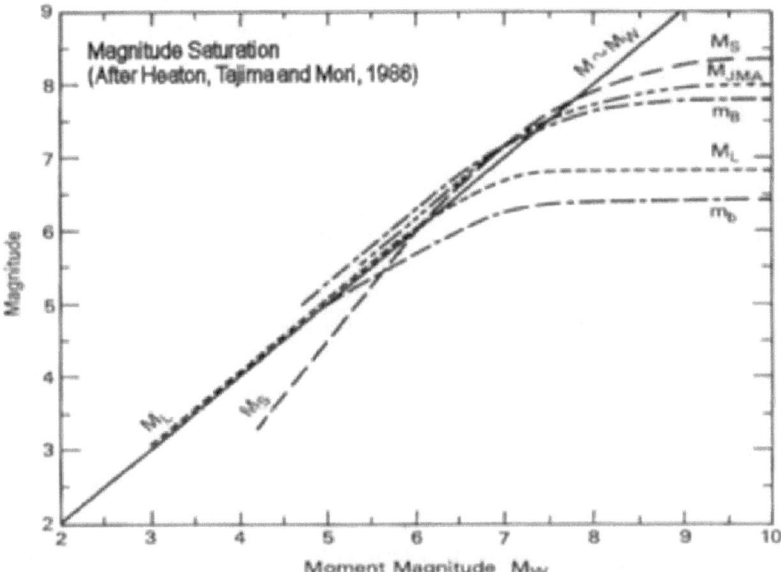

Fig. 3.1. Relationship of the moment magnitude M_w with other well established magnitude scales (Courtesy of the American Geophysical Union)

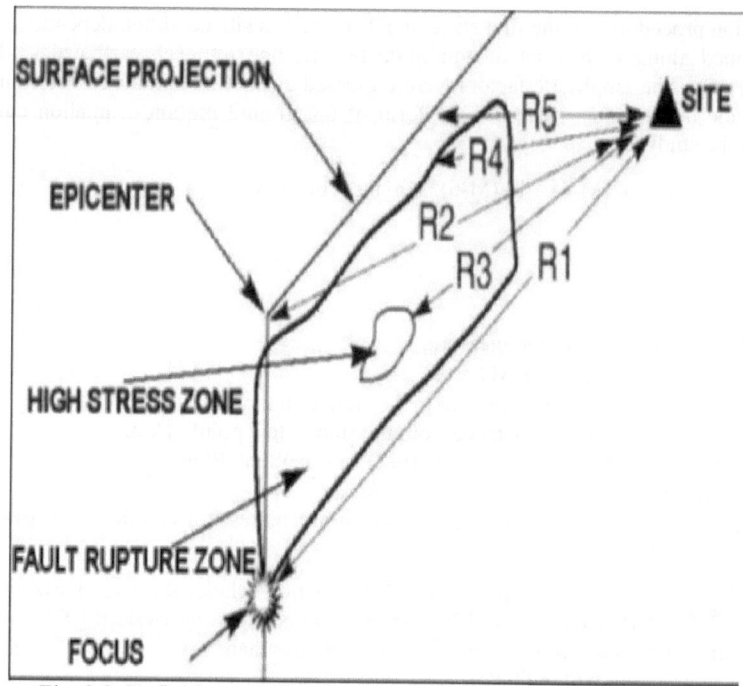

Fig. 3.2. Definitions of distance used in the attenuation relationships

For the estimation of PGA values Campbell (1997) uses the following expression:

$$\ln(A_H) = -3.512 + 0.904M - 1.328 \ln[R_{SEIS}^2 + (0\ 149e^{0.67M})^2]^{1/2}$$

$$+ [1.125 - 0.112\ln(R_{SEIS}) - 0.0957M]F$$

$$+ [0.440 - 0.171 \ln(R_{SEIS})] S_{SR} + [0.405 - 0.222 \ln(R_{SEIS})] S_{HR} + \varepsilon \qquad (3.4)$$

where:

A_H = median of the geometric mean of the two horizontal PGA (g)

M = moment magnitude,

R_{SEIS} = the closest distance to seismogenic rupture on the fault (km),

F = 0 for strike-slip and normal faulting earthquakes and 1 for reverse, reverse-oblique, and thrust faulting earthquakes,

S_{SR} = 1 for soft-rock sites,

S_{HR} = 1 for hard-rock sites,

$S_{SR} = S_{HR} = 0$ for alluvial sites,

ε = random error term with mean of zero and a standard deviation equal to the standard error of estimate of $\ln(A_H)$.

Sadigh et al. (1997) present attenuation relationships for shallow crustal earthquakes based on strong motion data primarily from California earthquakes. Relationships are presented for the geometric mean of the two horizontal components, strike-slip and reverse-faulting earthquakes, rock and deep firm soil deposits, earthquakes of moment magnitude M between 4 and 8+ and distances up to 100 km. The site conditions representative of rock attenuation models given here should be accepted as soft rock. The deep soil data are from sites with greater than 20 m of soil over bedrock. Attenuation relationships of horizontal Response Spectral Acceleration (5% damping) are given in two separate equations according to the soil condition. Relationship for reverse/thrust faulting are obtained by multiplying the given strike-slip amplitudes by 1.2. The general form of the equation for rock sites is as follows:

$$\ln(y)=C_1+C_2M+C_3(8,5-M)^{2.5}+C_4\ln[r_{rup}+\exp(C_5+C_6M)]+C_7\ln(r_{rup}+2) \qquad (3.5)$$

y = PGA or SA (in g) represented by the geometric mean of the two horizontal components,

C_1 to C_7 =amplitudes given in tabular form

M = moment magnitude,

r_{rup} =Minimum distance to the fault rupture surface (km).

Ambraseys et al. (1996) attenuation relationship is based on 422 strong motion records from 157 earthquakes in Europe and adjacent areas. The equations use the larger horizontal acceleration response ordinate fort 5 per cent damping and give ground motion in terms of surface wave magnitude, distance and site conditions. Site conditions are represented by soil classes as rock, stiff soil and soft soil. The ground motion estimation equation used is of the form:

$$\log(Y) = C'_1 + C_2 M + C_4 \log (r) + C_A S_A + C_S S_S \qquad (3.6)$$

where:

$$r = (d^2 + h_0^2)^{1/2}$$

In this equation;

Y = peak horizontal accelerations in g,

M= surface wave magnitude $4 \le M \le 7.5$,

d = shortest distance to the surface projection of the fault in km,

h_0 = a constant determined with C_1, C_2, C_3 and C_4,

S_A= 1 for stiff soils and 0 otherwise,

S_S= 1 for soft soils and 0 otherwise.

The period dependent coefficients C'_1, C_2, C_4, C_A, C_S and h_0 and the error term σ are provided in tabular form.

Figure 3.3 and Figure 3.4 provide comparison of above attenuation relationships with the strong motion data obtained from 1999 Kocaeli, Turkey Earthquake.

Spudich et al. (1997) collected ground motions from extensional regimes throughout the world and derived attenuation relationship for PGA and SA in extensional tectonic regimes using globally obtained data. In general, their values suggest that most other attenuation models will significantly overestimate ground motions from normal faulting earthquakes are smaller than for other tectonic regimes.

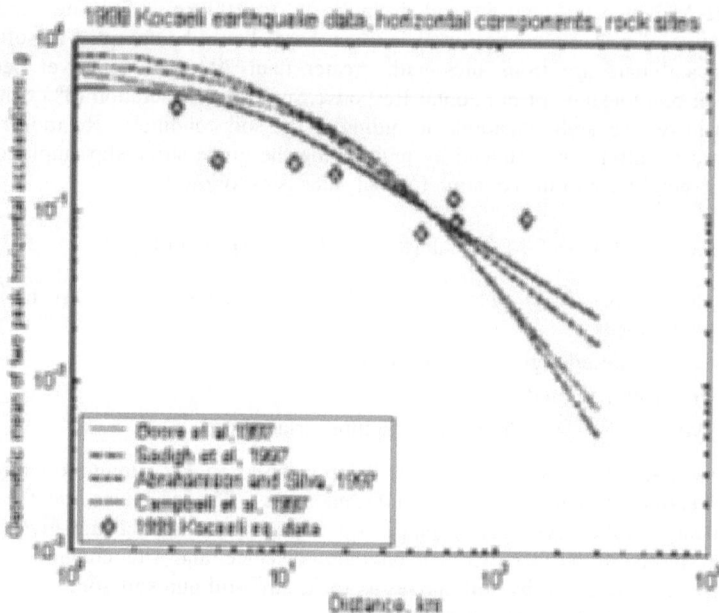

Fig. 3.3. Comparison of 1999 Kocaeli earthquake data with attenuation relationships, horizontal accelerations, rock sites (Durukal, 2002)

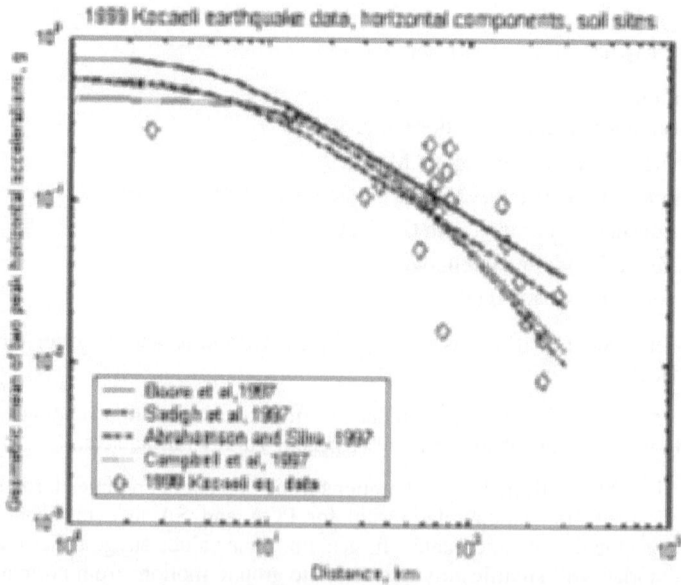

Fig. 3.4. Comparison of 1999 Kocaeli earthquake data with attenuation relationships, horizontal accelerations, soil sites (Durukal, 2002)

The predictions in these attenuation relationships at large magnitudes and short distances are based on rather limited data sets that do not incorporate the data from Kocaeli, Turkey earthquake (August 17, 1999, M_w=7.6) and ChiChi, Taiwan, earthquake (September 21, 1999, M_w=7.6). In both earthquakes, the peak ground accelerations on rock at near-fault distances were below the existing ground motion prediction equations. Among others, this discrepancy can be caused by low stress drop and smooth fault rupture with limited asperities. Such variations on the dynamics of fault rupture are currently treated as random uncertainties in the attenuation relationships.

3.3. Factors Affecting Earthquake Strong Ground Motions

Findings (i.e. Somerville, 2000) indicate that while the average ground motions from one large earthquake are similar to those of another, there are conditions that cause the ground motions to vary significantly from one location to another at the same distance from a given event. This variability is related to earthquake source process, propagation and site response.

It has been well recognized that earthquake ground motions are affected by earthquake source conditions, source- to-site transmission path properties, and site conditions. The source conditions include the stress drop, source depth, size of the rupture area, slip distribution, rise time, type of faulting, and rupture directivity. The transmission path properties include the crustal structure and the shear-wave velocity and damping characteristics of the crustal rock. The site conditions include the rock properties beneath the site to depths of up to about few kilometers, the local soil conditions, and the topography of the site.

3.3.1. EFFECTS OF THE EARTHQUAKE SOURCE

Recorded strong ground motion in the near field incorporates all the heterogeneities, complex arrivals and the high frequency content of the source process. Patches on the fault plane with higher slips are called asperities. Asperities are highly stressed regions surrounded by weak or zero stress zones. The fault plane is then composed of patches of high stress and low or zero stress, which leads no a non-uniform stress drop during an earthquake event. Barriers are identified as those portions of the fault plane that do not rupture. Simple source models assume that main fault rupture parameters (rupture velocity, rise time and stress parameters) are homogenous and coherent over the plane of the dislocation.

Seismic Moment, Stress Drop, Effective Stress and the Corner Frequency are the main parameters of the earthquake source that influences the strong ground motion characteristics. A short review of these parameters is provided below.

Seismic Moment, M_0, is the most recognized measure of the earthquake size given by the multiplication of the shear modulus (Lame's constant) of the medium, the average total dislocation (i.e. mean fault offset or slip) and the area of the dislocation surface (i.e. fault rupture surface). The seismic moment, M_0, is generally regarded as the best available single number to describe the size of an earthquake and can be estimated from the low frequency asymptote of the Fourier transform of the displacement seismogram. The moment magnitude (M_w) is derived from the seismic moment on the basis of the following equation (Kanamori, 1977)

$$M_w = (2/3) \log M_o -10.73 \qquad\qquad (3.7)$$

Figures 3.5, 3.6 and 3.7 show the accelerograms, corresponding response spectra and the Fourier amplitude spectra for six records provided by Anderson and Quaas (1988) for a wide range of magnitudes (3.1 to 8.1). All records have epicentre distances of about 25 km and are obtained on rock. As it can be seen in Figure 3.5 with increasing magnitude the amplitudes of ground motion generally increase, and the duration of the accelerogram rapidly increases. Figure 3.6 and 3.7 show that as the magnitude increases, the amplitudes of the low frequency waves increase dramatically, while the amplitudes of the high frequencies increase slowly. In other words, increasing magnitude results in greatly enriched relative frequency content (higher spectral shapes) at long periods with an approximately flat acceleration spectrum over a sizeable frequency band. This flat spectral shape has contributed to the development of models of strong motion as band-limited white noise.

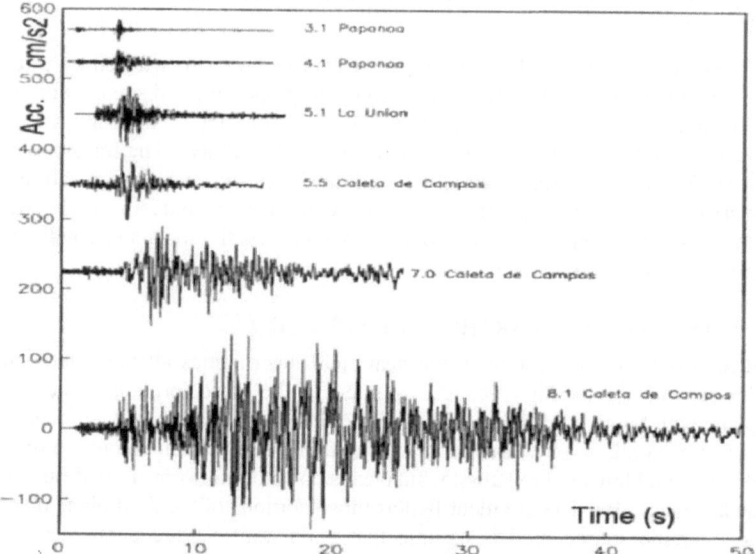

Fig. 3.5. Accelerograms of six records for a magnitude range of 3.1 to 8.1 (after Anderson and Quaas, 1988; Reproduced with permission from J.G. Anderson and the Earthquake Engineering Research Institute)

Stress Drop is the difference between the initial state of the shear stress (before the earthquake) and the final state of the shear stress (after the earthquake). Stress drop is about 3 MPa for interplate earthquakes and about 10 MPa for intraplate earthquakes of moderate and large magnitudes (Magnitude > 5).

Effective Stress is the difference between the initial static stress and frictional stress in existence during the rupture process.

Corner Frequency is the frequency where the high and low frequency trends of the Fourier Amplitude Spectrum. It is related to the inverse of the rise time (rate of growth in dislocation or roughly, time duration of rupture). By measuring the corner frequency from the Fourier Amplitude Spectrum the apparent duration of faulting at the source and hence the fault dimension can be estimated. A large magnitude earthquake will generally have a large fault dimension and hence a small corner frequency, implying a longer rupture and also strong motion duration.

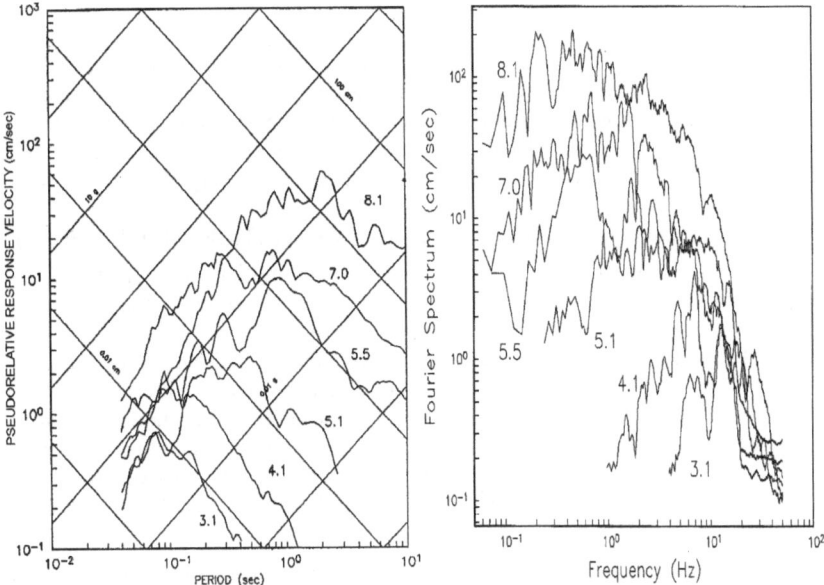

Fig. 3.6. PSRV of records shown in Figure 3.5 (Reproduced with permission from J.G. Anderson and the Earthquake Engineering Research Institute)

Fig. 3.7. Fourier spectra of records shown in Figure 3.5 (Reproduced with permission from J.G. Anderson and the Earthquake Engineering Research Institute)

3.3.2. SUBDUCTION ZONE AND SHALLOW CRUSTAL EARTHQUAKES

The collision of tectonic plates in subduction zones causes large and deep earthquakes. Ground motion data from subduction zone earthquakes are associated with slower rate of attenuation compared to those from shallow crustal earthquakes. Analysis of ground motion data also indicates that the response spectral shapes obtained from subduction zone earthquakes have smaller amplitudes in the long-period range than response spectral shapes from shallow crustal earthquakes.

3.3.3. EFFECTS OF DISTANCE

Attenuation is strongly influenced by distance for both the geometric spreading and the material damping. Excluding material damping and considering only geometric attenuation it can be observed that the cylindrical body waves attenuated with inverse of distance and spherical body waves attenuate with the inverse of the distance squared.

The material attenuation is generally given by the following expression.

$$\exp\left[-(\pi f/Qc)\,x\right] \tag{3.8}$$

Where f is the frequency, Q is the quality factor that accounts for the material damping, c is the shear wave propagation velocity and x is the distance to source. Assuming almost constant Q, it can be seen that the rate of attenuation increases exponentially with increasing frequency and distance.

The spectral shape of the strong ground motion on competent soil sites indicates reduction in the high-frequency regions and increase in the low frequency regions with increasing distance. However, within distances of about 50 km, the effect of distance on spectral shape is much smaller than the effect of magnitude. The duration of the accelerogram tends to increase with increasing distance (e.g., Dobry et al., 1978).

3.3.4. EFFECTS OF NEAR SURFACE WAVE PROPOGATION (SITE EFFECTS)

Variability that is introduced into the strong ground motions by effects of wave propagation in the source to site propagation media are comparable to the complexities introduced by the source dynamics. Site effects include modification of seismic waves by the local soil layers, the effect of alluvial basins and effect of local topography. It is well established that local soil conditions have a major effect on the amplitude and response spectral characteristics of earthquake ground motions depending on the type and depth of soil and on the level of ground motion. It was demonstrated by the dramatic differences in ground motions in Mexico City in the 1985 Mexico earthquake, in the San Francisco Bay Area in the 1989 Loma Prieta earthquake and in Adapazarı in the 1999 Kocaeli earthquake.

The soft soils that form low velocity layers near the Earth's surface trap energy, amplify all frequencies due to the decrease in seismic impedance, and preferentially amplify resonant frequencies. Several researchers have shown that for layers of given thickness, the relative shaking response will be greatest where the surface geologic units have the lowest impedance values and where the impedance contrast between the surface layer and the underlying one is the greatest.

For peak ground acceleration, this dependence of amplification on ground motion level is illustrated by the relationship for soft soil developed by Idriss (1991) shown in Figure 3.8. For peak rock accelerations less than about 0.4 g the ground motions are typically amplified in soft soils. However, for higher levels of ground motion, higher soil damping due to nonlinear soil behaviour tends to result in deamplification of peak ground accelerations (*or* high-frequency response spectral components). Nonlinearity in soil behaviour is generally recognizable in differences between site response (defined by spectral ratios) when peak accelerations exceed about 0.4 g, peak velocity exceeds 30 cm/s, or peak strain exceeds 0.1%. The effects of nonlinearity generally reduce the amplitudes by decreasing the effective shear stiffness of the sediments and increasing the hysteretic damping.

Amplification due to topography has been identified in theoretical as well as empirical studies. The top of isolated hills, elongated crests, edges of plateaus and cliffs are usually zones of amplification due to diffraction and focusing. The main results are that the topographic amplification is maximum at the top of the hill, and is maximum at the frequency at which one shear wavelength equals the width of the hill base. Motions on the hillsides are not amplified much, and motions around the base of the hill are usually deamplified with respect to motions far from the hill.

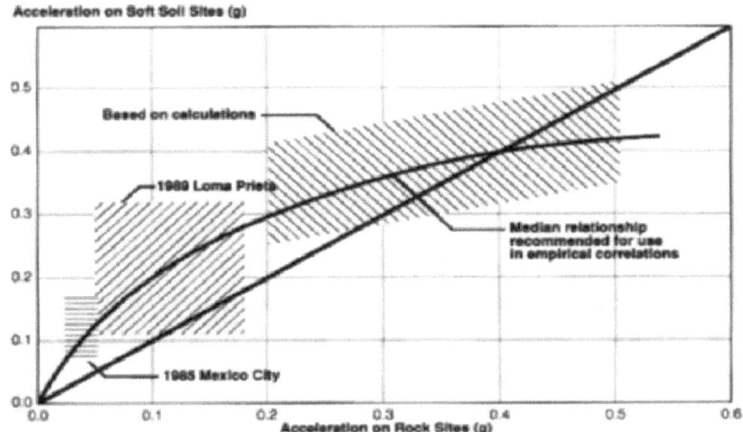

Fig. 3.8. Dependence of amplification-deamplification on peak ground acceleration for soft soil (After Idriss, 1991; Reproduced with permission from University of Missouri and Ed. S.Prakash)

3.3.5. BASIN RESPONSE EFFECTS

Surface waves generated by conversion of body waves at the boundaries of sedimentary basins dominate the ground motion amplitudes at long periods with much longer durations of strong shaking. In the Kobe earthquake, the zone with the highest damage is a linear band located at the zone of constructive interference of waves emanating directly from the fault through the sedimentary basin structure and the other one refracted into the basin. Other examples come from the Northridge earthquake, where there was an isolated zone of high damage in Santa Monica basin and from Kocaeli earthquake where a zone of heavy damage is located in the Adapazarı Basin (Beyen and Erdik, 2002).

3.4. Simple Earthquake Source Models

To model near field ground motion (Brune, 1970) considered a tangential stress pulse applied instantaneously to an interior of a dislocation surface (Figure 3.9). The fault propagation effects are neglected. This stress pulse generates a shear wave along the direction normal to the dislocation surface (i.e. fault surface). If x represents the perpendicular distance from the fault surface and H(t) is the Heaviside unit-step function and β is the shear-wave propagation velocity, the initial time function for shear stress pulse can be written as:

$$\sigma (x, t) = \sigma H(t - x/\beta) \qquad (3.9)$$

where σ is the effective shear stress (Figure 3.9).

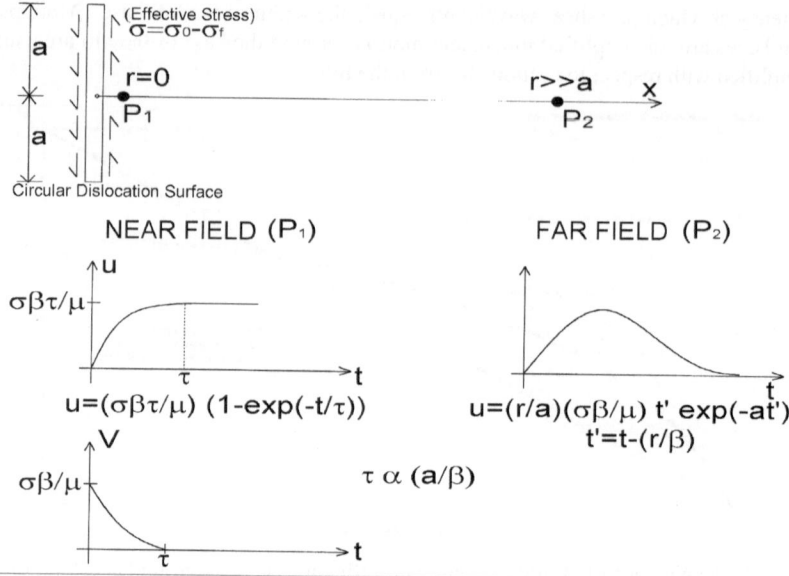

Fig. 3.9. Brune (1970) dislocation model for s-waves (Courtesy of the American Geophysical Union)

Ground displacement close to the fault centre in parallel direction to the fault surface, *u*, can be obtained through integration:

$$\sigma = \mu \, (\delta u / \delta x) \tag{3.10}$$
$$\text{at } x{=}0 \;\; u = (\sigma/\mu) \, \beta \, t \quad \text{for } 0 < t < T \tag{3.11}$$

where μ is the Lame's constant, T is the time required for the waves to propagate across the fault surface. The initial particle velocity parallel to the fault is:

$$v = (\sigma/\mu) \, \beta \tag{3.12}$$

For $\sigma{=}100$ bars (1 x 10^8 dyn/cm^2), $\mu = 3$ x 10^{11} dyn/cm^2 and $\beta = 3$ km/s the initial velocity takes a value of $v{=}1$m/s, in line with peak ground velocities experienced in large earthquakes around the world.

Brune (1970) has estimated the maximum ground acceleration as 2g, by assuming a stress drop of $\sigma{=}100$ bars and by considering the contribution of the band of frequencies between 0 and 10Hz. This value is also in line with the peak ground accelerations measured in earthquakes. Brune (1970) indicates that the near-field ground displacement parallel to the fault (Equation 3.11) increases linearly with time until the effects of the boundaries of dislocation reach to the observation point and then decrease gradually to zero. This effect is modelled by the following exponential factor:

$$v \, (x{=}0, t) = (\sigma/\mu) \, \beta \, \exp \, (\text{-}t \, / \tau \,) \tag{3.13}$$
$$u \, (x{=}0, t) = (\sigma/\mu) \, \beta \, \tau \, [(1{-}\exp \, (\text{-}t \, / \tau \,)] \tag{3.14}$$

where τ is the order of a/β, a being the appropriate fault dimension (dislocation surface) and, in reality, governs the speed of rise in dislocation displacement to its final value. For a large t, $u(x=0,t)$ tends to a constant level (final dislocation) given by:

$$u_{max} = (\sigma/\mu) \, \beta \, \tau \qquad (3.15)$$

The solutions of a static shear crack model with uniform stress drop on the fault plane can generally be given as (Keilis-Borok, 1957):

$$D = \xi \, \sigma \, a \, /\mu \qquad (3.16)$$

Where D is the average final dislocation, a is the critical dimension of the fault (radius for a circular fault) and ξ is a non-dimensional constant, equal to 1.37 for a circular fault. Through the consideration of Equations (3.15) and (3.16):

$$\tau_{av} = \xi \, (a/\beta) \qquad (3.17)$$

The Fourier amplitude spectrum of the near-field s-wave displacement, *UNF (ω)*, and acceleration, *ANF (ω)*, can be computed from Equation (3.14).

$$U_{NF} (\omega) = (\sigma/\mu) \, \beta \, \omega^{-1} \, (\omega^2 + \tau^{-2})^{-1/2} \qquad (3.18)$$
$$A_{NF} (\omega) = (\sigma/\mu) \, \beta \, \omega \, (\omega^2 + \tau^{-2})^{-1/2} \qquad (3.19)$$

The acceleration spectrum given by Equation (3.19) has a constant amplitude of $(\sigma\beta/\mu)$ in high frequency regions and and diminishes by $(\sigma\mu\tau/\beta)\omega$ in low frequency regions. The transition frequency (ω_c, corner frequency) between these two regions is given by $1/\tau$.

$$\omega_c = 2\pi f_c = 1/\tau \qquad (3.20)$$
$$A_{NF} (\omega) = (\sigma/\mu) \, \beta \, \omega \, (\omega^2 + \omega_c^2)^{-1/2} \qquad (3.21)$$

Figure 3.10 provides a plot of the Brune's near-field spectrum in log-log coordinates. As it can be seen the theoretical spectra has a flat high frequency amplitude that cannot represent the high frequency decay observed in empirical spectra. The dashed line in the high frequency region of Figure 3.10 illustrates the effect of high frequency diminution. The important difference is that in Brune's spectrum has only one corner frequency whereas a high-frequency corner frequency ($\omega h = 2\pi f h$) is generally evident from the empirical spectra. The high frequency decay can be considered to be a manifestation site effects, attenuation or source properties, such as, decay time of stress drop or size of the asperities. This high frequency diminution of the spectral amplitudes can be accounted for with the inclusion of a high frequency filter, with a high frequency cut-off frequency given by ωh, in Equation (3.21) (Trifunac, 1976).

$$A_{NF} (\omega) = (\sigma/\mu) \, \beta \, \omega \, (\omega^2 + \omega_c^2)^{-1/2} \, [\omega^2_h (\omega^2 + \omega^2_h)^{-1/2}] \qquad (3.22)$$

Far field shear wave displacement, *u(r,t)*, from a point shear dislocation in a homogenous elastic half space (with no energy loss and no surface effects) is given by (Aki and Richards, 1980):

$$u \, (r, \, t) = \lfloor (R\mu A) \, / \, (4\pi\rho\beta^3 r)\rfloor \, d'(t-r/\beta) \qquad (3.23)$$

where R is the scaling factor for the angular radiation pattern, ρ is the density of the medium, r is the hypocentral distance, $d'(t)$ is the time derivative of the average dislocation on the rupture surface (source time function).

By defining the source time function, d(t), as:

$$d(t) = D [1- (1+ t/\tau) \exp(t/\tau)] \tag{3.24}$$

and by using the definition of the Seismic Moment (Beresnev and Atkinson, 1997) has shown that the Fourier Amplitude Spectrum (FAS) of the far field shear wave displacement (Equation 3.23) becomes:

$$U(r,\omega) = [(RM_0) / (4\pi\rho\beta^3 r)] [1+(\omega/\omega_c)^2]^{-1} \tag{3.25}$$

Where $|U(r,\omega)|$ is the FAS of the far-field shear wave displacement and ω_c (or f_c) is the corner frequency equal to:

$$\omega_c = 2\pi f_c = 1/\tau \tag{3.26}$$

Using the definition of Seismic Moment, M_0, it can be shown (Brune, 1970) that the corner frequency becomes:

$$\omega_c = (7\pi/4)^{1/2} (\beta/a) = 2.34 (\beta/a) \tag{3.27}$$

Following (Brune, 1976) the following expression for the far-field shear wave RMS acceleration spectrum, $A_{FF} (r, \omega)$ can be given:

$$A(r,\omega)=R(\sigma\beta/\mu)(a/r)[\omega^2/(\omega^2 +(2.34\ \beta/a)^2)]=R(\sigma\beta/\mu)(a/r)[\omega^2/(\omega^2 +\omega_c^2)] \tag{3.28}$$

or as

$$A(r, \omega) = RM_0 (4\pi\rho r\beta^3)^{-1} \omega^2 [1+(\omega/\omega_c)^2]^{-1} \tag{3.29}$$

where R is the scaling factor for the RMS radiation pattern.

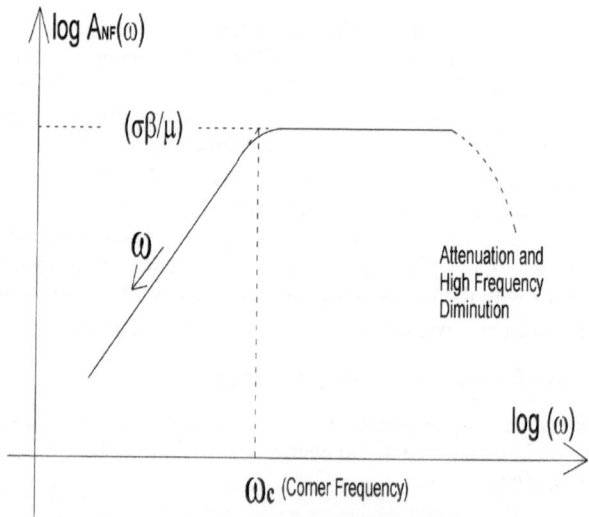

Fig. 3.10. Brune (1970) near-fault s-wave acceleration Fourier amplitude spectrum (Courtesy of the American Geophysical Union)

Figure 3.11 provides a plot of the Brune's far-field spectrum in log-log coordinates. As it can be seen for frequencies less then ωc the spectral amplitudes decay by ω^2. The spectral amplitudes reach asymptotically to a constant level equal to $R(\sigma\beta/\mu)$ (a/r). The effects of high frequency diminution, not accounted by Equation (3.28), are indicated by a dashed line in the high frequency regions.

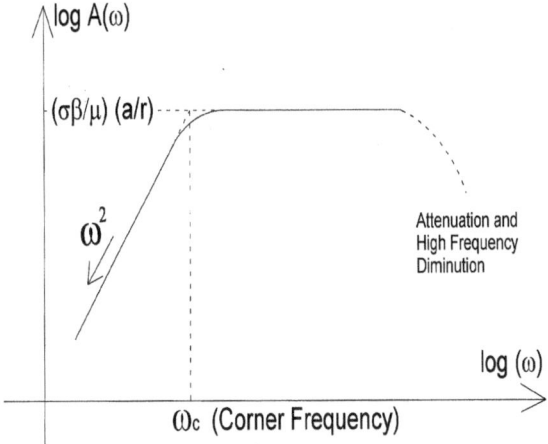

Fig. 3.11. Brune (1970) far-field s-wave acceleration Fourier amplitude spectrum
(Courtesy of the American Geophysical Union)

3.5. Time Domain Characteristics of Strong Ground Motion

Peak ground acceleration (PGA), velocity (PGV) and displacement (PGD) are the most common and easily recognizable time domain parameters of the strong ground motion. PGA, PGV and PGD are related to respectively high-, mid- and low-frequency ground motion components. Maximum recorded peak accelerations vary between 1g and 3g. Peak ground velocities reaching 4 m/s have been measured in 1999 ChiChi, Taiwan earthquake.

These parameters are indicated in Figure 3.12 where time history traces of acceleration, velocity and displacement of the E-W component of the YPT record obtained in the Aug. 17, 1999 Kocaeli (M_w=7.4) earthquake are illustrated (Erdik, 2001). This is a near fault record from a major strike-slip earthquake as evidenced by the pulse-like velocity and permanent displacement. This section will encompass the modelling of root-mean-square (RMS) acceleration, the duration of the strong ground motion and the time domain envelope functions.

3.5.1. MODELLING OF RMS-ACCELERATION

McGuire and Hanks (1980) have obtained an estimate of the root-mean-square (RMS) value of the ground acceleration associated with far-field shear waves through an operation of Parseval's theorem on the Brune (1970) source model. Hanks and McGuire (1981) provides this estimate as follows:

M. Erdik and E. Durukal

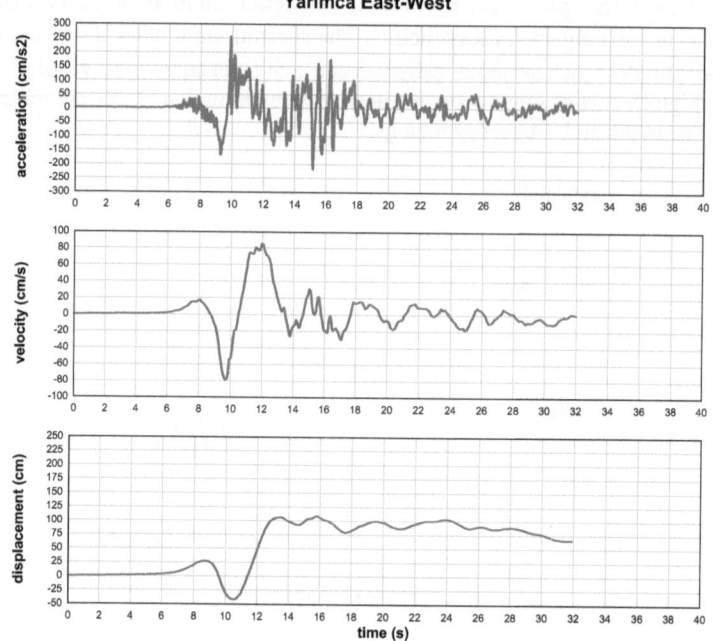

Fig. 3.12. Time history traces of acceleration, velocity and displacement of the E-W
component of the YPT record obtained in the Aug. 17th, 1999 Kocaeli (M_w=7.4)
(Durukal, 2002)

$$a_{rms} = 2 \text{ R } [(2\pi)^2/106] [\Delta\sigma/(\rho r^{2/3})] [(Q\beta)/(\pi f_c)]^{1/2} \qquad (3.30)$$

for cases where the anelastic attenuation of Fourier amplitude spectrum is of the form

$$\exp [(-\pi \text{ f r}) / (\text{Q }\beta)] \qquad (3.31)$$

In these expressions R is the radiation pattern factor, $\Delta\sigma$ is the stress drop, ρ is density, β is shear wave propagation velocity, r is the hypocentral distance, Q is the so-called quality factor and f_c is the corner frequency.

Defining $f_{max}=[(Q\beta)/(\pi r)]$ and assuming 2R=0.85, Equation (3.30) can be re-written as:

$$a_{rms} = 2 \text{ R } [(2\pi)^2/106] [\Delta\sigma/(\rho r)] [f_{max}/f_c)]^{1/2} \qquad (3.32)$$

Aki (1987) has developed a model for the estimation of the RMS value of the ground acceleration. For a circular dislocation surface the level of the acceleration power spectrum P_0 observed at a distance (r) from the dislocation can be given as (Aki, 1987):

$$P_0 = c \text{ W V } v_r^4 (\Delta\sigma/\mu)^2 (\beta r)^{-2} \qquad (3.33)$$

Where W is the fault width, V is the velocity of the rupture front, and v_r is the velocity of rupture spreading within a circular crack. If the acceleration is band-limited within f_c and f_{max}, the root-mean-square acceleration a_{rms} becomes (Aki and Richards, 1980):

$$a_{rms} = [P_0 2 (f_{max}-f_c)]^{1/2} \tag{3.34}$$

or approximately

$$a_{rms} = [P_0 2 f_{max}]^{1/2} \tag{3.35}$$

Since for California earthquakes of magnitude between 5.5 and 7.2, the cut-off frequency f_{max} is nearly constant at about 4-5 Hz (Aki, 1987), the RMS acceleration amplitudes are mainly controlled by the stress drop.

3.5.2. DURATION OF THE STRONG GROUND MOTION

Duration of a strong ground motion is a function of fault parameters (i.e. size of the rupturing part of the fault, rupture velocity), path from source to station, local site effects (soft soil, basin effects) and directivity. Duration of strong ground motion is also an important parameter playing a direct role in the destructiveness of an earthquake. A number of proposals exist in the literature for the identification of duration of the strongest part of shaking (Bommer and Martinez-Pereira, 2000). Perhaps the most widely used types of strong ground motion duration are the bracketed duration and the significant duration. The Bracketed Duration is the interval between the two points in time where the acceleration amplitude first and last exceeds a prescribed level such as 0.03 g (Ambraseys and Sarma, 1967) and 0.05g (Bolt, 1969). Significant Duration, defined as the time required to build up from 5 to 95 percent of the integral of $(\int a^2 \, dt)$ for the total duration of the record, where a is the acceleration (Trifunac and Brady, 1975). Arias (1970) showed that this integral is a measure of the energy in the ground motion acceleration.

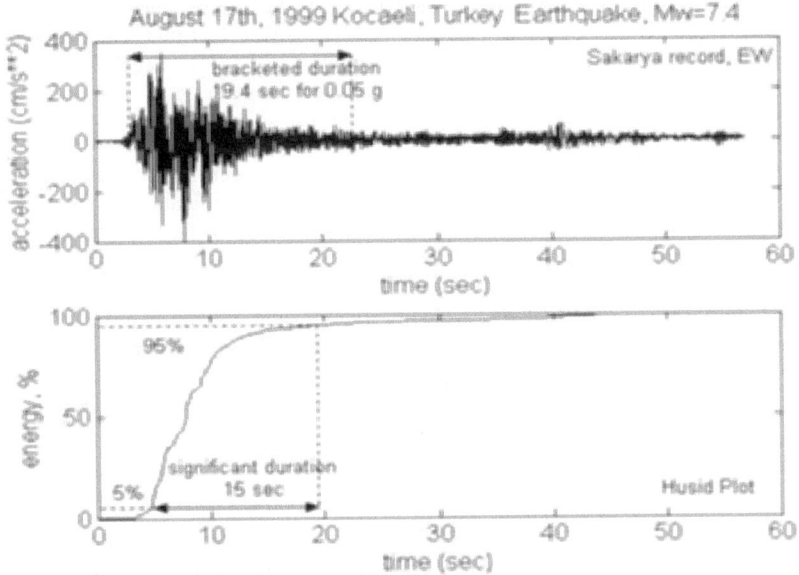

Fig. 3.13. Bracketed duration and significant duration using the August 17th 1999 Kocaeli, Turkey earthquake Sakarya accelerogram in the form of a Husid (1969) plot (Erdik and Durukal, 2003; Reproduced with permission from CRC Press LLC)

Dobry et al. (1978) have provided an empirical correlation of the "Significant Duration", T_s, of strong ground motion with magnitude.

$$\log (T_s) = 0.423\ M - 1.83 \qquad (3.36)$$

The correlation is valid for rock sites in Western US and for magnitude ranges $4.5 < M < 7.6$. Duration on soil sites may be up to twice the value for rock sites.

Bracketed duration and significant duration are shown in Figure 3.13 using the August 17 Kocaeli, Turkey earthquake Sakarya accelerogram in the form of a Husid (1969) plot. The definition of significant duration is based on energy. For records at large distances from an earthquake source the bracketed duration will have a smaller value than the significant duration.

Boore (2000) defines the duration of the strong ground motion, T_d, by:

$$T_d = T_s + T_p \qquad (3.37)$$

Where the first term (T_s) denotes the source duration and the second term (T_p) denotes the path duration. The source duration is given as the inverse of the corner frequency.

3.5.3. TIME DOMAIN ENVELOPE OF THE STRONG GROUND MOTION

The Gaussian white time series of duration T_w is windowed using the shape (coda) function, $w(t)$, of Saragoni and Hart (1974) as described in Boore (1983).

$$w(t) = a_n\ t^b \exp(-ct)\ H(t) \qquad (3.38)$$

where $H(t)$ is the Heaviside (unit step) function, a_n is the normalizing factor, b and c are the shape parameters. Saragoni and Hart (1974) showed that this window is a good representation of the averaged envelope of squared ground motion acceleration.

3.6. Frequency Domain Characteristics of Strong Ground Motion

In the frequency domain: Fourier amplitude and phase spectrum, power spectrum and several definitions of response spectra are used in the quantification of strong ground motion. Response Spectra ordinates present the amplitude of the response of a Single-Degree-of-Freedom system at each frequency (or period). Five types of response spectra are defined: relative displacement (S_d), relative velocity (S_v), absolute acceleration (Sa), pseudo-relative velocity (PSV), and pseudo-relative acceleration (PSA). Frequency content of the response spectrum has been described by Predominant Period and Mean Period. The predominant period is generally linked to the peak spectral acceleration at 5% damping. Rathje et al. (1998) defines mean period (T_m) of a Fourier Amplitude Spectrum as

$$T_m = \Sigma\ (C_i^2 / f_i) / \Sigma\ (C_i^2) \qquad (3.39)$$

where C_i is the spectral amplitude at frequency f_i.

The modelling of Fourier amplitude spectrum is of prime importance for simulation. Both empirical and theoretical models of Fourier amplitude spectra exist. Spectra of strong ground motion have been empirically estimated through. Trifunac and Lee (1989) provide empirical models for scaling Fourier amplitude spectra in terms of earthquake magnitude, source to site distance, site intensity and recording site conditions. These models are based on the regression of the empirical amplitudes at specific frequencies. Holistic theoretical models Fourier amplitude spectrum of the ground motion involves the elements of source, propagation path attenuation, high frequency diminution and site amplification is illustrated in Figure 3.14.

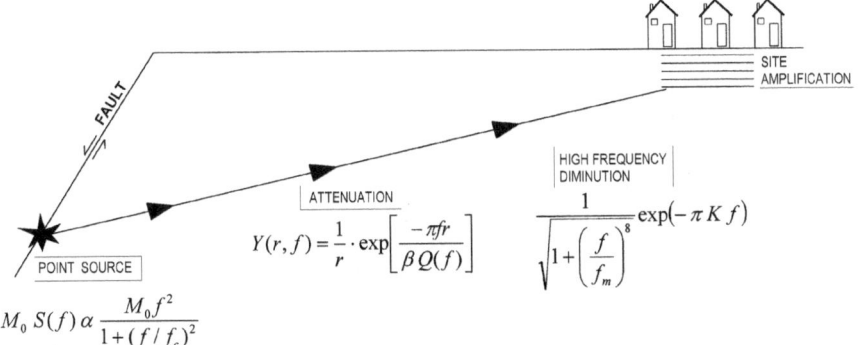

Fig. 3.14. Elements of the Fourier amplitude spectrum of the earthquake ground motion modelling (Erdik and Durukal, 2003; Reproduced with permission from CRC Press LLC)

3.6.1. THEORETICAL MODEL OF FOURIER AMPLITUDE SPECTRUM

Fourier amplitude spectrum of the free field acceleration of the horizontal ground motion at an epicentral distance (r) caused by the propagation of shear waves from an earthquake with a point source of slip model is given by (Boore 1983, Atkinson and Boore, 1998 and Atkinson and Silva, 2000):

$$A(f,r) = C_0 \, M_0 \, S(f) \, Y(f,r) \, P(f) \, Z(f) \qquad (3.40)$$

where C_0 is the frequency independent scaling factor, M_0 is the seismic moment, $S(f)$ is the source spectrum, $Y(f,r)$ is the attenuation factor, $P(f)$ is the high frequency decay factor and $Z(f)$ is the scaling factor that accounts for the site effects. For the ideal cases where $Y(f,r)$, $P(f)$ and $Z(f)$ are not considered $A(f,r)$ will be given by:

$$A(f,r) = C_0 \, M_0 \, S(f) \qquad (3.41)$$

Equation (3.41) has the same form and is essentially identical to the Fourier Amplitude Spectrum of the far-field shear wave acceleration given in Equations (3.28) and (3.29).

$S(f)$ is the "Source Spectrum" that accounts for the spectral model of the radiated waves from the source. It consists of two parts: the spectral shape and the scaling law (the relationship between the seismic moment and the corner frequency). One of the simplest and most commonly used source spectrum with a single corner frequency, f_c, is called the "omega-squared" spectrum, similar to the Brune's spectral shape

$$S(f) = (2\pi f)^2 / (1 + (f / f_c)^2) \qquad (3.42)$$

The corner frequency and the seismic moment are related (Brune 1970) by the so-called spectral scaling law:

$$f_c = 4.9 \times 10^6 \beta \left(\Delta\sigma / M_0 \right)^{1/3} \qquad\qquad (3.43)$$

where $\Delta\sigma$ is in bars (1 bar = 10^5 Pa) and M_0 is the seismic moment in (dyn-cm) and β is in km/s. The variation of the corner frequency f_c with respect to earthquake size can be seen in Figure 3.15 using records of earthquakes with magnitudes changing between 3.1 and 8.1 and in Figure 3.16 using data from the 1999 Kocaeli, Turkey earthquake.

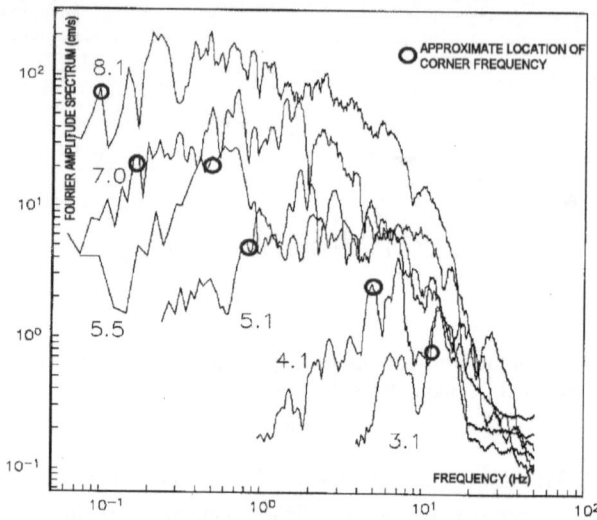

Fig. 3.15. Variation of the corner frequency fc with respect to earthquake size (after Anderson and Quaas, 1988; Reproduced with permission from J.G. Anderson and the Earthquake Engineering Research Institute)

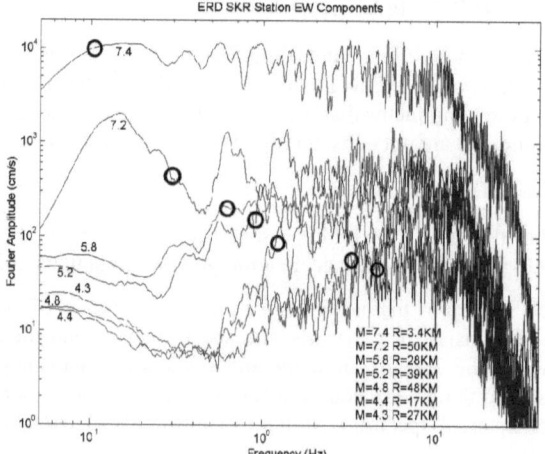

Fig. 3.16. Variation of the corner frequency f_c with respect to earthquake size, using data from the August 17, 1999 Kocaeli earthquake sequence recorded at station Sakarya.

Y *(f, r)* is called the attenuation factor.

$$Y(f, r) = Y_G (r) \, Y_A(f, r) \tag{3.44}$$

Y_G *(r)* is the geometric attenuation factor due to geometric spreading of the seismic energy. At epicentral distances *(r)* less than about 100km, empirical evidence indicates a geometric attenuation by *(1/r)*. Atkinson and Silva (2000) states that the geometric attenuation is proportional to *(1/r)* at epicentral distances less than 40km but to *(1/r)*$^{1/2}$ at epicentral distances greater than 40km.

Y_A*(f,r)* is the anelastic attenuation (or whole-path attenuation) factor given by the following expression:

$$Y_A(f, r) = \exp\left[(-\pi \, f \, r) / (Q \, \beta)\right] \tag{3.45}$$

Where Q is the so-called "quality factor" and, at its simplest definition, can be taken as a constant $(Q=Q_0)$.

P(f) in Equation (3.40) serves as the high frequency diminution factor that accounts for the decay of spectral amplitudes at high frequencies, believed to be caused by the weathering in the upper layers of the medium. Boore (1983) assigns a fourth order Butterworth filter for *P(f)*.

$$P_2(f) = [1 + (f/f_m)^8]^{-1/2} \tag{3.46}$$

Anderson and Hough (1984) models P(f) by the spectral decay factor κ as:

$$P_1(f) = \exp(-\pi \, \kappa \, f) \tag{3.47}$$

Z(f) represents the scaling factor to account for the site effects. Boore and Joyner (1997) provides amplification values as a function of frequency towards the assessment of *Z(f)* in terms of typical soil profiles associated with NEHRP (1997) site classes.

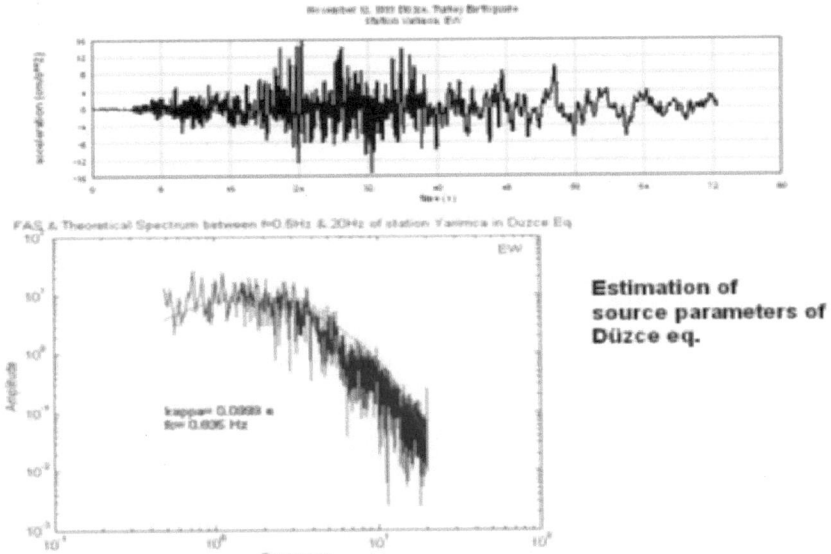

Fig. 3.17. Estimation of corner frequency fc and near surface attenuation factor kappa, κ using data from the Nov. 12 Düzce, Turkey earthquake recorded at station Yarımca

An estimation of kappa, κ and corner frequency f_c, is presented in Figure 3.17 using data from the November 12, 1999 Düzce, Turkey earthquake, as it is recorded at station Yarımca.

3.7. Radiation Pattern and Directivity

Several phenomena observed in strong motion can only be understood in the context of finite source models involving directivity and near-source pulse motions. The amplitude and polarity of a seismic wave radiated from an earthquake source change with the orientation of the source and the receiver. This dependence is called as the radiation pattern. There is a difference in the radiation patterns of P and S waves for a point source (Figure 3.18, from Das, 1997). The effect of an extended fault on the radiation patterns of p- and s- waves can be seen in the same figure as well.

Directivity is the effect of rupture propagation along the fault on the ground motion. It impacts both the high frequency and the low-frequency accelerograms. Let us imagine a rupture propagating at a certain velocity along a fault plane. Stations located in the direction of rupture propagation experience shorter duration ground motions than the ones located in the direction opposite to the direction of rupture. This is called directivity. Associated ground motion amplitudes are larger for stations in the forward directivity region than the ones in backward directivity region due to conservation of energy. At high frequencies, directivity shows up as a short, intense accelerogram at the far end of the fault, in contrast with a lower-amplitude, long-duration accelerogram near the origin of rupture. At high period ranges forward directivity effects at near-fault locations result in high amplitude velocity pulses.

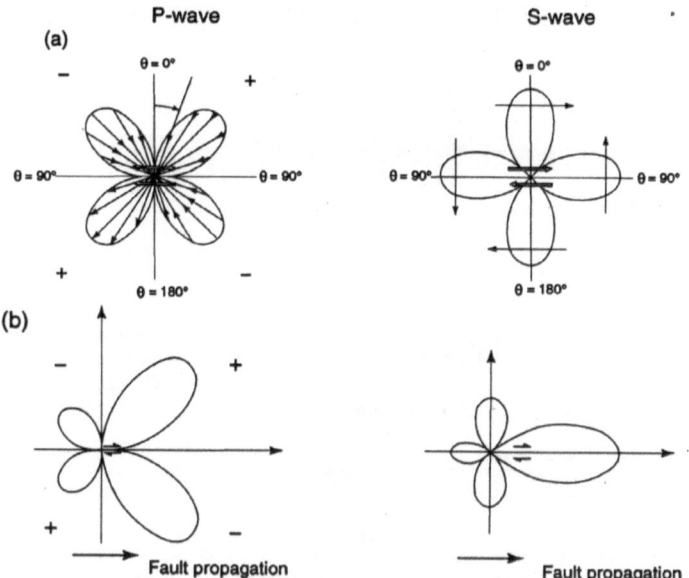

Fig. 3.18. Radiation patterns for p- and s-waves for a point source (from Das, 1997; Reproduced with permission from Institute of Engineering Seismology and Earthquake Engineering, ITSAK)

The fault normal component of the ground velocity will generally consist of a full cycle velocity pulse, which upon integration will not create a permanent displacement. Whereas, the fault parallel components will generally have half-cycle velocity pulse, which creates a permanent absolute displacement equal to the fault offset. These effects can be clearly seen in the acceleration, velocity and displacement time history traces of Sakarya record of 17.8.1999 Kocaeli earthquake given in Figure 3.19. The station is located at about 3 km from the fault trace. The rise time of this displacement is about 3s.

1999 ChiChi, Taiwan, earthquake has confirmed that the hanging walls of thrust faults move much more, and have greater high-frequency ground motions (peak accelerations) than the footwalls during earthquakes.

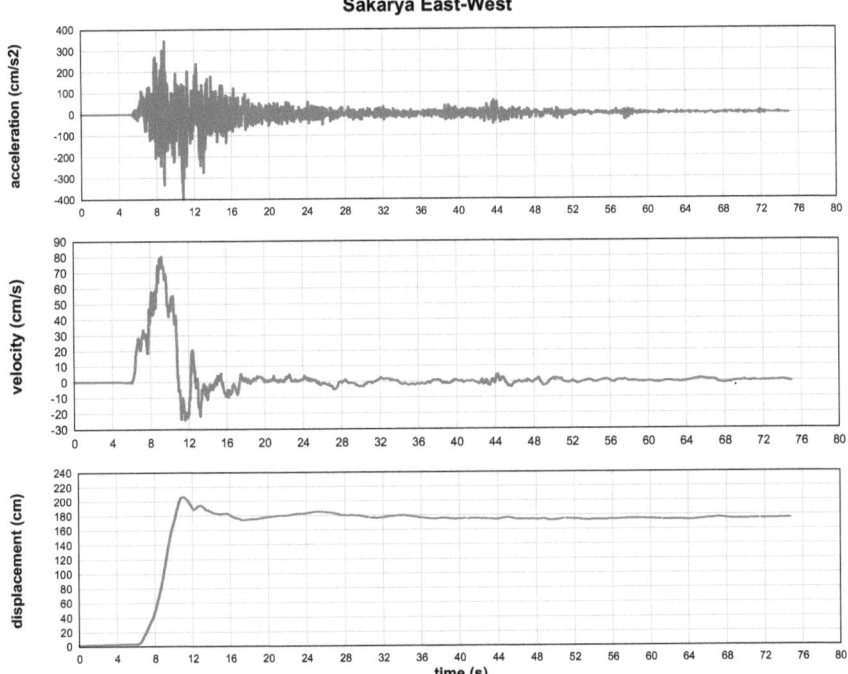

Fig. 3.19. Acceleration, velocity and displacement time history traces of the Sakarya record of 17.8.1999 Kocaeli, Turkey earthquake (Durukal, 2002)

Luco and Anderson (1983) have studied the near-fault ground motions using a simple theoretical model. The fault is modelled by a finite width, infinitely long vertical strike slip dislocation. The fault, buried in a homogenous half space, extends from a depth of z_u=2km to z_d=10km (as illustrated in the top of Figure 3.20). The p- and s-wave propagation velocities of the medium are 6 km/s and 3.464 km/s. A step-type dislocation of amplitude 100 cm propagates horizontally with a rupture velocity of 3.184 km/s along the fault. The rupture front is vertical. Fault-parallel, fault-normal and vertical acceleration, velocity and displacement time histories at any observation point along the fault for various distances to the surface projection of the fault are shown in Figure 3.20 (Anderson and Luca, 1983).

M. Erdik and E. Durukal

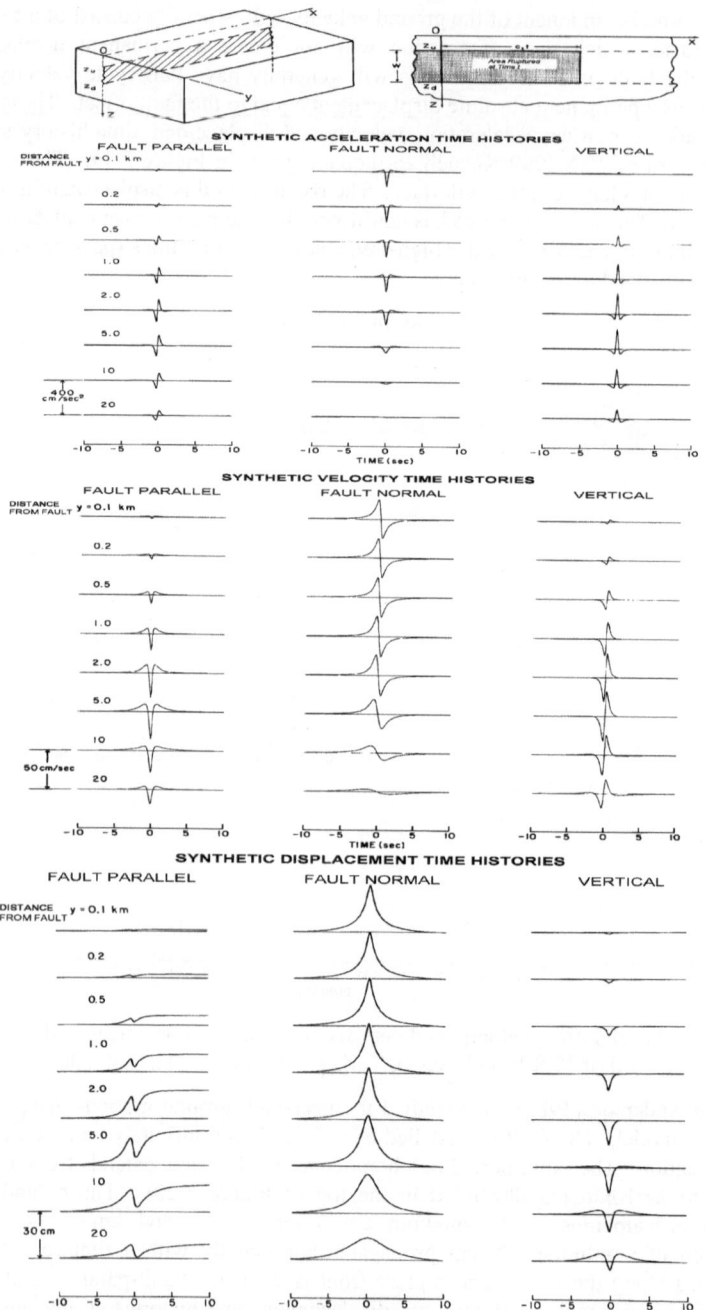

Fig. 3.20. Characterization of near-fault ground motion (after Anderson and Luca, 1983; Courtesy of the Seismological Society of America)

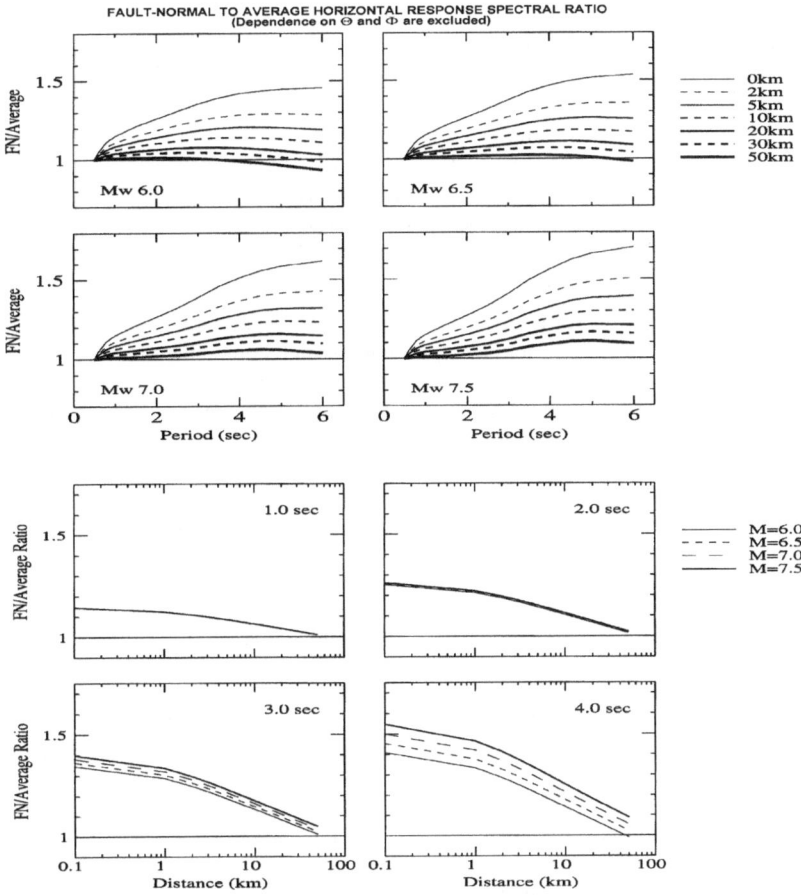

Fig. 3.21. An empirical model of the ratio of the fault-normal spectral component amplitudes to average spectral amplitudes (Somerville et al., 1997; Courtesy of the Seismological Society of America)

As it can be seen the peak amplitudes occur at distances (about 2km) comparable to the depth of the top of the fault. Figure 3.20 shows prominent velocity pulses in the fault-normal direction. The durations of the acceleration and velocity pulses increase, especially for fault-normal components, with distance from fault. Fault-normal displacement and velocities stay approximately constant for fault distances less than the vertical distance to the top of the fault. The pulse amplitudes are sensitive to the location of the top of the fault relative to the observation point and inclusion if softer layers at the top of the half-space can lead to substantial amplifications ranging from 2 to 8 (Bouchon, 1987).

Quantification of rupture directivity effects is an emerging component of attenuation studies. Somerville et al (1997) have used empirical recordings from shallow crustal earthquakes in active tectonic regions to illustrate how directivity causes spatial variation in ground motion amplitude and duration around faults, leading to different strike-normal and strike-parallel components of horizontal ground motion and to develop modifications to ground motion attenuation relationships due to rupture directivity. These variations appear to be significant at a period of 0.6 second, and Figure 3.19 Acceleration, velocity and displacement time history traces of the Sakarya record of 17.8.1999 Kocaeli, Turkey earthquake generally grow in size with increasing period. The spectral acceleration is larger for periods longer than 0.6 second, and the duration is smaller, when the rupture propagates toward a site. For sites located close to faults, the strike-normal spectral acceleration is larger than the strike-parallel spectral acceleration at periods longer than 0.6 second in a manner that depends on magnitude, distance, and angle.

For design purposes, the variation of the average horizontal response spectra and the difference between the fault-normal and fault-parallel components of the response spectra in near-fault conditions becomes an important consideration. Somervillle et al. (1997) has presented a procedure for the modification of response spectra of near-fault strong ground motion to account for the rupture directivity. Figure 3.21 (Somerville et al., 1997) provides an empirical model of the ratio of the fault-normal spectral component amplitudes to average spectral amplitudes. Models of this ratio are given for different magnitudes and epicentral distances against period, and for different magnitudes and periods against distance. As it can be assessed, forward directivity caused larger spectral amplitudes at periods larger than 0.6s and the ratio of the fault spectral amplitudes to the average spectral amplitudes can be as high as 1.6 at periods in the vicinity of 6 s under favourable forward directivity conditions.

3.8. Simulation of Strong Ground Motion

A major goal of strong motion studies is to be able to synthesize strong motion seismograms suitable for use in engineering analyses. The simulation process is generally expressed mathematically using a representation theorem. The ground motion at the site is computed as the integral over space of the contributions (Green's Function) from each point on the fault surface. The integration over time incorporates the effect of the rupture at each point taking a finite amount of time to reach its final value. Green's function characterizes the response of the earth to a point source earthquake. In the summation process of the representation theorem it is used as a building block to simulate ground motion from a more general source.

Forward modelling in strong ground motion seismology deals with the estimation of ground motion at the ground surface by modelling the earthquake faulting process, the earth medium between the earthquake source and the station, and local site effects near the station, such as modelling of topography, basin structure and soft soil conditions.

There are two types of source models: kinematic and dynamic. In kinematic source models the slip over the rupturing portion of a fault as a function of fault plane coordinates and of time is known or given a-priori and it is not a function of stresses causing it. In dynamic source models on the other hand, slip over the rupturing segment of a fault is a function of tectonic stresses acting on the region.

Theoretical Green's Function Models generate synthetic Green's functions based on Earth structure models of varying complexity, and combine these with a range of models of the earthquake source. Empirical Green's Function Models use records of small earthquakes as empirical Green's functions. Seismograms used as empirical Green's functions incorporate all the complexities of wave propagation. The limitation is that the empirical Green's functions may not have an adequate signal-to-noise ratio at all the frequencies of interest, may not be available for the desired source-station pairs, and may originate from sources with a focal mechanism different from the desired mechanism. One relatively simple model that has been used increasingly to simulate earthquake rupture and source-to-site wave propagation is the Band Limited White Noise/Random Vibration Theory. The broad flat portion of the Fourier amplitude spectrum provides support to model strong motion as band-limited white noise. The process of constructing a seismogram with this model begins with generating a white-noise time series, applying a shaping taper in the time domain to match the envelope of the expected strong motion, and then applying a band-pass filter in the frequency domain to mould the Fourier spectrum to the expected spectral shape.

For simulation of ground motion deterministic, empirical (e.g. Hartzell, 1978), semi-empirical (e.g. Irikura, 1983; Somerville et al., 1991), stochastic (e.g Boore, 1983; Silva et al., 1990) and hybrid methods have been proposed and are being widely utilized. The recent trend and need in the earthquake engineering community is towards a simulation technique that will incorporate broadband ground motions of longer period, directivity effects and also higher frequencies.

3.8.1. STOCHASTIC SIMULATIONS

Stochastic approaches to the simulation of strong ground motion filter and window the white-noise time series according to seismologically determined average spectra and duration (Boore, 1983). Stochastic simulations for point source earthquakes are an "Engineering" approach to the simulation of strong ground motion. Ground acceleration modelled as a filtered Gaussian white noise modulated by a deterministic envelope function (Safak, 1988). The filter parameters are determined by either matching the empirical properties of the spectrum of the strong ground motion (e,g, Trifunac and Lee, 1989), theoretical spectral shapes (i.e. Kanai-Tajimi Spectrum, Housner and Jennings, 1964) or are determined on basis of reliable physical characteristics of the earthquake source and propagation media (e,g. Hanks and McGuire, 1981; Boore, 1983). The Fourier amplitude spectrum model used in the latter stochastic simulations is essentially S-wave ground motion spectra based on the far-field model of Brune (1970). It has been found to satisfy main parameters of high frequency ground motion for earthquakes within a wide magnitude range (McGuire and Hanks, 1980).

The stochastic simulation of strong ground motions that relies on seismic source physics has found substantial applications in Earthquake Engineering with successful comparisons of predicted and recorded data. Boore (1983) developed a so-called Band Limited White Noise model for stochastic simulation of strong ground motion with seismological constraints. In the time-domain procedure, elaborated in Boore (1983), a Gaussian white noise is windowed with a shaping function having a prescribed duration. The window is chosen such that the mean level of the spectrum of the windowed white noise is unity. The windowed time series is transformed into the frequency domain. Its Fourier amplitude spectrum is scaled to the square root of the mean squared absolute spectra and multiplied by the site-specific shape of the theoretical Fourier amplitude spectrum of the free field acceleration of the horizontal ground motion at the site, $A(f,r)$, given by Equation (3.40). Transformation back into time domain results in the simulated (synthetic) time history of the horizontal component of the ground motion.

Software developed originally by Dr. Erdal Safak of USGS (Pasadena, California) with modifications by the authors of this paper will be used to provide some examples for the simulation of strong ground motion from point sources. The software is written in MATLAB (http://www.mathworks.com) language and follows, almost exactly, the time-domain procedure developed by Boore (1983) and coded in Boore (2000).

For the example simulations the spectral constants are taken as: Soil density = 2.8 g/cm3; Shear wave velocity = 3.6 km/s; Partition factor for energy = 0.707; Factor for radiation pattern = 0.55; Factor for free surface amplification = 2

For geometric spreading a simple r^{-1} model is assumed. A single corner frequency ω^2-model is used for the spectral shape (Section 3.6.1, Equation 3.42). The frequency-dependent Q model for the whole-path attenuation is taken identical to the model used by Boore (2000) for Western US. Source duration is taken equal to the inverse of the corner frequency and the path duration is modeled as 5% of the epicentral distance. An exponential time windowing function is used. The high frequency diminution function (Section 3.6.1) is modeled by a fourth order Butterworth filter with a cut-off frequency of f_m =50Hz and a spectral decay factor κ=0.035 (Anderson and Hough, 1984).

One of the simulations of ground acceleration for a M_w=7.6 earthquake at 20km epicentral distance is provided in Figure 3.22 together with the match of the averaged Fourier amplitude spectra of 20 simulations with the target spectra. An example simulation and the match of simulated and target spectra are provided in Figure 3.23 for the same earthquake (M_w=7.6) but at an epicentral distance of 100km.

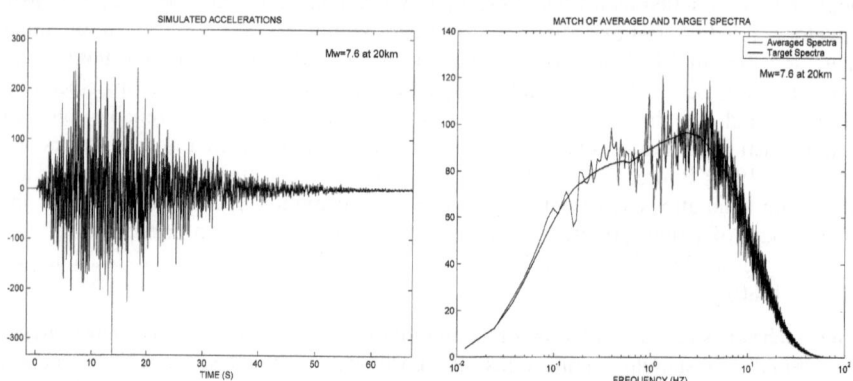

Fig. 3.22. A stochastic simulation of horizontal ground acceleration and the comparison of average simulated and theoretical Fourier amplitude spectra (M_w=7.6 earthquake at 20km epicentral distance) (Erdik and Durukal, 2003; Reproduced with permission from CRC Press LLC)

Beresnev and Atkinson (1997) have developed a procedure (FINSIM) for the stochastic simulation of strong ground motion from finite-fault ruptures. The fault rupture plane is modelled with an array of sub-faults. The radiation from each sub-fault is modelled as a point source with a ω^2-spectrum, similar to Boore (1983). Fault rupture initiates at the hypocenter and spreads uniformly along the fault plane with a constant rupture velocity triggering radiation from sub-faults in succession. Simulations from each sub-fault, properly lagged and summed at the observation point, provide the simulation of ground motion from the modelled finite fault rupture. The size of the sub-faults controls the overall spectral shape at medium frequencies. The total number of sub-faults is controlled by the constraint that the total seismic moment of the sub-faults must be equal to the target seismic moment.

To provide an example for stochastic simulation of strong ground motion from finite fault ruptures, the accelerations recorded by near-field stations during the Aug. 17, 1999 (M_w=7.4) Kocaeli (Turkey) earthquake will be simulated using the code FINSIM developed by Beresnev and Atkinson (1997).

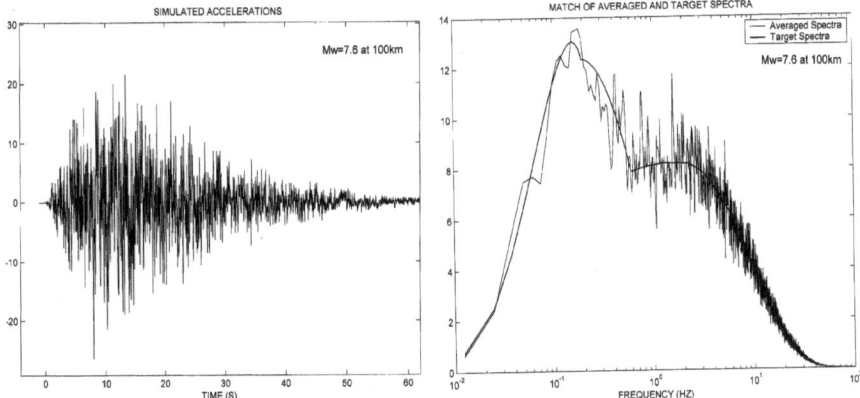

Fig. 3.23. A stochastic simulation of horizontal ground acceleration and the comparison of average simulated and theoretical Fourier amplitude spectra (M_w=7.6 earthquake at 100km epicentral distance) (Erdik and Durukal, 2003; Reproduced with permission from CRC Press LLC)

An isometric view of the fault rupture plane, location of the epicenter and the recording stations Yarımca, İzmit and Adapazarı are indicated in Figure 3.24. The geometry of the fault rupture plane and the relative distribution of slip have been adopted from the slip model developed by Yagi and Kikuchi (2000) for the Kocaeli earthquake. The fault dislocation is modelled as a 110 km long and 20 km deep vertical plane with sub-faults of size 5km x 5km. Rupture velocity is 2.7 km/s.

For the whole-path attenuation a crustal Q model of $Q(f) = 180 f^{0.45}$ is adopted as reported by Atkinson and Silva (2000) for Western US. For site amplification a frequency-dependent amplification function values provided by Boore and Joyner (1997) using the NEHRP site class for each station is used. As for the near surface attenuation we used the kappa (κ) values of 0.07, 0.05 and 0.03 respectively for the Yarımca, İzmit and Sakarya stations. The factor that controls the strength of subfault radiation (sfact) is 1.5 for simulations in İzmit and Sakarya stations and 2.0 in Yarımca. A Saragoni and Hart (1974) based envelope function is used for radiation from sub-faults similar to Boore (1983).

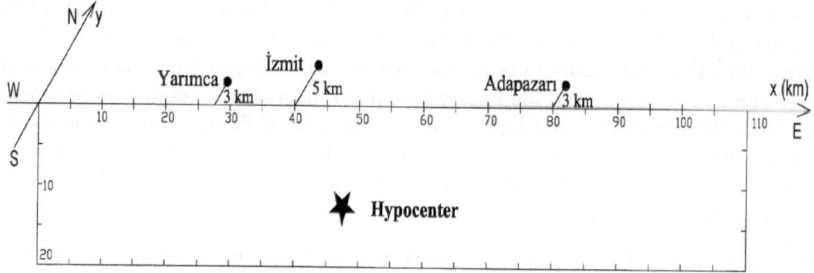

Fig. 3.24. Fault rupture geometry, hypocenter and the location of the recording stations
in 17.8.1999 Kocaeli (M_w=7.4) earthquake (Durukal, 2002)

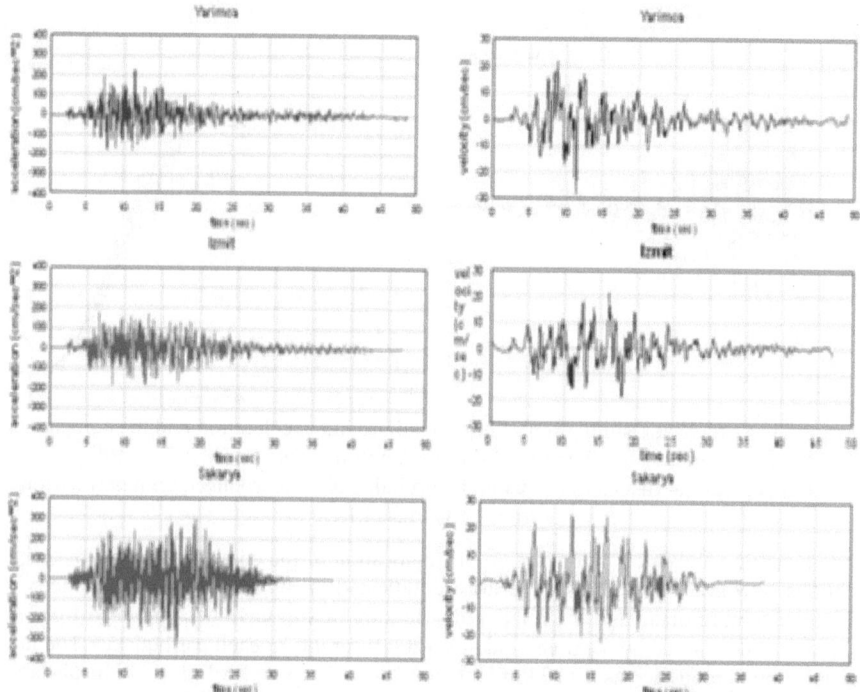

Fig. 3.25. Simulated accelerations and velocities at stations Yarımca, İzmit and Sakarya
for Kocaeli earthquake using code FINSIM (Durukal, 2002)

Simulated accelerations for stations Yarımca, İzmit and Sakarya are presented in Figure
3.25. Spectral accelerations averaged over five simulations can be seen in Figure 3.26.
For comparison, we also show empirical spectral accelerations from (August 17, 1999
earthquake) in the same figure. With the model it was possible to get a very satisfactory
fit of spectral accelerations with the recorded data (Figure 3.26).

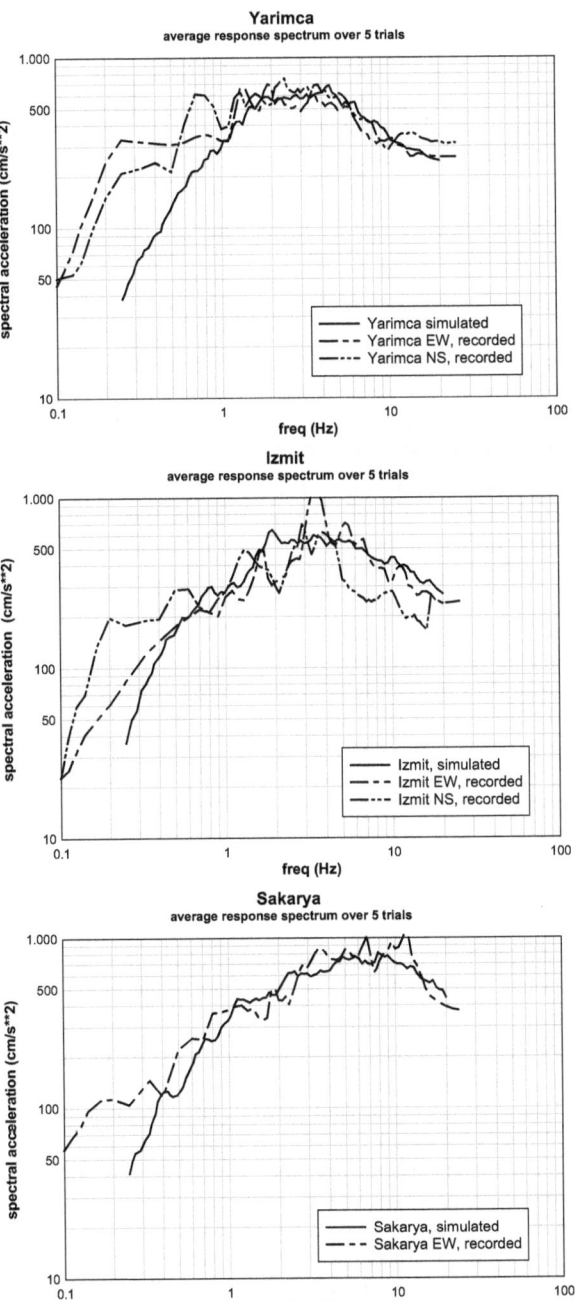

Fig. 3.26. Comparison of simulated and recorded acceleration response spectra (Durukal, 2002)

Peak accelerations are somewhat lower than the recorded ones. Simulated peak horizontal accelerations are 226 cm/s^2, 186 cm/s^2 and 334 cm/s^2 for Yarımca, İzmit and Sakarya stations respectively. The code FINSIM inherently cannot account for the differences in the ground motion as a result of polarity. Calculated peak accelerations are to be considered horizontal accelerations. Therefore it is believed that the calculated accelerations are satisfactory. Respective velocities calculated from the accelerations using a polynomial baseline correction are 28 cm/s, 21 cm/s and 25 cm/s for stations Yarımca, İzmit and Sakarya.

3.8.2. HYBRID SIMULATIONS

3-D finite difference method, discrete wave-number method, indirect boundary element method, modal summation method, ray theory, 2.5-D discrete wave number-boundary integral equation method, 2.5-D pseudo spectral method and 2.5-D finite difference method have been used for deterministic simulation of strong ground motion. Essentially deterministic methods convolve the source function with synthetic Green's functions to produce the motion at ground surface.

Deterministic theoretical predictions of the ground motion can be achieved by convolution of the Green's Functions and the slip function. Green's functions can be calculated through empirical and synthetic means. Although certain predictions can be made for the total slip and the mode of faulting no prediction can be made regarding the rupture characteristics. This necessitates the consideration of different rupture models. Such deterministic predictions cannot be extended into the frequency regions above around 1Hz, since, high frequency ground motions are controlled by the heterogeneities in the fault rupture, which cannot be a-priori accounted for in a deterministic manner. This requires either the use of stochastic source models or the stochastic treatment of the high frequency components in the ground motions. Thus hybrid procedures are developed for simulation of strong ground motion, which address the low and high frequency components of the ground motion separately and than combine the two motions.

In the hybrid broadband simulation procedure adopted by Somerville et al. (2000) the source is represented by an empirical source time function. For simulations of ground motion for frequencies below 1Hz a theoretically rigorous representation of radiation pattern, rupture directivity and wave propagation effects are incorporated in Green's function computations. At higher frequencies stochastic simulation techniques that includes source radiation pattern and scattering in the path and site is utilized.

In a recent paper Erdik and Durukal (2001) applied a hybrid method to simulate strong ground motion for a container port facility near the Sapanca segment of the North Anatolian Fault. Low-frequency ground motion (DC-1Hz) calculated with the help of the discrete wave number method is combined with the stochastically simulated high frequency components using the methodology proposed by Boore (1983). The basic tenets of the simulation procedure were chosen to: (1) Preserve the deterministic displacement shape; (2) Satisfy the corresponding theoretical Fourier Amplitude Spectrum and; (3) Yield a coda shape in conformance with applicable empirical findings. Furthermore, the peak ground acceleration (PGA) and the pseudo spectral relative velocity (PSRV) values should be favourably compared against those obtained from empirical attenuation relationships for conformity. The essential elements of the procedure were:

1. Assessment of the source parameters of the DBE motion associated with the corresponding return period for specific conditions of site and seismicity.
2. Deterministic assessment of the low frequency (DC-1Hz) ground motion, at the outcrop of a reference soil layer, due to rupture of seismic faults.
3. Use of a Boore (1983)-type stochastic simulation method to complement the deterministic low frequency ground motion with high frequency (1Hz-50Hz) components.
4. Combination of the two parts of ground motion to yield a site-specific simulation for a frequency range of DC-50Hz.
5. Site response analysis, if required, to include the local wave propagation effects in the soil media above the reference soil layer.

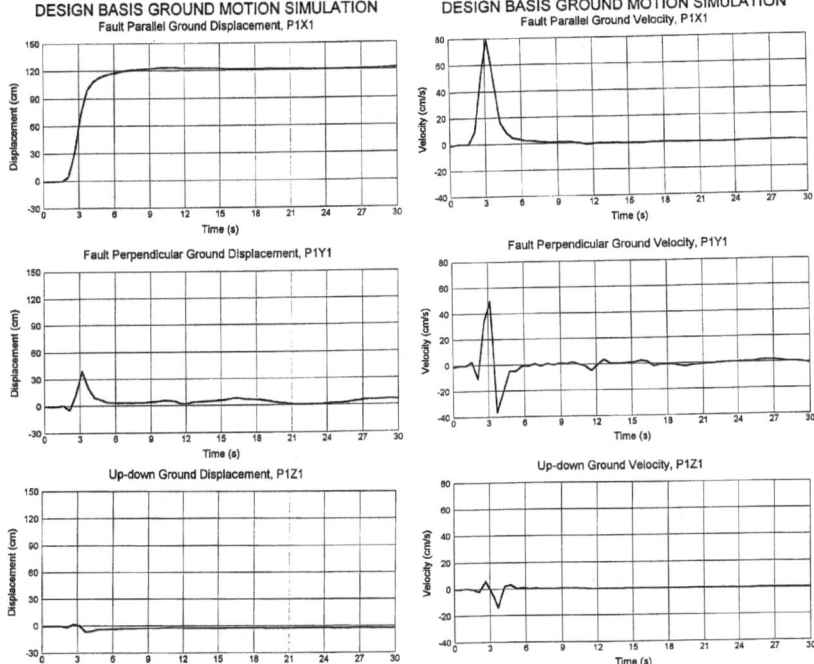

Fig. 3.27. Simulated low-frequency ground motion on competent soil (Bi-lateral rupture on Sapanca segment) (Erdik and Durukal, 2001)

The design basis simulation is assumed to be controlled by the bi lateral rupture of one of the segments of the North Anatolian Fault (the 70km x 15km Sapanca Segment). Haskell type ramp function is considered as the slip function with a rupture velocity of 2.8 km/s, 1s rise time, 1.5m final strike slip and 0.05m final dip slip. Figure 3.27 provides the simulated low frequency (less than 1Hz) fault-parallel, fault normal and vertical velocity and displacements. The final broad-band hybrid simulations that involve the combination of low and high frequency (stochastic) simulations at the competent soil outcrop are given in Figure 3.28. These simulations were performed prior to the 17.8.1999 Kocaeli earthquake that essentially hit the same region and site with same design earthquake level. Considering that the simulations also represent a "blind-test", comparison of simulated and recorded accelerations, velocities and displacements, as well as of spectral accelerations, provide a remarkable fit to the simulations (Erdik and Durukal, 2001).

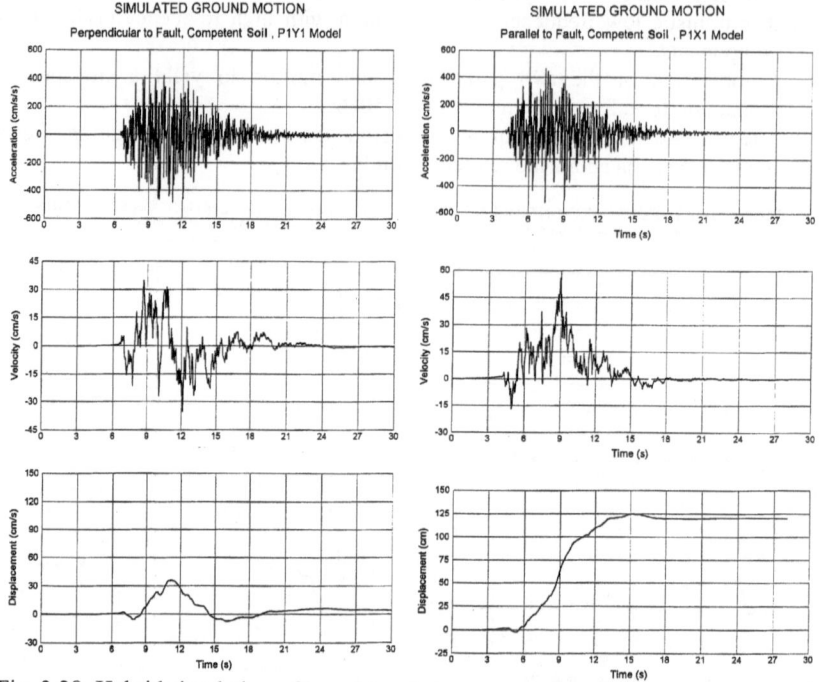

Fig. 3.28. Hybrid simulation of broad-band strong ground motion on competent soil (Bilateral rupture on Sapanca segment) (Erdik and Durukal, 2001)

3.9. Conclusions

The availability of empirical strong ground motion data will always be less that what would be needed to meet the needs of a variety of ever demanding engineering problems. A set of strong ground motions, either recorded or theoretically simulated, is the necessary database for the civil engineering design, regarding both new construction and performance assessment of the existing built environment. The future of performance based earthquake resistant design and sophisticated non-linear dynamic analysis will rely on the development of analytical tools that can simulate realistic ground motions in terms of tectonic structure, earthquake physics, local geological and geotechnical conditions. This need is more acute for large magnitude earthquakes in near-field conditions. The state-of-the art and success in strong ground motion simulation is developing at a fast rate and we all hope that it meets the demand in foreseeable future.

CHAPTER 4
GEOPHYSICAL AND GEOTECHNICAL INVESTIGATIONS FOR GROUND RESPONSE ANALYSES

Diego Lo Presti, Carlo Lai, and Sebastiano Foti
Department of Structural and Geotechnical Engineering
Politecnico di Torino, Italy

4.1. Introduction

Seismic response analyses require thorough geophysical and geotechnical investigations in order to assess the mechanical and geometrical parameters of the system. Planning and selecting appropriate geophysical and geotechnical investigations are delicate tasks that require a deep knowledge of available techniques. This chapter discusses in situ and laboratory tests with particular focus on the following aspects:

- comparison of different laboratory tests (monotonic and cyclic triaxial tests, resonant column tests, cyclic torsional shear tests);
- comparison of different geophysical testing methods (invasive and non-invasive tests);
- interplay between geological, geophysical and geotechnical investigations.

Main objectives of the chapter are i) to point out the multi-disciplinary character of seismic response studies and ii) to highlight capabilities and limitations of different in situ and laboratory testing methods. The chapter ends with a brief description of the lesson learnt from a case history.

Ground response analysis is a multi-disciplinary task involving various types of professional competencies including engineering seismology, structural geology, geophysics, geotechnical earthquake engineering. In fact, the whole process requires the implementation of the following activities: i) definition of the expected ground motion at the outcropping rock, ii) assessment of the geo-morphological features of the area under study and iii) determination of the mechanical properties of the soil deposit and of the underlying bedrock by means of geophysical and geotechnical investigation campaigns. Ground response analysis is used to predict the design ground motion at a site for seismic microzoning or alternatively to evaluate the seismic stability of slopes and earth structures. The main purpose of microzonation is to provide the local authorities with tools for assessing the seismic risk associated with the use of lands as well as to estimate the seismic motion to be used in the design of new structures and/or retrofitting existing ones.

The main focus of this chapter is on the geological, geophysical and geotechnical investigations that need to be carried out when performing studies of microzonation. Planning of such investigations depends primarily on the definition of following aspects of the problem:

- Geology and geo-morphology of the area under study

A. Ansal (ed.), *Recent Advances in Earthquake Geotechnical Engineering and Microzonation*, 101–137.
© 2004 *Kluwer Academic Publishers.*

- Kinematics of the wave-field and type of analysis concerning the geometry (1D, 2D, 3D)
- Constitutive modelling of the subsurface

Concerning the first aspect and the associated geometry of the problem, in the absence of topographic irregularities and of deep geological structures, 1D-solutions yield reasonable results and therefore are often used. In other cases 2D or 3D solutions are required. The shape of the boundaries of non-consolidated sediment valleys as well as of deeper geologic structures introduce additional wave field effects such as generation of surface waves which tends to increase the amplitude as well as the duration of ground motion (Dobry and Iai, 2000). These effects cannot be taken into account by simple 1D or 2D modelling which considers only vertically incident plane wave (Riepl et al., 2000). In these cases or when the kinematics of the problem cannot be represented by vertically incident shear waves, a more complex wave field need to be considered which includes both non-vertical impinging body waves rays and surface waves. As far as constitutive modelling of geomaterials is concerned, current soil models can be divided in three main families: equivalent linear models, simplified cyclic non-linear models, advanced cyclic non-linear models. Of these families, equivalent linear and simplified cyclic non-linear models are the most commonly used because of their simplicity of implementation and ease of determination of the constitutive parameters. On the other hand, advanced cyclic constitutive models use fundamental principles of continuum mechanics to describe complex aspects of soil behaviour such as strain localization and instability. Furthermore these models are applicable to a wide variety of initial stress and drainage conditions, stress/strain paths, stress/strain magnitude etc. Unfortunately the implementation of these models is generally complex and the calibration of their constitutive parameters is not simple. For this reason, the use of advanced constitutive models in geotechnical earthquake engineering practice is still limited. Some key aspects of equivalent linear and simplified cyclic non-linear models will be illustrated in the next section together with a brief illustration of the techniques used to solve the dynamic equations of motion.

4.2. Mechanical Behaviour of Geomaterials

The mechanical response of geomaterials to earthquake loading can be modelled using a large variety of constitutive models. A detailed description of the capabilities and limitations of each of these models is beyond the scope of this work. Instead it will be made reference for simplicity to the framework of soil behaviour proposed by Jardine (1985, 1992, and 1995). Jardine's qualitative model considers in addition to the State Boundary Surface (SBS, Y_3), two other kinematic sub-yield surfaces (Y_1 and Y_2) which are located inside the SBS and are always dragged with the current stress point. On the contrary the SBS is relatively immobile so that any sharp change of the Effective Stress Path (ESP) from the Y_3 inwards leaves its position unchanged except in soils with highly developed fabric in which the collapse of the structure can cause the SBS to contract.

The essential features of this model are shown in the p'-q plane (Figure 4.1), that can be divided into three distinct zones in order to match the behaviour of a large variety of soils and soft rocks, which has mainly been observed in monotonic and cyclic triaxial (TXT) and torsional shear (TST) tests.

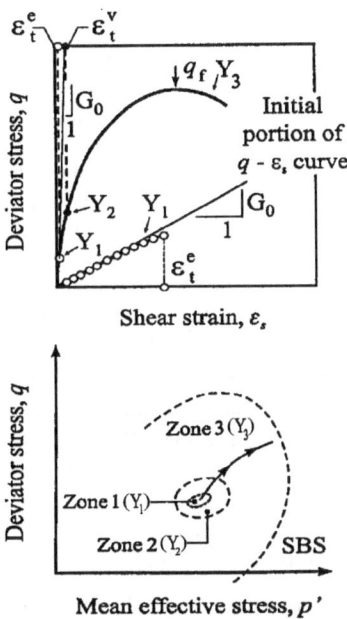

Fig. 4.1. Qualitative stress-strain behaviour of soils (Jardine 1985, 1992, 1995; Reproduced with permission from Balkema)

Zone 1: within this zone, for all practical purposes, the soil exhibits a linear stress-strain response. Soil deformability mainly depends on soil structure and state. The latter can be conveniently expressed by the void ratio and effective geostatic stresses (Jamiolkowski et al., 1995; Tatsuoka and Shibuya, 1992; Tatsuoka et al., 1995a). Therefore, the Young's modulus E_0 and shear modulus G_0, within this zone, can be regarded as the initial stiffness of the relevant stress-strain curves for a given soil. Both these moduli, if properly normalised with respect to the void ratio and effective stresses, are independent of the type of loading (monotonic or cyclic), applied shear strain level, number of loading cycles N and exhibit a negligible dependence on the strain rate and the stress/strain history. Energy dissipation, within this zone, is mainly due to frictional losses between soil particles and fluid flow losses due to the relative movement between the solid and fluid phases, and exhibits a strong dependence on the loading rate or loading frequency (Tatsuoka et al., 1995b; 1997). The limits of the "linear" zone are defined by the so-called linear threshold strain ε_t^l (Vucetic, 1994; Jamiolkowski et al., 1995). For young uncemented soils the linear threshold strain ranges in between $\varepsilon_t^l \square 0.0007 \div 0.002\%$ (Jamiolkowski et al., 1995). Aging, cementation and other diagenetic processes increase ε_t^l up to one order of magnitude (Tatsuoka and Kohata, 1995; Tatsuoka et al., 1997). The linear threshold strain also increases with strain rate or frequency (Isenhower and Stokoe 1981, Tatsuoka and Shibuya, 1992; Lo Presti et al., 1996; 1997). Such an increase is due to the fact that the size of the "linear domain" in the q-p' space (see Figure 4.1) is to a certain extent rate dependent, especially in the case of fine granular soils. With regard to constitutive modelling, it is reasonably to assume that within Zone 1 soils behave like linear viscoelastic solids.

Zone 2: When soil is strained beyond ε_t^l the ESP penetrates into Zone 2 (Figure 4.1). In this zone the stress-strain response becomes non-linear. Consequently the secant stiffness G and E depend not only on the current state of the soil but also on the imposed shear stress or strain level. Moreover G and E are influenced by many other factors such as strain rate, ageing, OCR, recent stress history, direction of the applied ESP, etc. The decay of E and G for increasing strains generally does not exceed 20-30% of their initial value. Moreover, in cyclic tests, the stiffness is only moderately affected by the number of loading cycles and after few cycles the stress-strain response becomes stable. This indicates that the plastic strains inside Zone 2 are negligible and therefore the soil behaviour can be modelled as non-linear viscoelastic. The boundary between Zones 2 and 3 can again be defined in terms of the strain level, which is called the volumetric threshold strain ε_t^v (Dobry et al., 1982). The volumetric threshold strain coincides with the onset of important permanent volumetric strains (ε_v^p) and residual excess pore pressure (Δu) in drained and undrained tests respectively. The values of ε_t^v are at least one order of magnitude higher than those of ε_t^l (Vucetic, 1994; Stokoe et al., 1995; Dobry et al., 1982; Chung et al., 1984; Lo Presti, 1989). Moreover, the values of ε_t^v are influenced by the following factors and phenomena: creep, moderate cyclic loading at strains larger than ε_t^v, overconsolidation ratio or prestressing, direction of the stress path and strain rate (Tatsuoka et al., 1997).

Zone 3: When the deformation process engages Y_2, the soil starts to yield and the plastic deformations become important. As the ESP proceeds towards the State Boundary Surface (SBS) that coincides with Y_3, the ratio of the plastic shear strain to the total shear strain increases approaching values close to unity at Y_3. In Zone 3 the stress-strain response of soils becomes highly non-linear. G, E and D depend to a great extent on the shear stress and strain level. Factors such as strain rate, creep and OCR greatly influence the magnitudes of these parameters. Moreover, the stress-strain response to cyclic loading is no longer stable and a continuous degradation of the mechanical properties of soil is observed. In the case of undrained cyclic loading of granular soils, this leads to a continuous accumulation of the pore pressure and eventually to liquefaction. Within this zone appropriate soil modelling requires the adoption of constitutive models based on viscoplasticity theories.

Figure 4.2 summarises some of the concepts that have been previously discussed, showing, for different strain intervals, the essential features of the stress-strain behaviour, the main influential factors, the constitutive model and the method of analysis.

The soil parameters needed for the models considered in the present chapter are the following:

- the small strain shear modulus G_0. It can be obtained from i) the velocity of propagation of shear waves V_S according to the well-known formula $G_0 = \rho V_s^2$ with ρ = mass density; ii) the slope of unload-reload loops at very small strains (less than 0.001%); iii) the initial slope of the stress-strain curve obtained for monotonic loading at very small strains (less than 0.001%);
- the small strain Young modulus E_0 or the small strain bulk modulus (K_0) or the small strain constrained modulus (M_0) or the Poisson ratio (v_0);

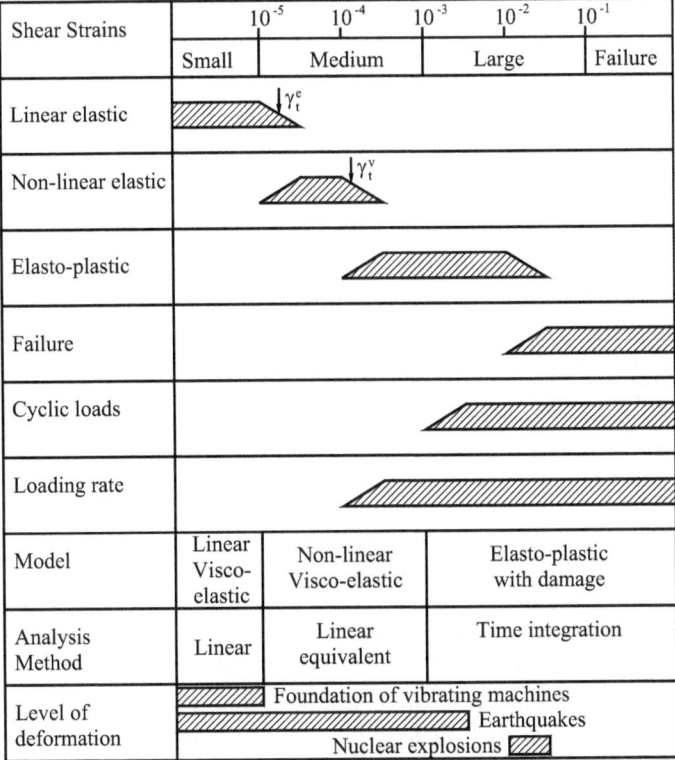

Shear Strains	10^{-5}	10^{-4}	10^{-3}	10^{-2}	10^{-1}
	Small	Medium		Large	Failure
Linear elastic	γ_t^e				
Non-linear elastic		γ_t^v			
Elasto-plastic					
Failure					
Cyclic loads					
Loading rate					
Model	Linear Visco-elastic	Non-linear Visco-elastic		Elasto-plastic with damage	
Analysis Method	Linear	Linear equivalent		Time integration	
Level of deformation	Foundation of vibrating machines	Earthquakes		Nuclear explosions	

Fig. 4.2. Soil behaviour - Constitutive models and methods of analysis - Strain levels (modified after Ishihara, 1996; Reproduced with permission from Oxford University Press)

- the small strain damping ratio D_0. In principle, two values of the damping ratio, from compression and shear tests, should be determined. Typically these two values results experimentally almost coincident (Tatsuoka et al., 1995b);
- if adopting equivalent linear models the $(G - \gamma)$ and $(D_S - \gamma)$ and $(E - \varepsilon)$ and $(D_E - \varepsilon)$ curves should be determined to characterise the variation of stiffness and damping ratio with strain level. If a constant Poisson ratio is assumed, only the $(G - \gamma)$ and $(D_S - \gamma)$ degradation curves are required.
- if adopting simplified cyclic non-linear models the following pieces of information are required: i) the so-called "backbone" curve needed to describe the stress-strain curve for monotonic loading; ii) a "rule" to simulate the unloading-reloading behaviour and stiffness degradation as the seismic excitation progresses (Masing, 1926; Idriss et al., 1978; Tatsuoka et al., 1993). For 1D problems both in geometry and kinematics, the definition of simplified cyclic non-linear models is straightforward.

The parameters previously introduced can be determined from undrained tests and typically they are used in total stress analyses. Effective stress analyses require the experimental evaluation of pore pressure build-up with strain level and number of loading cycles. Alternatively, it is possible to implicitly account for the effects of pore-pressure build-up by determining experimentally the corresponding degradation of soil properties.

In order to complete the model for ground response analysis, it is finally necessary to i) specify the spatial variation, within the domain considered in the analysis, of the soil parameters previously defined, and ii) prescribe the location of the bedrock where the ground motion is specified together with its mechanical properties.

4.3. Laboratory Tests

Assessment of soil non-linearity is mainly achieved with laboratory tests on undisturbed or reconstituted samples. The influence of disturbance or destructuration on the normalised G/G_0-γ and D -γ curves is not fully understood and contradictory results have been published in the literature. The use of high quality samples and the improvement of sample quality are therefore the only possibilities of obtaining more reliable parameters.

4.3.1. TRIAXIAL TESTS

Triaxial tests are typically performed on cylindrical specimens with height to diameter ratio H/D ranging between 2.0 and 2.5. The total stress-path during cyclic or monotonic compression loading triaxial tests is shown in Figure 4.3 together with the scheme of the apparatuses in use at the geotechnical laboratory of Politecnico di Torino.

The difference between the total stress-path of a triaxial test and that induced by an earthquake is well recognised. Indeed, the loading history induced by an earthquake can be decomposed into harmonic shear stresses. Hence the stress-path experienced by the specimen during a resonant column or torsional shear test is more representative. Indeed, due to the anisotropic behaviour of soils, it is expected a different stress-strain response under different stress-paths. Nevertheless, the triaxial test offers some advantages in comparison to other laboratory tests. In particular, triaxial tests can be easily performed under stress or strain control and there is no limit in the maximum achievable strain.

In the last 15 years the triaxial apparatus has been deeply innovated and enhanced. Some of these innovations are fundamentals in order to obtain accurate stress-strain measurements. In particular, the following requirements are extremely important:
- local strain measurements are always preferable and are strongly recommended for any kind of soil. In particular, local strain measurements are imperative when testing hard soils or soft rocks that usually exhibit very small strains during the reconsolidation to the in situ geostatic stress;
- end capping is strongly recommended for hard soils and soft rocks;
- LDTs (Goto et al., 1991) have proved to be very effective in the measurement of local axial strains of hard soils (gravels, sands) and soft rocks. In the case of clay samples, the use of submersible LVDTs or proximity transducers seems to be preferable. The ability to re-setting the sensor position from outside the cell (Fioravante et al., 1994) can be very useful;

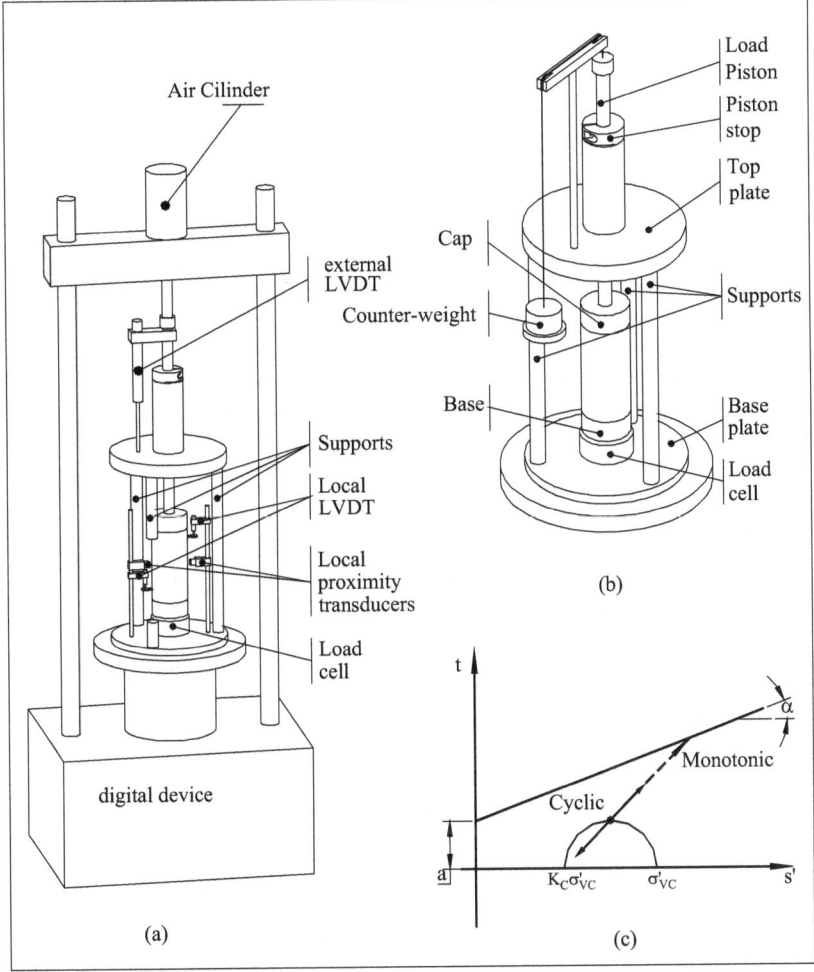

Fig. 4.3.Triaxial apparatus: (a) scheme; (b) cell structure; (c) stress path

- a cell structure with very low compliance and a loading ram virtually frictionless are also critical (Tatsuoka, 1988);
- the actuator resolution and its ability to apply a given constant strain rate is another essential feature of laboratory testing that should be carefully considered. The actuator should also be able to apply small and large cyclic loading under displacement control without backlash. Two examples of system having these characteristics can be found in the literature. Tatsuoka et al. (1994) used an analogous motor with Electro-magnetic clutches to change the direction of loading ram motion without backlash. Shibuya and Mitachi (1997) used a digital servomotor to control a minimum axial displacement of 0.00015 micrometer spanning over several orders of the rate of axial straining.

4.3.2. RESONANT COLUMN AND TORSIONAL SHEAR TEST

The Resonant Column (RCT) gives a very accurate and repeatable estimate of the small strain shear modulus. Recently, the measurement of shear wave velocity in the laboratory by means of piezo-ceramics, called bender elements, has become popular. This measurement is quite inexpensive and can be performed repeatedly during a triaxial or torsional shear test; however, the results do not have the same repeatability and accuracy of the RCT. The main drawbacks of RCT are the following:

- RCT uses very high frequencies and consequently the sample is subjected to very high strain rates. Equivalent strain rates in cyclic tests can be computed as $\dot{\gamma} = 4\,\gamma_{SA}\,f$ [%/s] where: $\dot{\gamma}$ = shear strain rate; γ_{SA} = single amplitude shear strain [%]; f = frequency [Hz]. According to this equation the equivalent strain rates in a RCT increase from several %/min to several thousands %/min as the strain increases from 0.001 % to 0.1 % (Lo Presti et al., 1996). The influence of strain rate on the small strain shear modulus is quite negligible for a great variety of geomaterials (Tatsuoka et al., 1997) but it becomes increasingly important for higher strain level (see as an example Figure 4.4). The influence of strain rate on G as a function of the strain level can be evaluated by means of the following empirical parameter (Tatsuoka et al., 1997; Lo Presti et al., 1996):

$$\alpha(\gamma) = \frac{\Delta G(\gamma)}{\Delta(\log \dot{\gamma}) \cdot G(\gamma, \dot{\gamma}_{REF})} \tag{4.1}$$

i.e. the increase in shear modulus for one logarithmic cycle of strain rate normalised with respect to the shear modulus correspondent to a reference strain rate.

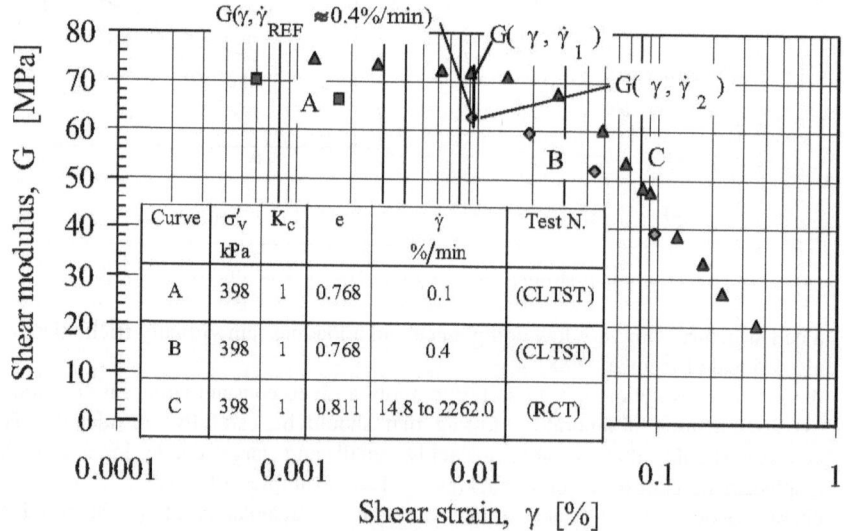

Fig. 4.4. G-γ curves from CLTST and RCT tests of Augusta clay (Lo Presti et al., 1996. "Rate and creep effect on the stiffness of soils", GSP No.61, ASCE; Reproduced with permission from ASCE)

Figure 4.5 shows, for a variety of geomaterials, that $\alpha(\gamma)$ increases with PI and, for a given soil, increases with γ. Consequently, the shear modulus decay curve $G/G0$ is rate dependent and very different results can be obtained depending on the loading frequency. Therefore, in the case of static problems or even for seismic problems it is preferable to obtain the $G/G0$ curve from TST with constant $\dot\gamma$ and frequencies not larger than 10Hz;

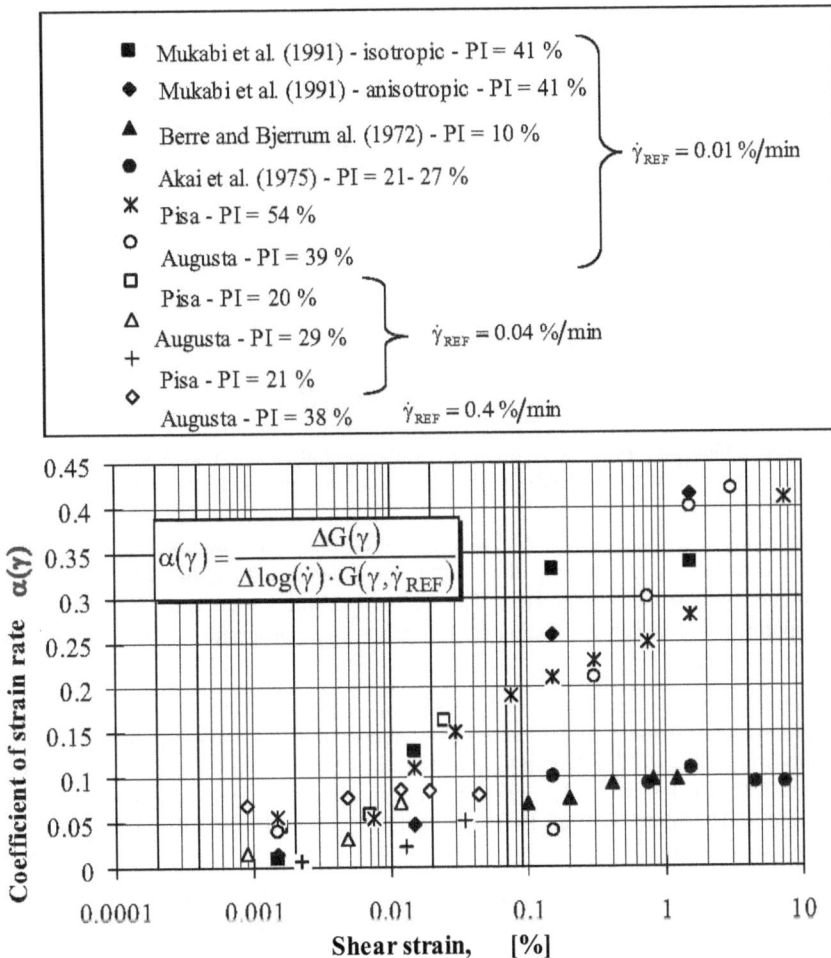

Fig. 4.5. Coefficient of strain rate vs. shear strain (Lo Presti et al., 1996. "Rate and creep effect on the stiffness of soils", GSP No.61, ASCE; Reproduced with permission from ASCE)

- damping ratio (D) is much more dependent on frequency or strain rate than G and even at very small strains the frequency dependency of D has been observed (Papa et al., 1988; Tatsuoka and Kohata, 1995; Stokoe et al., 1995; Lo Presti et al., 1997;

Cavallaro et al., 1998; d'Onofrio et al., 1999). In particular, the damping ratio values obtained from RCT are markedly greater than those inferred from cyclic tests at frequency from 0.1 to 1.0Hz as shown in Figure 4.6;

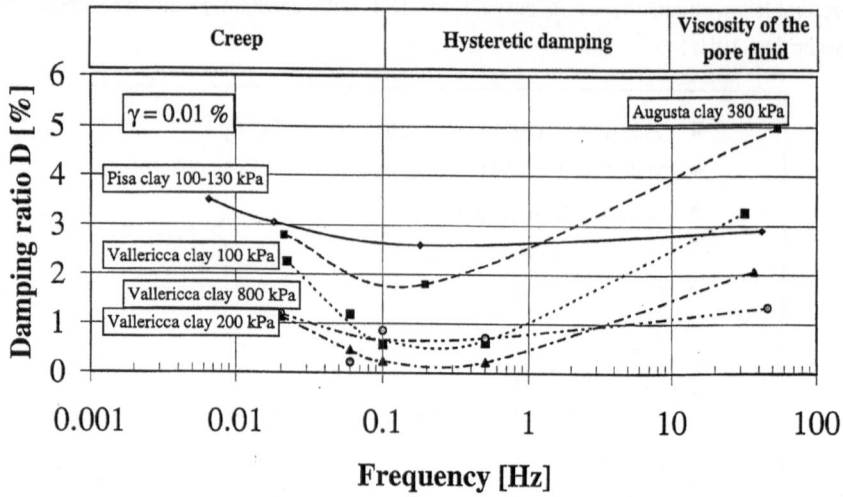

Fig. 4.6. Damping ratio vs. frequency (Lo Presti et al., 1997, Cavallaro et al., 1998, D'Onofrio et al., 1999; Reproduced with permission from Balkema)

- according to Stokoe et al., (1995) the D values obtained with a RC apparatus should be corrected by subtracting the equipment-generated damping (D_{app}) which is a frequency dependent parameter. The $D_{app} - f$ calibration curve is different for each apparatus and it can be evaluated following an appropriate calibration procedure (Stokoe et al., 1995).

The use of the Torsional Shear Tests (TST), instead of the Resonant Column, seems to be preferable and many researchers have adapted the RC apparatus to perform TST (Isenhower et al., 1987; Alarcon-Guzman et al., 1986; Lo Presti et al., 1993; Kim and Stokoe, 1994; d'Onofrio et al., 1999). The advantage of this hybrid apparatus is that it is possible to determine G_0 from RCT and perform cyclic loading torsional shear tests at various frequencies.

On the other hand, several hollow cylinder torsional shear apparatuses have been developed in various research laboratories (Hight et al., 1983; Miura et al., 1986; Pradhan et al., 1988; Teachavorasinskun, 1989; Alarcon-Guzman et al., 1986; Vaid et al., 1990; Yasuda and Matsumoto, 1993; Ampadu and Tatsuoka, 1993; Cazacliu, 1996; Di Benedetto et al., 1997; Ionescu, 1999; Yamashita and Suzuki, 1999). The main advantage of these devices is that they can operate at constant strain rate and over a very wide strain interval. Some of these devices have been developed to study the stress-strain relationship of geomaterials under a more general stress state and stress-path. Other devices have been developed as an alternative to triaxial tests. An effort to standardise this test has been undertaken in Japan (Toki et al., 1995).

4.4. Field Tests

4.4.1. GEOPHYSICAL TESTS

Seismic tests are conventionally classified into borehole (invasive) and surface (non-invasive) methods. They are based on the propagation of body waves [compressional (P) and/or shear (S)] and surface waves [Rayleigh (R)], which are associated to very small strain levels (i.e. less than 0.001 %) (Woods, 1978). Assuming a linear elastic response, the following relationships allow computing the small-strain deformation characteristics of the soil from the measured body wave phase velocities:

$$G_o = \rho V_s^2 \qquad (4.2)$$

$$M_o = \rho V_p^2 \qquad (4.3)$$

$$v = (V_p^2 - 2V_s^2)/2(V_p^2 - V_s^2) \qquad (4.4)$$

where: G_0, M_0 = small strain shear and constrained modulus respectively; ρ = mass density; V_S, V_P = velocity of shear and compressional waves respectively; v = Poisson ratio.

In saturated porous media the measured P wave velocity corresponds to the compression wave of the first kind (Biot, 1956a, 1956b) that is strongly influenced by the pore fluid. In this case the above equations are no longer valid and must be replaced with the corresponding ones of poroelasticity theory.

Seismic tests may also be used to determine the material damping ratio by measuring the spatial attenuation of body or surface waves:

$$D_o = \frac{\alpha V}{2\pi f} \quad (D_o < 10\%) \qquad (4.5)$$

where D_0 = small-strain material damping ratio; α, V = attenuation coefficient and velocity, respectively, of P, S or R waves and f = frequency.

Material damping measurements are difficult because they require accurate measurements of seismic wave amplitude and accurate accounting of the effects of geometric (radiation) attenuation (Rix et al., 2000).

Even at strains less than the linear threshold strain, soils have the capability not only of storing strain energy (elastic behaviour) but also of dissipating it over a finite period of time (viscous behaviour) (Ishihara, 1996). This type of behaviour can accurately be modelled by the theory of linear viscoelasticity. An important result predicted by this theory is that soil stiffness and material damping are not two independent parameters, but they are coupled due to the phenomenon of material dispersion (Aki and Richard, 1980). Assuming that material-damping ratio at small strains is rate-independent over the frequency range of interest (i.e. the viscoelastic response is assumed hysteretic), the following dispersion relation describes the coupling between the material damping ratio and the velocity of propagation of seismic waves:

$$V(f) = \frac{V(f_{ref})}{\left[1 + \dfrac{2D_o}{\pi} \ln\left(\dfrac{f_{ref}}{f}\right)\right]}$$

(4.6)

where: f_{ref} = reference frequency (usually 1Hz).

This implies that a vigorous procedure to estimate the small-strain dynamic properties of geomaterials should determine velocity of propagation of seismic waves and material damping ratio simultaneously rather than separately as it is done in the current practice (Lai and Rix, 1998; Lai et al., 2001; Rix et al., 2001; Lai et al., 2002).

Borehole methods

The most widely used borehole methods in geotechnical engineering are Cross Hole (CH), Down Hole (DH), Suspension PS logging (PS) (Nigbor and Imai, 1994) tests. Strictly speaking, the Seismic Cone (SCPT) test is not a borehole method, but it is based on the same principle. Their popularity is due to the conceptual simplicity. The measurement of the travel time of P and/or S waves, travelling between a source and one or more receivers is determined from the first arrival of each type of wave. Current practice and recent innovations of borehole methods are covered by many comprehensive works (Auld, 1977; Stokoe and Hoar, 1978; Woods, 1978; Woods and Stokoe, 1985; Woods, 1991, 1994). In the following, only some aspects of the borehole methods are briefly summarized. In particular, the focus is placed on emphasising the importance of respecting these testing procedures:

- good mechanical coupling between receiver, borehole casing (if used) and surrounding soil must be guaranteed. A distinct advantage of the SCPT is that good coupling is virtually assured. With conventional cased and grouted boreholes, good coupling is less certain and, more importantly, is difficult to verify. The need for good coupling is particularly important for attenuation measurements, which require accurate amplitude data;
- a check of the borehole verticality with an inclinometer is also highly recommended in order to determine accurately the length of wave travel path in CH tests;
- it is important to generate repeatable waveforms with the desired polarity and directivity. This allows receivers to be oriented in such a way to optimise the measurement of a particular wave type, the use of reversal polarity to make the identification of wave arrivals easier, and measurements along different directions to infer structural and stress-induced anisotropy as explained below;
- in down-hole measurement, the use of two of receivers located at a fixed distance apart (Patel, 1981) can increase the accuracy and the resolution because the *true interval method* for data interpretation can be implemented;
- dedicated portable dynamic signal analysers and computer-based data acquisition systems allow more sophisticated data processing methods. Thanks to these enhancements, it is now possible to routinely use cross correlation (time domain) or cross power spectrum (frequency domain) techniques to estimate travel times instead of subjective identification of the first arrivals in the time histories. In addition, as multi-channel data acquisition systems become more common, the logical extension will be to use arrays of receivers and array-based signal processing (seismic tomography).

Generally, the shear wave velocity profiles inferred from various borehole tests are in good agreement (see the example in Figure 4.7). However, SCPTs generally provide values of the shear wave velocity slightly higher than those inferred from down-hole or cross-hole tests.

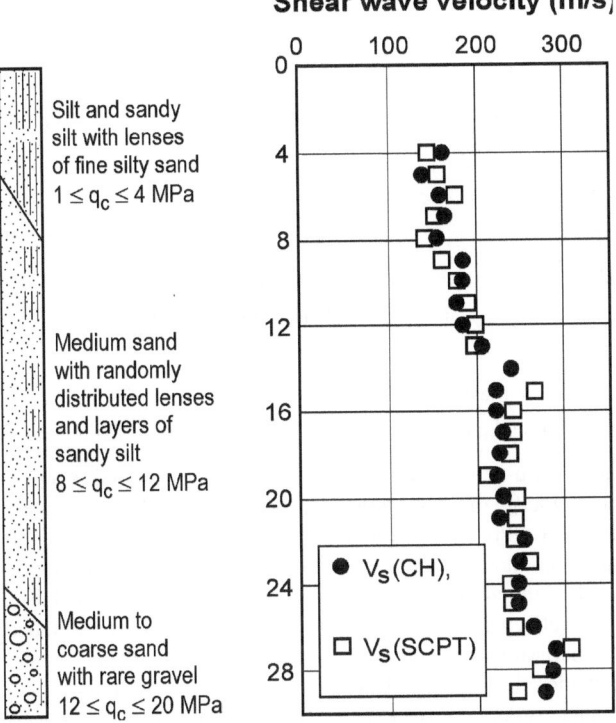

Fig. 4.7. Shear wave velocity in Po river sand (Jamiolkowski et al., 1998; Reproduced with permission from Balkema)

In the past, differences between CH and DH velocities have been attributed to soil heterogeneity and anisotropy. The following considerations explain why anisotropy is not responsible for these differences. In CH tests, S waves propagate in the horizontal direction with vertical particle motion (S_{hv}). In DH tests, propagation of the S wave is sub vertical with horizontal particle motion (S_{vh}). In a continuous medium, the V_s^{vh} and V_s^{hv} shear wave velocities are the same and a unique value of the shear modulus ($G_{vh} \equiv G_{hv}$) is expected. Figure 4.8 illustrates that $V_s^{vh} = V_s^{hv}$ or $G_{vh} \equiv G_{hv}$ using BE measurements of S_{vh} and S_{hv} waves on a reconstituted sample of Fujinomori clay (Lo Presti et al., 1999). Similar results have been obtained in the case of reconstituted sands by Stokoe et al. (1991), Lo Presti and O'Neill (1991) and Bellotti et al. (1996). Hence, different values of shear wave velocity from CH and DH tests are most likely due to soil heterogeneity (Stokoe and Hoar, 1978).

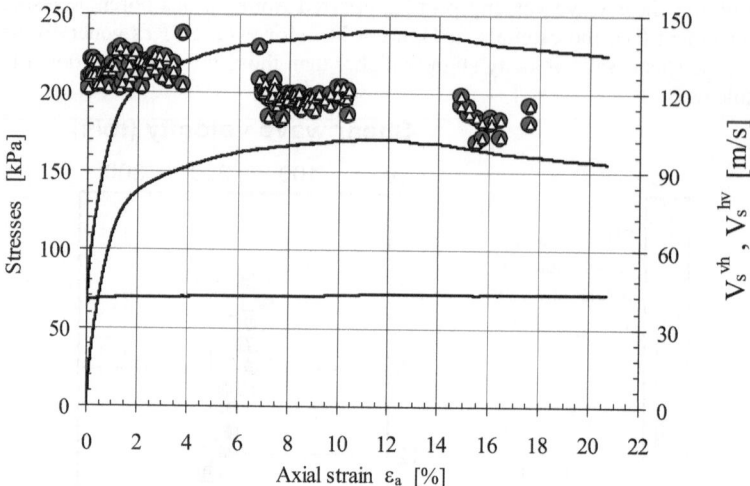

Fig. 4.8. V_s^{vh} and V_s^{hv} measured with BE during drained CLTX test on Fujinomori clay (Lo Presti et al., 1999; Reproduced with permission from Balkema)

The assessment in situ of inherent and stress induced elastic anisotropy is possible by measuring the velocity of propagation of both S_{hv} and S_{hh} waves in CH tests (Jamiolkowski and Lo Presti, 1991; Mitchell et al., 1994; Fioravante et al., 1998). S_{hh} waves propagate in the horizontal direction with particle motion polarized in the complementary horizontal direction. This additional information enables the evaluation of the G_{hh}/G_{vh} ratio, which is a function of inherent and stress-induced anisotropy. Figure 4.9 summarises some field and calibration chamber data. Figure 4.9 indicates that, for the considered granular soils, the inherent anisotropy (inferred at $K_C=1$) causes a 20% to 25% increase in G_{hh} over G_{vh}. The influence of stress induced anisotropy is apparent for other values of K_C.

Recently researchers devoted more attention to inferring the small-strain damping ratio, D_0 from borehole tests. The current methods are based on measures of the spatial attenuation between two or more receivers. The most widely used methods include:

a) *The spectral ratio method* (Mok, 1987; Fuhriman, 1993) is based on the following assumptions which hold only in the far field: i) the amplitude of the body waves decreases in proportion to r^{-1}, where r is the distance from the source, due to geometric attenuation and ii) the soil-receiver transfer function can be considered identical for both receivers. Based on the above assumptions, the damping ratio can be computed by means of the following equation:

$$D(f) = \frac{\ln\left[A_1(f) \cdot r_1 / A_2(f) \cdot r_2\right]}{\Phi(f)} \tag{4.7}$$

where: r_1 and r_2 are the distances from the source of a pair of receivers, $A_1(f)$ and $A_2(f)$ are the amplitude spectra at the two receivers and $\Phi(f)$ is the phase difference between the two receivers.

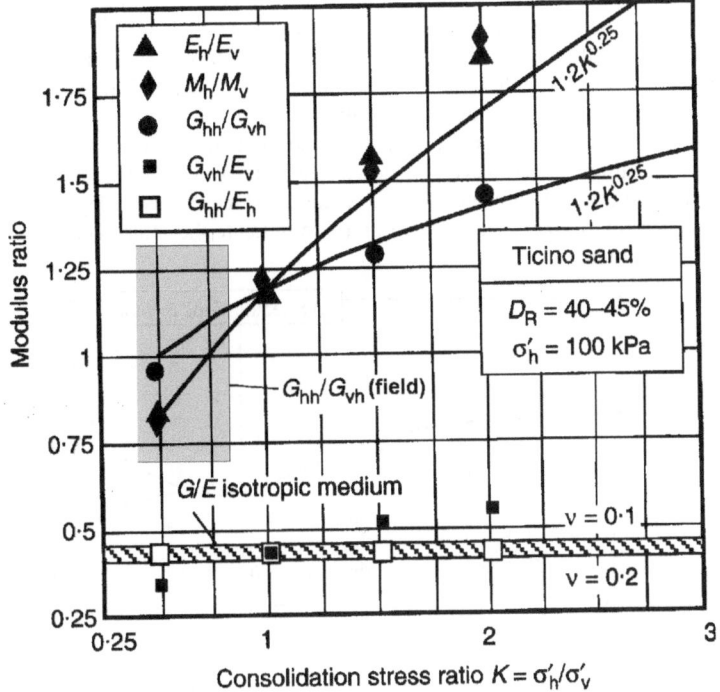

Fig. 4.9. Small strain stiffness anisotropy: field versus laboratory data (modified after Bellotti et al., 1996; Reproduced with permission from the Institute of Civil Engineers)

b) *The spectral slope method*, originally developed for downhole measurements (Redpath et al., 1982; Redpath and Lee, 1986), differs from the spectral ratio method because it assumes that material damping is frequency independent and that it is not necessary to define the law for geometric attenuation. The attenuation constant, defined as the ratio of attenuation coefficient to frequency $k = \alpha/f$, represents the spectral slope, i.e. the slope of the spectral ratio vs. frequency curve:

$$k = \frac{-\Delta\{\ln[A_1(f)/A_2(f)]\}}{\Delta f(r2-r1)} \tag{4.8}$$

therefore the material damping can be computed using the following expression:

$$D(f) = \frac{-\Delta\{\ln[A_1(f)/A_2(f)]\}}{\Delta f \cdot 2\pi \cdot \Delta t(f)} \tag{4.9}$$

Both methods require signal processing prior to interpretation to isolate direct arrivals and frequency ranges. They provide damping values in the bandpass range of the filter. Khawaja (1993) and Fuhriman (1993) recommend performing crosshole tests with four boreholes, in order to obtain stable values of damping with the spectral ratio method.

D. Lo Presti, C. Lai, and S. Foti

They suggested placing the source in the outer boreholes, in order to propagate waves in both forward and reverse directions, and the receivers in the two central boreholes. The spectral ratio method with combined directions provides stable values of damping and avoids the extreme case of negative damping values (Campanella and Stewart, 1990). Campanella and Stewart (1990) studied the applicability of the above methods to the downhole SCPT's. They found that the spectral slope method provides more realistic values of material damping. However, in downhole tests, wave amplitudes are also affected by reflection/transmission phenomena at the interfaces between layers and by ray path divergence: these phenomena make more complicate the interpretation of the particle motion amplitude.

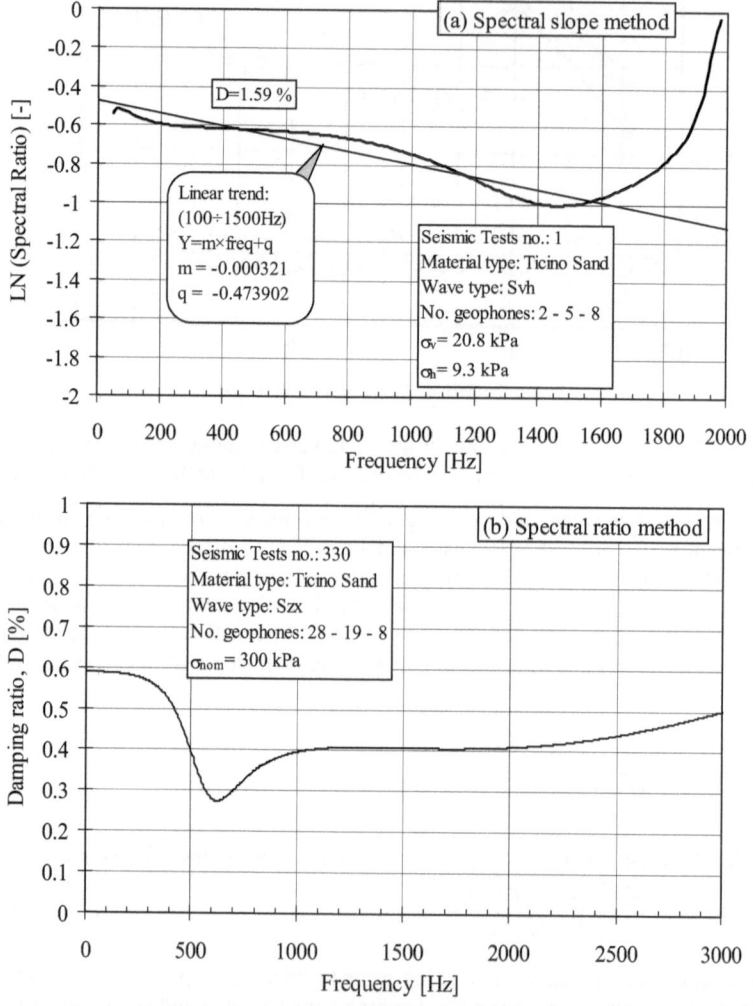

Fig. 4.10. Damping ratio from geophysical tests in calibration chamber (Puci and Lo Presti, 1998; Reproduced with permission from Balkema)

Examples of damping measurements with the spectral ratio and spectral slope methods are given in Figures 4.10a and 4.10b for reconstituted Ticino sand (Puci and Lo Presti, 1998). The results are from seismic tests, performed with miniature geophones embedded in large-size calibration chamber specimens. Figure 4.11 (Puci and Lo Presti, 1998) compares the damping ratio values obtained in the case of reconstituted Ticino sand from laboratory tests (RCT) and those inferred from the spectral ratio and spectral slope methods applied to calibration chamber seismic tests. In this very controlled experiment, the seismic methods yield values of the material damping ratio that generally agree with laboratory values. Measured damping ratios are plotted vs. the corresponding consolidation stresses which have a great influence on the results.

Other approaches to measure the material damping ratio include the rise time method, based on the experimental evidence that a seismic wave signal broaden with distance because of material damping, and the waveform matching method. However, at the present time, none of the available borehole methods to measure material damping ratio appears to be robust enough for routine use in geotechnical engineering practice.

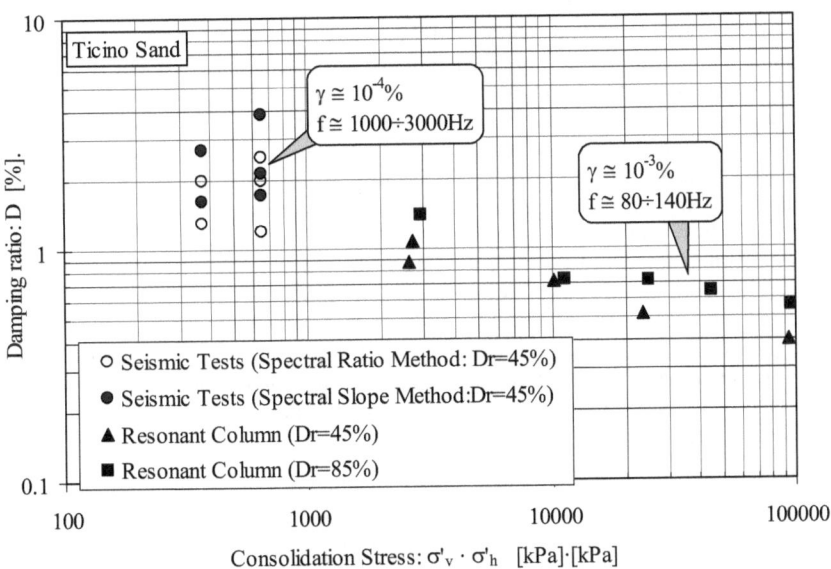

Fig. 4.11. Damping ratio from laboratory and geophysical seismic tests on reconstituted sands (Puci and Lo Presti, 1998; Reproduced with permission from Balkema)

Surface methods

Surface methods are non-invasive field techniques that are executed from the ground surface of a soil deposit; hence they do not require drilling of boreholes or insertion of probes. They include seismic refraction, high-resolution reflection and surface wave methods. Seismic refraction and reflection methods are based on the analysis of body wave propagation and can be performed considering either compressional (P) or shear waves (S) waves. P-wave refraction is often used to locate underlying bedrock formations, while S-wave refraction can be used to obtain the small strain stiffness profile. However particular care must be taken in planning such investigations because there are situations (stiffer-over-softer layers; hidden layers) where the seismic refraction method is not reliable (Reynolds, 1997). High-resolution reflection, on the other hand, does not suffer such limitations; however it requires very intensive data processing.

Advantages of surface methods are mainly related to their non-invasive nature. They are more economical and can be performed more rapidly than borehole methods. Furthermore, in sites like solid waste disposals and landfills, due to environmental concerns, surface methods can be the only choice for geotechnical investigations. Another peculiar aspect of surface methods is related to the volume of soil involved in the test, which is much larger than in borehole methods. As a result, surface methods are particularly useful if the average properties of a soil deposit are to be assessed as in the case of ground response analyses.

In the following, the discussion on surface methods will focus exclusively on surface wave methods mainly because of their relevance in near-surface site characterization.

Early surface wave methods employed laborious field procedures to measure the dispersion curve (i.e. a plot of Rayleigh phase velocity vs. frequency) and crude inversion techniques to obtain the S-wave profile from the experimental dispersion curve (Jones, 1958). Stokoe and his co-workers (i.e. Nazarian, 1984; Stokoe et al., 1989) re-invented engineering surface wave testing by taking advantage of portable dynamic signal analysers, to efficiently measure the dispersion curve, and of the widespread availability of high-speed computers, to implement theoretically-based robust inversion algorithms. In recent years, the Spectral Analysis of Surface Waves (SASW) method has been further improved by incorporating i) the results of a better understanding of surface wave propagation in stratified media, ii) robust and efficient procedures for phase velocity measurements including those associated with passive methods, iii) techniques to determine the shear damping ratio profile from attenuation measurements, and iv) efficient algorithms capable of performing the coupled and uncoupled inversion of surface wave data. Later in this section some of these recent innovations will be briefly described.

The traditional SASW method uses either impulsive sources such as hammers or steady-state sources like vertically oscillating hydraulic or electro-mechanical vibrators that sweep through a pre-selected range of frequencies, typically between 5 and 200Hz (Rix, 1988). R-waves are detected by a pair of transducers located at distances D and D+X from the source (Figure 4.12). The signals at the receivers are digitised and recorded by a dynamic signal analyser. The Fast Fourier Transform is computed for each signal and the cross power spectrum between the two receivers is calculated. Multiple signals are averaged to improve the estimate of the cross power spectrum. The phase angle of the cross power spectrum is used to determine the travel time between the two receivers:

$$t(f) = \phi(f) / 2\pi f \qquad\qquad (4.10)$$

where $\phi(f)$ is the phase difference in radians and f is the frequency in Hz. The Rayleigh phase velocity is then computed as:

$$V_R = \frac{X}{t(f)} \qquad (4.11)$$

The wavelengths corresponding to the frequency dependent Rayleigh phase velocities are computed as $\lambda_R = V_R(f)/f$. The result is the experimental dispersion curve for the considered receiver spacing. The calculations are then repeated reversing the source position and with different receiver spacing (Figure 4.13): the results are combined to form the composite dispersion curve of the site (Figure 4.14a).

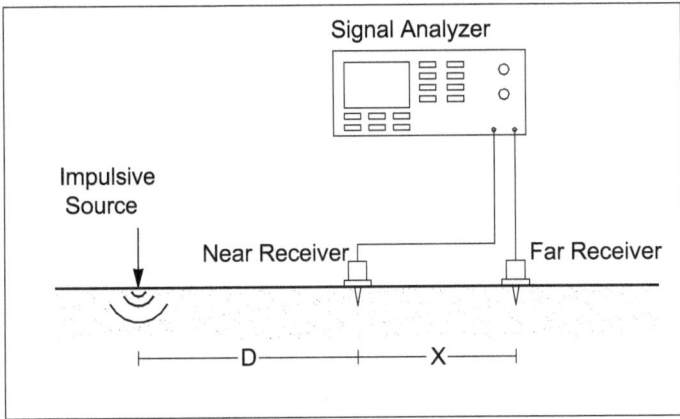

Fig. 4.12. Typical configuration for SASW test

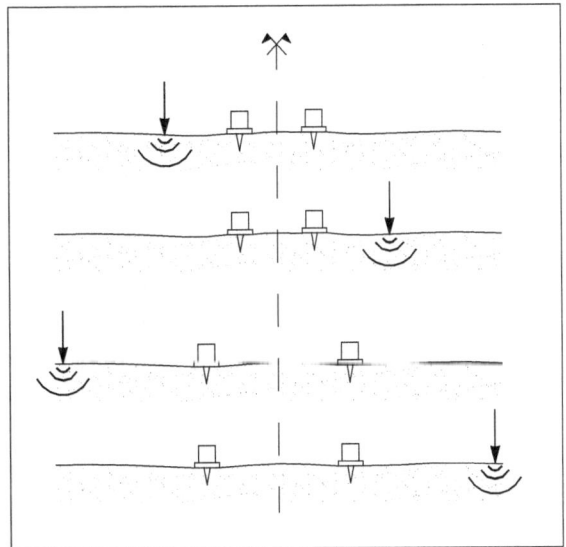

Fig. 4.13. Common receiver midpoint geometry used for SASW test

D. Lo Presti, C. Lai, and S. Foti

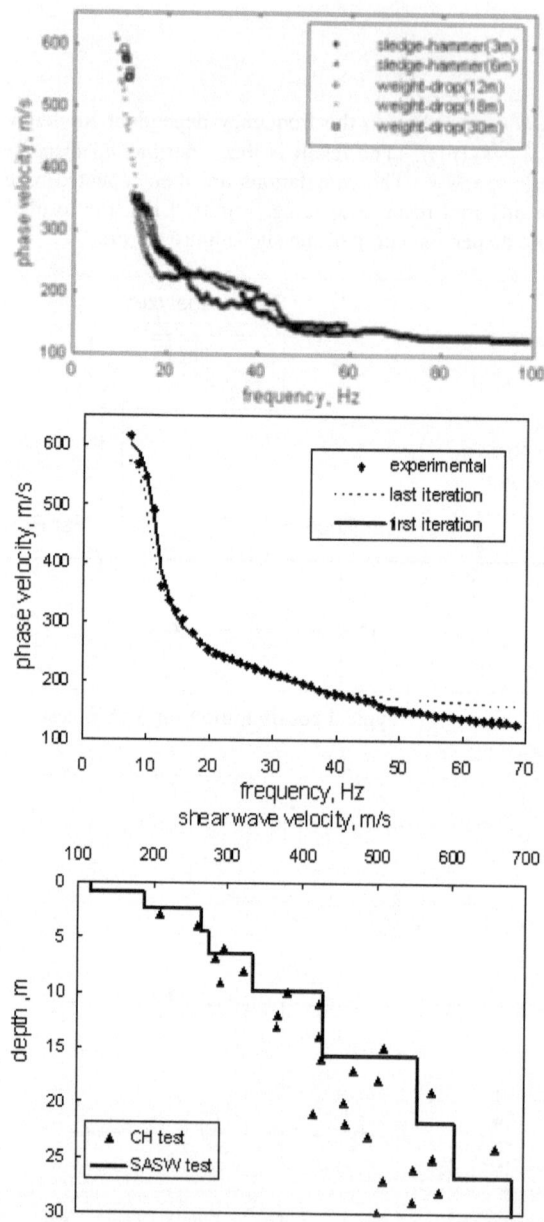

Fig. 4.14. Example of SASW test: a) experimental dispersion curve from multiple receiver spacings, b) Comparison between experimental and theoretical dispersion curves, c) Shear wave velocity profile obtained with the inversion process compared to cross-hole test results (Foti, 2000)

The use of a multi-station testing setup (Figure 4.15) can introduce several advantages in surface wave testing. In this case, the motion generated by an impact source is detected simultaneously at several receiver locations and the corresponding signals are analysed as a whole (i.e. in both the time and space domains) using a double Fourier Transform. It can be shown (Tselentis and Delis, 1998) that the dispersion curve can be easily extracted from the location of the spectral maxima in the frequency-wavenumber domain in which the original data are transformed. Using this technique, the evaluation of the experimental dispersion curve becomes straightforward; furthermore, the procedure can be easily automated (Foti, 2000). With the adoption in SASW testing of the multi-station method, the need of several testing configurations required by the conventional two-station procedure is avoided with the result that testing and interpretation times are considerably reduced.

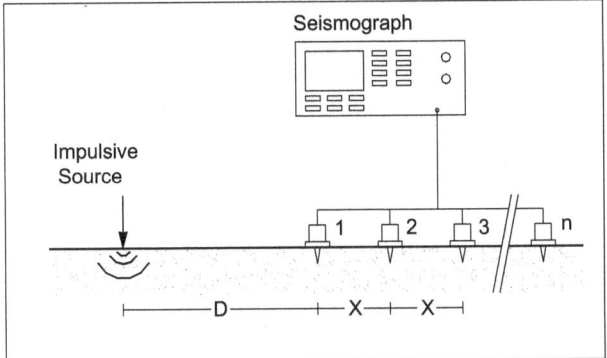

Fig. 4.15. Multistation configuration for SASW test

The experimental dispersion curve is used to obtain the shear wave velocity profile via a process called *inversion*. A theoretical dispersion curve is calculated for an assumed vertically heterogeneous layered soil profile using one of several available algorithms (Haskell, 1953; Thomson, 1950; Kausel and Roësset, 1981; Chen, 1993; Hisada, 1994, 1995). The theoretical dispersion curve is then compared with the corresponding experimental curve and the "distance" between the two curves is used as a basis of an iterative process consisting of updating the current soil profile until the match between the two curves is considered satisfactory. The soil profile may be updated manually by *trial and error* or using an automated minimisation scheme based on an unconstrained or constrained inversion algorithm (Lai and Rix, 1998). When a satisfactory agreement between theoretical and experimental dispersion curves is attained (Figure 4.14b), the final shear wave velocity profile (Figure 4.14c) is taken as representative of the site conditions.

For a successful application of SASW testing, it is recommended to observe the following guidelines:

- in choosing the relative spacing between source and receivers, attention should be placed to minimize near-field effects and spatial aliasing. In this context, the near-field is defined as a region close to the source where the magnitudes of the body wave components of the wave field are of comparable magnitude to the surface wave components. Efforts should be made to eliminate or minimize near-field effects unless they are explicitly accounted for during the inversion process (Roësset et al., 1991; Ganji et al., 1998). In normally dispersive media, the body wave field is significant until D/λ exceeds about 0.5, hence the nearest receiver should be located at least one-half wavelength from the source:

$$D \geq \lambda / 2 \qquad\qquad\qquad (4.12)$$

This recommendation is consistent with other studies of the influence of near-field effects, but more strict requirements are necessary for inversely dispersive stratigraphies (Sanchez-Salinero, 1987; Tokimatsu, 1995). It is also important to limit the distance between receivers to avoid spatial aliasing; a simple criterion is given by:

$$X \leq \lambda / 2 \qquad\qquad\qquad (4.13)$$

- the length of the receiver array must be sufficiently large, if the stiffness profile at great depth has to be estimated. A rule of thumb is that the survey length must be as long as about 3 times the maximum depth of interest. This requirement may not be compatible with the space available at the site. Moreover, massive sources are needed to get good quality signals with long testing arrays, causing an increase of testing time and cost;
- it is important to account for multiple modes of surface wave propagation, especially in irregular, inversely dispersive soil profiles (Gucunski and Woods, 1991; Tokimatsu, 1995). Currently several approaches are used to account for multiple modes. Individual, modal dispersion curves can be calculated and compared with the experimental dispersion curve during the inversion process. Unfortunately, the use of only two receivers in the traditional SASW method prohibits resolving individual modes in the experimental dispersion curve; only the effective velocity representing the combination of several modes can be determined. Also using a multi-station approach the individual modes cannot be separated if a relatively short receiver array is used, as required by engineering practice (Foti et al., 2000). Thus, it must be assumed that the experimental curve represents an individual mode, usually the fundamental mode. This approach is satisfactory only in normally dispersive profiles. Another approach is to calculate the effective velocity directly and use it as the basis of the inversion. Lai and Rix (1998) have developed an efficient procedure based on the normal mode solution to calculate the effective velocity as well as closed-form partial derivatives required for inversion. Finally, it is possible to numerically simulate the SASW test using Green's functions that calculate the complete wave field (Roësset et al., 1991). This approach is computationally expensive, in part because the partial derivatives must be calculated numerically, but it accurately models the actual field procedure used in SASW tests;
- for the inversion of the experimental dispersion curve, it is essential to use theoretically-based inversion algorithms. Prior to the widespread availability of high-speed computers, simple empirical inversion techniques were used. Furthermore, in recent years, there have been attempts to develop simple methods based on parametric studies and regression equations. These methods have limited usefulness and are

likely to yield erroneous results. It is remarked that the rapidly increasing power of personal computers makes it possible to use theoretically-based inversion methods routinely;

- the non-linear inversion of the experimental dispersion curve is inherently ill-posed with the consequence that the solution (i.e. the S-wave profile) is not unique. This problem can be overcome with the recourse of two strategies (Lai and Rix, 1998). First, a priori information about the soil profile can be used to limit the range of possible solutions. Second, additional constraints such as smoothness and regularity (e.g., Constable et al., 1987) may be imposed on the solution.

Among the most recent advances in surface wave testing, there are the determinations of shear damping ratio profile from attenuation measurements and the combined use of active and passive sources to determine the experimental dispersion curve. In the following, some aspects of these two topics will be briefly discussed.

Damping ratio profile. As a surface wave travels away from the source, its amplitude decreases. This decay is due in part to the geometrical spreading of energy over wider wavefronts (geometrical attenuation) and to the energy dissipation (material attenuation). If the material attenuation is estimated experimentally from field measurements, it is possible to evaluate the damping parameters of the soil deposit by using an inversion process similar to that used to infer the stiffness profile. The difference is that, in the case of the damping ratio profile, the experimental curve to be inverted is the attenuation rather than the dispersion curve.

Rix et al. (2000) proposed a method to determine the shear damping ratio profile based on the following assumptions: a) the soil can be modelled as a layered weakly dissipative medium, and b) the propagation of Rayleigh waves is governed by the fundamental mode. The experimental attenuation curve can be estimated using the multi-station method based on measuring the amplitude of the particle motion at several receiver offsets placed on the ground surface at increasing distance from a frequency-controlled source (Rix et al., 2000). One of the most delicate aspects of this procedure is related to a correct estimation of the geometrical attenuation, which is needed to separate the material attenuation. For a homogenous medium, the geometrical attenuation is inversely proportional to the square root of the distance from the source, but in a layered medium it becomes a non-trivial function of the stiffness profile.

Lai and Rix (1998) have proposed a technique where the shear damping ratio profile is determined simultaneously with the shear wave velocity profile in a joint inversion of experimental dispersion and attenuation curves. The method is based on combining results from linear viscoelasticity with some theorems of complex analysis. A coupled simultaneous inversion of both dispersion and attenuation curves is a mathematically better-posed problem, if compared with the corresponding uncoupled inversion. In this view a new multi-station scheme for the simultaneous measurement of dispersion and attenuation curves has also been proposed, based on the transfer function concept (Rix et al., 2001). The experimental transfer function is obtained measuring the acceleration of a controlled harmonic source and the induced motion at several receivers placed along the ground surface. Once the experimental transfer functions have been determined, they are used to compute, in a complex-valued non-linear regression, the frequency dependent complex wavenumbers, which contain information about both the phase velocity and the attenuation. After the experimental dispersion and attenuation curves have been simultaneously measured, the process is completed by using the simultaneous inversion algorithm described above to yield the stiffness and damping ratio profiles (Lai et al., 2002).

Passive measurements. A limitation of surface wave methods based on the use of active sources is related to the depth of investigation. Stiffness characteristics of deep strata can be obtained only using very massive sources and very long receiver arrays. A possible alternative is the measurement of short-period microtremors (T<1s), which are caused by natural events or by human activities in the nearby of the site. Because of the absence of a specific source, such methods are often called passive methods. Microtremors can be associated to Rayleigh waves if the measurements are conducted in favourable weather condition, i.e. in absence of strong winds (Horike, 1985) and if the presumable sources of noise (e.g. heavy traffic on a highway) are far enough from the testing site.

A satisfactory characterisation of a site can be obtained with a hybrid method using short period microtremors for the investigation of deeper layers jointly with an active method to cover the need of high resolution at shallow depth, because in the high frequency range microtremors are strongly affected by noise.

The basic steps of soil characterization using microtremors (Horike, 1985; Tokimatsu, 1995; Zwycki and Rix, 1999) are essentially the same of the SASW test: i) field measurements, ii) determination of the experimental dispersion curve, and iii) inversion process. Several sensors are required because there is no restriction to a single mode or a single direction of propagation, since the actual position and geometry of the source are unknown. Usually the receivers are deployed in a circular array either with or without a receiver in the centre of the array. Using three-dimensional receivers, it is possible to analyse both vertical and horizontal particle motions related to microtremors (Tokimatsu, 1995). As for active surface wave measurements, more than one receiver configuration is needed because spatial aliasing, wavenumber resolution and leakage in the wavenumber domain limit the spectrum of frequency that can be obtained from a single receiver lay-out (Zywicki and Rix, 1999).

Data analysis is performed using high-resolution frequency-wave number spectral estimation techniques, steering the data array with a trial wavenumber array in many directions. The resulting spectrum can be represented as a 2D wavenumber (kx-ky) contour plot for each analysed frequency. The peak of this spatial plot is used to evaluate the wavenumber that is associated to the dominant surface wave and its direction of propagation. The experimental dispersion curve is determined by repeating the above procedure for different frequencies. Finally the dispersion curve is inverted using the same algorithms used in active surface wave testing.

4.4.2. IN SITU LARGE STRAIN TESTS: PRESSURIMETER AND PLATE LOAD TESTS

Pressuremeter and Plate Load Tests (PLT) are used to determine the large strain stiffness of soils in situ.

Several different pressuremeter devices are currently used. The differences mainly concern the way in which the probe is inserted into the soil. In the following, only the

Self Boring Pressurimeter (SBPT) is considered because it causes a limited soil disturbance. On the contrary, the insertion of kinds of probes causes an unavoidable destructuration of the soil fabric and a consequent underestimation of the stiffness especially in the case of aged deposits. The SBPT has been developed in France and UK at the beginning of the 1970s (Baguelin and Jezequel, 1973; Wroth and Hughes, 1973).

A pressuremeter test consists of the expansion of a cylindrical cavity which has a finite length L and diameter D. During the test, the applied cavity pressure (p) and the corresponding circumferential strain at the cavity wall ε are measured. The test yields an expansion curve of the type shown in Figure 4.16 which allows, at least in principle, the direct determination of the following shear moduli:

- G_0 from the initial slope of the expansion curve;
- G_{ur} from small unload-reload cycles which can be performed during the expansion curve or even during the contraction phase of the test.

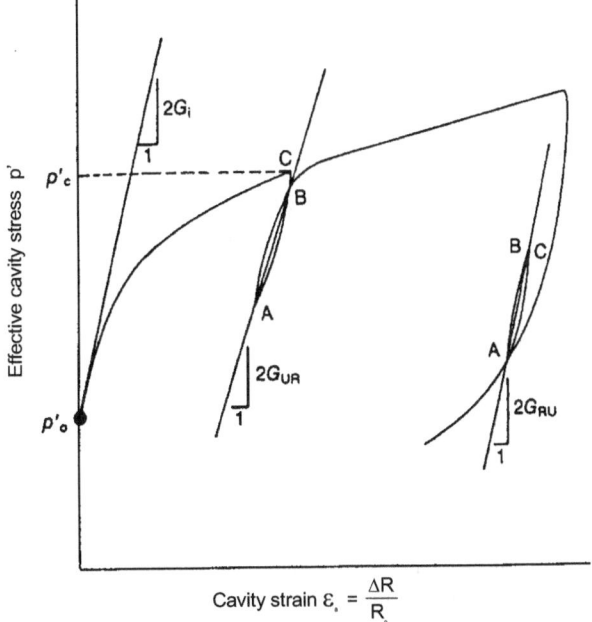

Fig. 4.16. Shear moduli from SBP tests

These moduli can be directly determined from the expansion curve because in both cases it is possible to neglect soil non-linearity as a first approximation. However, the initial shape of the expansion curve is sensitive to soil disturbance and to the compliance of the measuring system even in the case of SBP tests. Consequently, the direct assessment of G_0 from the expansion curve is not possible in practice, while the determination of a pseudo-elastic stiffness from small unload-reload cycles seems quite suitable and reliable if appropriate assumptions are made (Bellotti et al., 1986; 1989; Byrne et al., 1990; Fahey, 1991; Fahey and Carter, 1993; Ghionna et al., 1994).

The interpretations of small unload reload loops (*u-r* loops) in sand have been accomplished by several researchers. By referring to the SBP expansion curve obtained at the Viadana site (Bruzzi et al., 1986; Bellotti et al., 1989) in Holocene Po river sand (Figure 4.17), the G_{ur}value of the loop can be computed using the following expression:

$$G_{ur} = \frac{1}{2}\frac{(p_B - u_o) - (p_A - u_o)}{\varepsilon_B - \varepsilon_A} \tag{4.18}$$

where: p_A, p_B = total cavity stress applied at points A and B respectively; ε_A, ε_B = corresponding circumferential strains at the cavity wall; u_o= hydrostatic pore pressure.

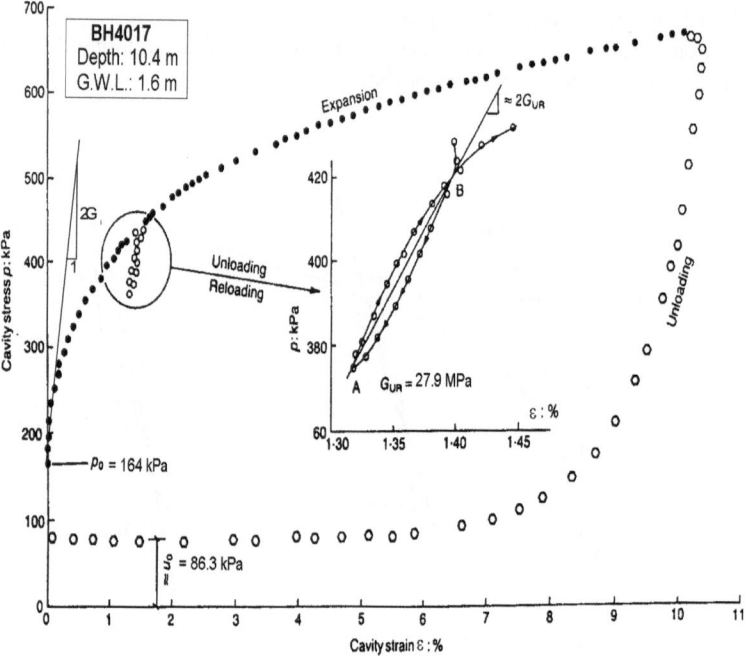

Fig. 4.17. Example of SBPT in Po river sand (Bellotti et al., 1989; Reproduced with permission from the Institute of Civil Engineers)

Considering the pressure-dependency and non-linearity of soil stiffness, it is evident that any rational interpretation of the ur-loop should attempt to link the G_{ur} value to the following factors:

- the average plane strain effective stress $s' = \frac{1}{2}(\sigma_1' + \sigma_3') = \frac{1}{2}(\sigma_r' + \sigma_\theta')$ existing at

  the start of the loop which is a function of p_C' (σ_r'= radial effective stress; σ_θ'= circumferential effective stress, p_C' = effective cavity stress at the start of unloading);

- the relative depth of the loop $R = \dfrac{p'_C - p'_A}{2 \cdot \sin \varphi'_{PS} \cdot p'_C /(1 + \sin \varphi'_{PS})}$ (Fahey, 1991) (φ_{PS}=

  plane strain angle of shear resistance which can be inferred from the expansion curve).

The R relates the current depth of the unload loop $(p'_C - p'_A)$ to the amount of stress unload at which the reverse plasticity can occur. In practice, R plays the same role of the shear stress ratio τ/τ_{max} that describes the non-linearity of soil stiffness measured in laboratory tests (τ/τ_{max} = mobilisation factor, i.e. the ratio of the current shear stress to the failure value).

Plate load tests (PLT) can provide an average (operational) stiffness linked to the load vs. displacement characteristics of the considered boundary value problem. Usually the values of E from these tests are evaluated using the formulae of the theory of elasticity of an isotropic medium. In this case, the soil elements beneath the test foundation not only experience very different strain levels, as in the SBPT, but also follow different stress-paths. To extend such investigations at depth it is necessary to run the test inside a pit excavation.

4.4.3. EMPIRICAL CORRELATIONS FROM PENETRATION TESTS

In spite of the fact that the penetration resistance represents the soil response to very large deformations, it is possible to establish reliable correlations between small strain moduli and penetration parameters. Indeed, both of them are mainly dependent on the soil state, although to a different degree.

Many empirical correlations between the penetration resistance from Standard Penetration Test (*SPT*) (Ohta and Goto, 1978; Imai and Tonouchi, 1982) or Cone Penetration Test (*CPT*) (Sykora and Stokoe, 1983; Robertson and Campanella, 1983; Rix, 1984; Baldi et al., 1986; 1989a; 1989b; Bellotti et al., 1986; Lo Presti and Lai, 1989; Rix and Stokoe, 1991; Mayne and Rix, 1993) and the small strain shear modulus G_0 have been established using different databases. Among the many available correlations, it is worthwhile to mention the following:

a) Ohta and Goto (1978), adapted by Seed et al. (1986)

$$V_s = 69 \cdot N_{60}^{0.17} \cdot Z^{0.2} \cdot F_A \cdot F_G \qquad (4.19)$$

where: V_s = shear wave velocity (m/s), N_{60} = number of blow/feet from *SPT* with an Energy Ratio of 60%, Z=depth (m), F_G =geological factor (clays=1.000, sands=1.086), F_A = age factor (Holocene=1.000, Pleistocene=1.303)

b) Jamiolkowski et al. (1988) have shown that the ratio G_0/q_c mainly depends on relative density and is only moderately influenced by overburden stress. Figure 4.18, which is based on field and CC data, can be used to infer the small strain shear modulus from *CPT* for sands.

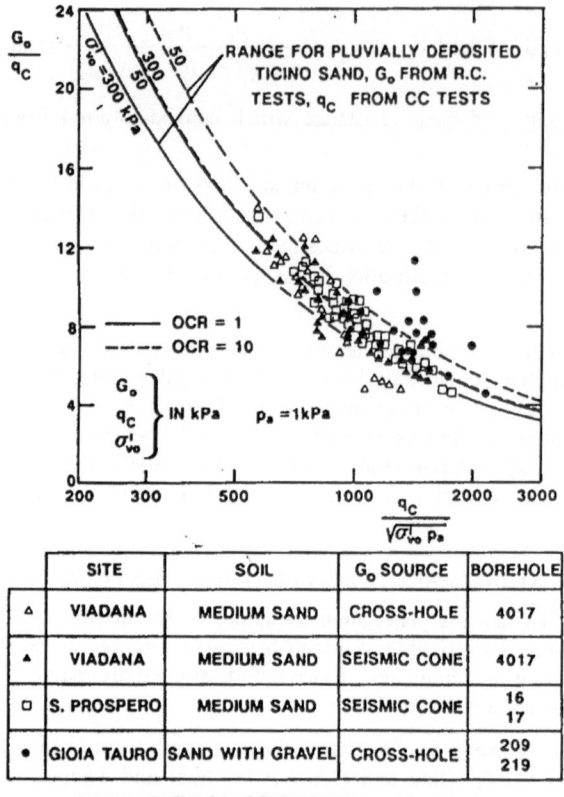

DEPTH BELOW G.L. CONSIDERED: 5.5 TO 43.5 m

Fig. 4.18. q_c vs. G_0 correlation for uncemented predominantly quartz sand (Jamiolkowski et al., 1988; Reproduced with permission from Balkema)

Mayne and Rix (1993) have proposed the following empirical correlation between the shear wave velocity and the cone penetration resistance:

$$G_o = 99.5(p_a)^{0.305}(q_c)^{0.695}/e_o^{1.130}$$

(4.20)

This correlation is based on database from 31 clay sites and takes into account the in situ void ratio which significantly influences G_0, whilst it has much smaller impact on q_c.

Site specific empirical correlations can be profitably used to assess the spatial variability of soil properties in a very cost effective way. However, due to their purely empirical nature, these correlations, when applied to sites which are different from those considered in the original database, can provide just an approximate estimate of G_0 which, in many cases, can be quite far from the actual value.

Empirical correlations between G_0 and the Flat Dilatometer Test results (DMT) (Marchetti, 1980; 1997) have also been established. A review of the many existing correlations has been provided by Mayne and Martin (1998). For a given soil, relatively good correlations have been found between G_0 and the following DMT measurements: the conventional lift-off pressure (p_0), the pressure corresponding to a displacement of 1.1 mm of the central part of the steel membrane (p_1), the wedge resistance (q_D) and the horizontal stress index $(K_D=(p_o-u_o)/\sigma'_{vo})$. On the other hand, poor correlations have been found between G_0 and the conventional dilatometer modulus $[E_D=34.7\ (p_1-p_0)]$.

4.5. Case History

This case history is related to a extensive project for the evaluation of site effects in about 60 municipalities, located in the territories of Garfagnana and Lunigiana (Ferrini et al., 2000). Castelnuovo Garfagnana is the first site in which the comprehensive program of field and laboratory investigations has been completed (Calosi et al., 2001).

The preliminary step for field investigation is the reconstruction of the geology of the whole representative area. This delicate task has to be accomplished accurately because it leads to the choice of significant sections, which will be investigated with geotechnical tests. The location of boreholes and sections for detailed investigations must be chosen carefully, so that the main geological formations are adequately characterised.

The geology of Castelnuovo has been investigated by classification of existing information, field inspection and some seismic refraction tests, aimed at locating the main discontinuities in the formations. According to the simplified geological cross sections reported in Figure 4.19, the main geological formations are (Ferrini et al., 2000):

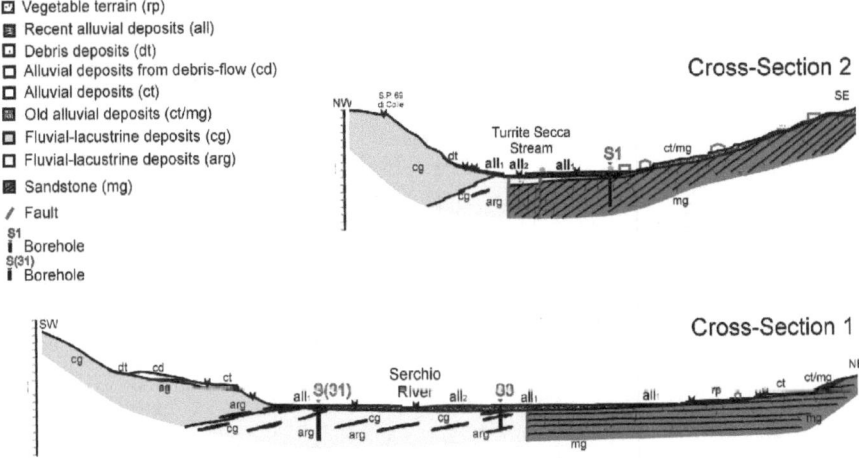

Fig. 4.19. Geological cross-sections of Castelnuovo Garfagnana

- Quaternary deposits that mainly consist of: i) Holocene alluvial deposits of well-graded gravels, sands and silts of variable thickness (ALL), ii) Holocene alluvial terrace of well graded gravels sands and silts (CT), iii) Pleistocene alluvial terrace of uniform pebbles of sandstone in a sand matrix (CT/MG) and iv) Other Holocene deposits (CD, DT);

- fluvial-lacustrine formations from Plio-Pleistocene epoch (i.e. 1.5 to 3 million year old) that mainly consists of i) clay and grey sand, sandy clay and clayey sand of thickness up to 90m (ARG) and ii) gravel and conglomerate in a clayey sandy matrix and pebbles of sandstone (CG, C/MG) of thickness even greater than 50m;
- Macigno sandstone (MG) from Oligocene epoch (i.e. 20 to 30 million year old) of thickness up to 2000m. This formation is strongly weathered at the top and becomes intact after a transition zone. The thickness of the weathered zone ranges from 10 to 20m while that of the transition zone is much more variable.

4.5.1. FIELD TESTS

Once the main geological formations were recognised, a series of geophysical seismic tests have been planned to assess their shear wave velocity and hence their small strain shear modulus. Three down-hole (DH) tests have been executed using a pair of receivers and interpreting the data both with the direct travel time and the visual true interval methods, without significant differences. During the drilling of the boreholes, *SPT* data have been collected.

Seismic refraction tests using horizontally polarised shear waves (SH) and multistation SASW tests have been performed at the same locations of the DHTs. An array of 24 geophones has been used for multistation SASW measurements that have been interpreted in the frequency-wave number domain (fk).

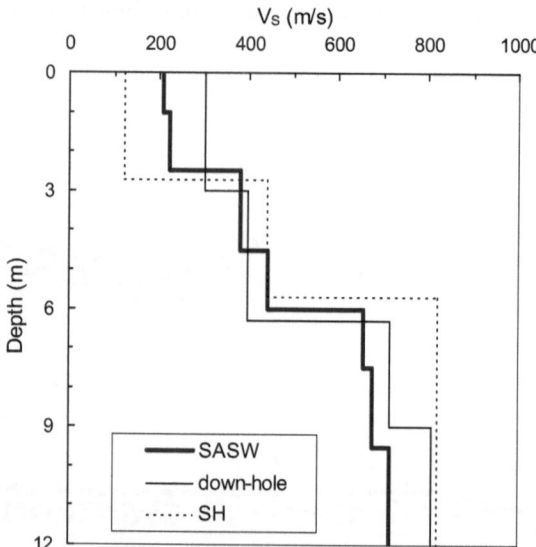

Fig. 4.20. V_S– profiles from different geophysical methods at Castelnuovo

Figure 4.20 shows a comparison between the shear wave velocities obtained with the different methods at the S3 borehole location. A very high initial shear modulus has been found for the ARG formation, leading to the idea that this formation could be locally considered as bed-rock. This initial statement was later modified on the basis of laboratory test results.

Table 4.1 reports the average values of the shear wave velocity for each formation, which have been estimated on the basis of the field test results.

Table 4.1. Average V_s values at Castelnuovo Garfagnana

Geological Formation	Shear wave velocity V_s [m/s]
Holocene Alluvial Deposits (ALL)	250
Holocene Alluvial Deposits (CT, DT, CD)	200
Pleistocene Alluvial Terrace (CT/MG)	380
Fluvial-Lacustrine Deposit (ARG)	800
Fluvial-Lacustrine Deposit (CG)	500
Weathered Sandstone (MG1)	600
Sandstone - transition zone (MG2)	950
Intact Sandstone (MG3)	1200 – 1500

4.5.2. LABORATORY TESTS

Laboratory tests have been performed on samples retrieved using a Shelby tube sampler for the ARG formation and a double core sampler for the MG (Ferrini et al., 2000). On the basis of classification test results the ARG samples are mainly classified as CL-ML and rarely as SM-SC. ARG can be defined well-graded and very dense soils of low plasticity. On the undisturbed ARG samples several Resonant Column (RC) and Cyclic Loading Torsional Shear tests (CLTS) have been performed. Moreover one specimen was tested in the triaxial apparatus in Cyclic Loading Triaxial conditions (CLTX), at constant strain rate and, after a rest period of 24 hrs with open drainage, in Monotonic Loading Triaxial conditions (MLTX). During both CLTX and MLTX tests, the shear wave velocity was measured using bender elements (BE), in order to assess the "elastic" damage due to the progressive straining imposed to the specimen. The details of the testing operations can be found in Calosi et al. (2001).

Normalised shear modulus G/G_0 and damping ratio (D) obtained with CLTX, RC and CLTS tests are compared in Figure 4.21. Because of the high non-linearity exhibited by the ARG specimens in the laboratory tests, it was decided that it would have been better to consider the Macigno formation as bedrock.

Another important aspect to be assessed was the effect of loading cycles. Figure 4.22 shows the degradation parameter ($t = -\log \delta / \log N$) (Idriss et al., 1978) and Figure 4.23 reports a comparison of CLTX and MLTX results. At large strains the stiffness from CLTX becomes even smaller than that from MLTX, because of cyclic degradation. The influence of the number of cycles on the stiffness and damping ratio of ARG specimen is quite negligible, as confirmed by the shear wave velocity (V_s) measurements before and after each loading cycle of the CLTX test, which gave almost the same results.

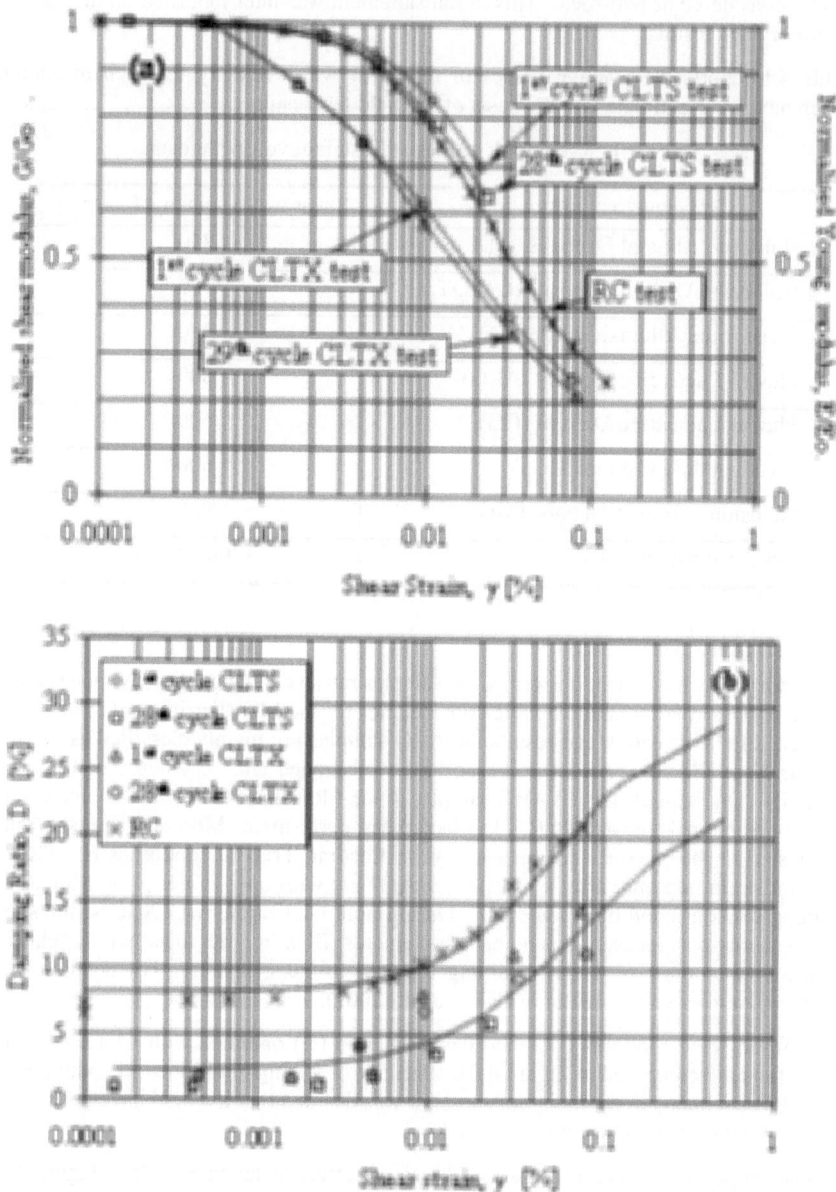

Fig. 4.21. Normalised shear modulus and damping ratio from RC, CLTS and CLTX tests

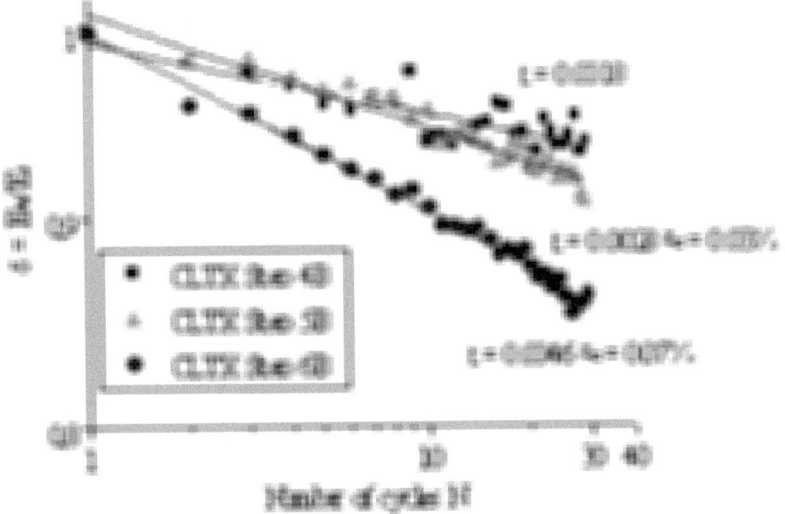

Fig. 4.22. Degradation parameter from CLTX tests

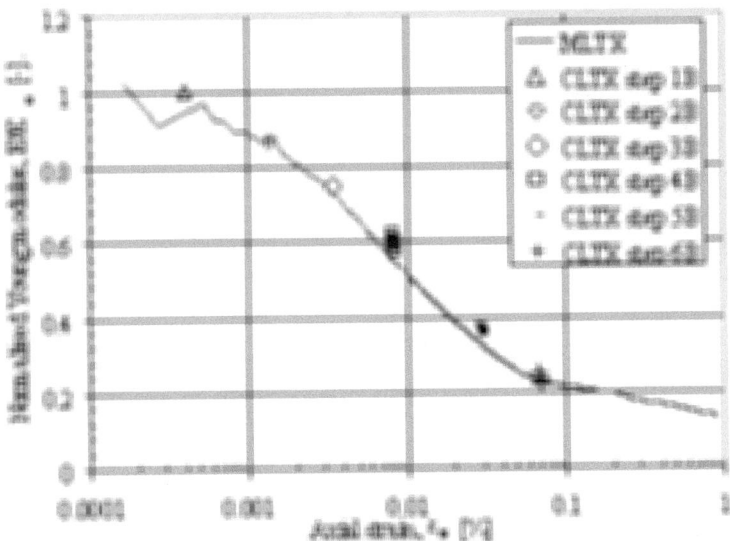

Fig. 4.23. Young modulus from MLTX and CLTX

One specimen of weathered MG (MG1) and another of intact MG (MG3) have been firstly subjected to unconfined cyclic compression loading test UCCL and after to unconfined monotonic compression loading test UMCL (Calosi et al., 2001). The normalised stiffness decay and damping ratio are shown in Figure 4.24.

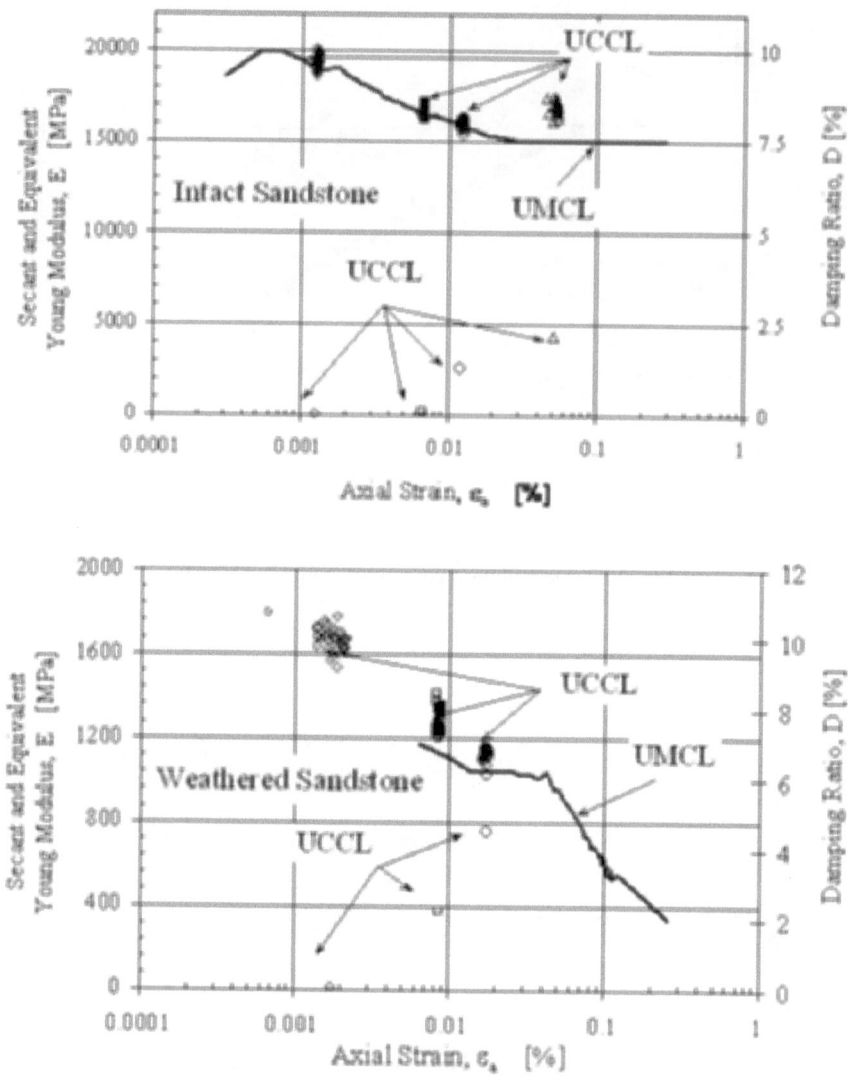

Fig. 4.24. Stiffness and damping ratio from UCCL and UMCL

4.5.3. LABORATORY VS. FIELD TESTS

The small strain shear modulus of ARG inferred from RC and CLTX tests is plotted vs. depth in Figure 4.25. The G_0 values inferred from CLTS and MLTX tests are almost the same. Figure 4.25 also shows the small strain shear modulus inferred from DH tests at S3 location. Laboratory tests largely underestimate the initial stiffness probably because of sampling disturbance, which can be very important if Shelby samplers are used in very dense low plasticity soils. The poor sample quality should not influence the shear modulus decay curves if the sampling disturbance influences the soil stiffness at small and at large strains. Yet, high quality samples, retrieved with in-situ freezing should be taken in order to clarify this aspect. Laboratory tests on high quality samples could definitely confirm the non-linearity of ARG specimens.

As the MG formation is concerned, the values of shear wave velocity inferred from in situ and laboratory tests are shown in Table 4.2.

The shear wave velocities inferred from in situ and laboratory tests show very similar values. In the case of intact sandstone, those measured in the laboratory are even greater than those determined in situ. It is possible to assume that the effect of sampling disturbance is not very important in the case of rock specimens. On the other hand, possible scale effects could explain why the shear wave velocity of intact sandstone determined from laboratory tests is larger than that obtained from in situ tests.

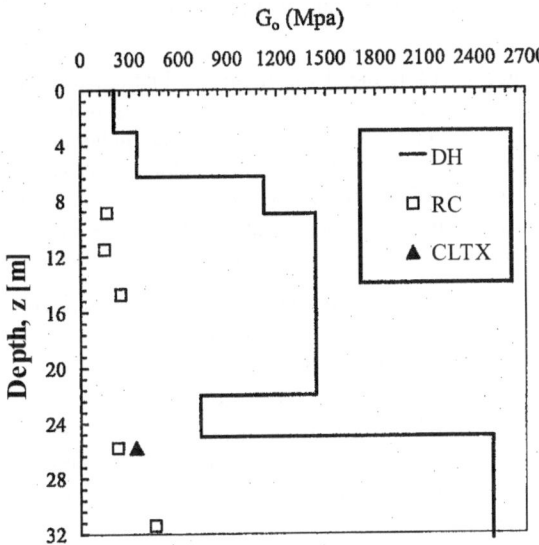

Fig. 4.25. Comparison between G_0 from in situ and laboratory tests

Table 4.2. MG Shear wave velocity from in situ and laboratory tests

Type of soil	$V_s[m/s]$ Situ	$V_s[m/s]$ Laboratory
Weathered sandstone (MG1)	600	530 - 550
Transition zone (MG2)	900	-
Intact sandstone (MG3)	1200-1500	1700-1800

4.5.4. DEFINITION OF SOIL PARAMETERS FOR SEISMIC ANALYSIS

Two different sets of parameters had to be selected in the reported case history, indeed the site response was analysed by mean of two different approaches: linear equivalent analysis and non-linear analysis.

1D and 2D linear equivalent analyses

Laboratory test results on ARG specimens have been used to analytically describe the G-γ and D-γ curves, adopting the following equations, proposed by Yokota et al. (1981):

$$\frac{G}{G_o} = \frac{1}{1 + \alpha \cdot \gamma^\beta} \tag{4.21}$$

$$D = D_{max} \cdot \exp(\lambda \frac{G}{G_o}) \tag{4.22}$$

where the shear strain γ is expressed in percent [%]. Note that the knowledge of the shear strength is not required for the evaluation of the curves using the above formulation.

The shear modulus decay was defined by a single set of parameters inferred from CLTS and RC tests ($\alpha = 65$ $\beta = 1.15$). The triaxial test results were disregarded for this purpose, because the effect on the seismic response of the higher non-linearity exhibited in triaxial tests was considered negligible. On the contrary, two sets of parameters were defined to describe the damping ratio decay ($D_{max} = 34.8$ $\lambda = -1.9$) from RC tests and ($D_{max} = 23.3$ $\lambda = -2.2$) from cyclic quasi-static tests, CLTS and CLTX). Distinct analyses have been performed with the two sets of parameters.

The G-γ and D-γ curves, obtained according to the method described above, have been used to characterise the stiffness and damping non-linearity of both ARG, CG and CT/MG formations in the framework of linear equivalent analyses.

For the MG formation, shear modulus and damping ratio decay curves have been defined numerically on the basis of experimental results. Finally, for quaternary deposits (ALL, CT, CD and DT), for which no undisturbed sampling was possible, the decay curves proposed by Rollins et al. (1998) with their range of variation have been adopted.

1D Non-linear analysis

The Ramberg-Osgood (1943) equation was used for the stress-strain backbone curve:

$$\frac{E}{E_o} = \frac{1}{1 + \alpha \cdot \left| \frac{E}{E_o} \cdot \frac{\varepsilon_a}{\varepsilon_{rif}} \right|^R} \tag{4.23}$$

where: $\varepsilon_{rif} = q_{max} / E_o$; $\frac{E}{E_o} \equiv \frac{G}{G_o}$; $\gamma = (1 + v) \cdot \varepsilon_a$

In the case of ARG, CG and CT/MG formations the Ramberg-Osgood (RO) parameters were obtained from the MLTX test. The RO parameters were obtained from UMCL tests, in the case of MG formation. For the alluvial deposits, the RO parameters were obtained fitting a numerical curve derived from the Rollins et al. (1998) equation.

The unload-reload branches were simulated by means of the second Masing (1926) rule as modified by Tatsuoka et al. (1993). In particular they proposed to assume a scale amplification factor "n" not necessarily equal to 2 as postulated by the second Masing rule. For the case under consideration, the only available experimental data concerning the n parameter where deducted by the comparison of CLTX and MLTX results on ARG specimen. It was decided to assume n=2.6 for ARG, CG, CT/MG and ALL and n=2 for MG.

4.6. Conclusions

A seismic ground response analysis must always be preceded by an accurate dynamic characterization of the site whose main objective is the identification of:

- shallow subsurface geology;
- lithostratigraphic-geotechnical units;
- hydrogeological regime;
- physical properties and state parameters of the formations;
- parameters of mechanical and hydraulic behavior of the formations under earthquake loading.

Typically, the dynamic characterization of the site is carried out implementing an adequate program of geophysical and geotechnical field-lab investigation testing campaign. This Chapter presented an overview of the capabilities and limitations of some of the most common geophysical and geotechnical investigation techniques as well as a detailed case history.

The starting point for planning the investigation is an accurate geological and geo-morphological model constructed on the basis of geological maps and cross-sections at an adequate scale (at least 1:5000).

Near-surface geophysical investigations are especially important for the definition of the dynamic soil properties at the site (i.e. shear wave velocity and damping ratio fields) and of the shallow subsurface geological structure. Often, the results of the geophysical investigations are used to re-define the original geological model. On the other hand, laboratory testing methods are important to determine the soil non-linear stress-strain relationships and their damping ratio degradation curves. A still open question, only partially discussed in the Chapter, concerns with undisturbed sampling, sample quality and the effect of sample disturbance on the normalized shear modulus and damping ratio degradation curves. In-situ large strain techniques should be used for those soils that cannot be successfully retrieved and tested in the laboratory. However the costs of these methods can be very high and not affordable in many cases.

CHAPTER 5
SITE EFFECTS

Kyriazis Pitilakis
Aristotle University, Thessaloniki, Greece

5.1. Introduction

Surface geology and geotechnical characteristics of soil deposits have a paramount importance on seismic ground shaking. The variations of ground shaking in space, amplitude, frequency content and duration are called "site effects". Site effects include primarily the effects of impedance contrast of surface soil deposits to the underlined bedrock, or firm soil considered as rock, which is rather well modelled using 1D ground models (i.e. linear elastic, equivalent linear or non-linear). They also include deep basin effects, and basin edge effects, produced from strong lateral geological discontinuities (i.e. geological anomalies, faults etc). These effects which are dominated by the presence of surface waves additionally to body waves can only be studied using 2D and 3D models. Finally, site effects are also dealing with spatial variation of ground shaking characteristics due to surface topography.

The physics and the importance of site effects is more and more understood and quantified with the increasing number of strong motion measurements in dense accelerometric arrays all over the world. Advanced numerical models using powerful computer facilities have also contributed significantly to the progress during the last two decades. Mexico City (1985) and Loma Prieta (1989) earthquakes, recorded in many stations located in different and well constrained ground conditions, relieved for the first time in a very precise experimentally documented way, the importance of the impedance contrast. Additional evidence of the significance of the more complex site effects on seismic ground motions have been brought from recent destructive earthquakes (Armenia 1988, Philippines 1990, Northridge 1994, Kobe 1995, Kozani 1995, Aegion 1995, Kocaeli and Duzce, Turkey 1999, Athens 1999, Ji-Ji Taiwan 1999 etc).

Specific experimental sites, operating the last ten years, such as Euroseistest (http://euroseis.civil.auth.gr) and Ashigara valley as well as other accelerometric arrays, mainly in Japan (i.e. K-net) and in California, are continuously producing high quality experimental data in densely instrumented sites, often sedimentary valleys, which allowed very detailed experimental and theoretical analyses of complex site effects, and revealing their complexity due to deep basin and basin edge effects, additionally to the pronounced role of the impedance contrast and the role of soil non-linear behaviour.

However, there is not yet a wide-spread agreement as regards to what could be the best way to estimate the amplification or de-amplification or the spatial variability caused by site effects. There are also different approaches to model and account for site effects in seismic risk studies. A typical example is the very different approach to model site effects, which range from 1D to 3D models, using linear or non-linear material behaviour. Probably this may be attributed to the fact that, still, very few site effect studies have been performed both involving a detailed study of subsurface structure and

A. Ansal (ed.), Recent Advances in Earthquake Geotechnical Engineering and Microzonation, 139–197.
© 2004 *Kluwer Academic Publishers.*

numerous high-quality recordings and/or observations of earthquake ground motion.

In general, site effects may be defined as the modification of the characteristics (amplitude, frequency content and duration) of the incoming wave field, due to the specific characteristics and geometrical features of the soil deposits and the surface topography. The modification is manifested as an amplification or de-amplification of ground motion amplitudes at all frequencies, which is dependent on many parameters. Some of them are inherent of the dynamic soil behaviour and its physical properties (i.e. D_r, PI, V_s, V_p, G_o, shear modulus degradation with shear strain increase, soil internal damping, soil non-linearity, etc), others are related to the characteristics and the intensity of the incoming wave-field and others are related to purely geometrical features like surface/bedrock topography, lateral geological discontinuities etc.

In order to understand the physics and the spatial variation of ground motion in each particular case and particularly to be able to quantify the phenomenon, it is necessary to have an accurate description of the above characteristics for the specific site. As a result, site and soil characterization is an important and indispensable parameter for site effect analyses.

In the present chapter we are discussing few basic topics outlined previously, using mainly experimental data and results from studies performed during the last few years in Greece. More precisely, after a brief overview of the physical concepts and the definitions of site effects, we first present a critical evaluation of the most usual methods to assess site effects both experimentally and theoretically and then we present and discuss site effects in case of (a) horizontally layered soil deposits, (b) basin edge effects and (c) topographical relief. Finally, the last part of this chapter is devoted to (a) discuss the way that modern seismic codes, like NEHRP, IBC 2000 and EC8, take in to account site effects, and (b) propose a new site categorization and a new set of amplification factors and normalized response spectra to specify input design earthquake motion for engineering purposes.

5.2. Basic Physical Concepts and Definitions

Earthquake recordings at soil surface include "information" that is related to three stages of the earthquake phenomenon evolution: a) the source activation (fault rupture), b) the propagation path of seismic energy and c) the effect of local geology on the wave-field at the recording site (Figure 5.1). The physical amplitude $r(t)$, potentially representing acceleration, velocity or displacement, which is recorded at a site, can be written in the time domain in the form of the convolution of three factors:

$$r(t) = e(t) * p(t) * s(t) \qquad (5.1)$$

where $e(t)$ is the source signal, $p(t)$ is the function that characterize the propagation from the source to the site and $s(t)$ expresses the effect of local soil conditions on ground motion (which from now on will be denoted as site effects). In the frequency domain, Equation (5.1) is written with the form of a product

$$R(f) = E(f) \cdot P(f) \cdot S(f) \qquad (5.2)$$

where $R(f)$, $E(f)$, $P(f)$, and $S(f)$ are the Fourier transform of the time depended functions $r(t)$, $e(t)$, $p(t)$, and $s(t)$ respectively. All of the above mentioned factors contribute to overall site response, either independently or in combination with the others. However, in the framework of this chapter, only the "site effects" factor is discussed. The other two factors are simply considered in the presentation of different models that are used to estimate ground response.

The term "site effects" introduces the effect of local geology in the modulation of seismic wavefield at a recording site; where local geology consists of surface sedimentary sites and surface topography. The main parameters that characterize a site are the geometry of the soil stratigraphy (thickness and lateral discontinuities), the shape of the topographic relief and the dynamic, physical and mechanical properties of soil and rock materials.

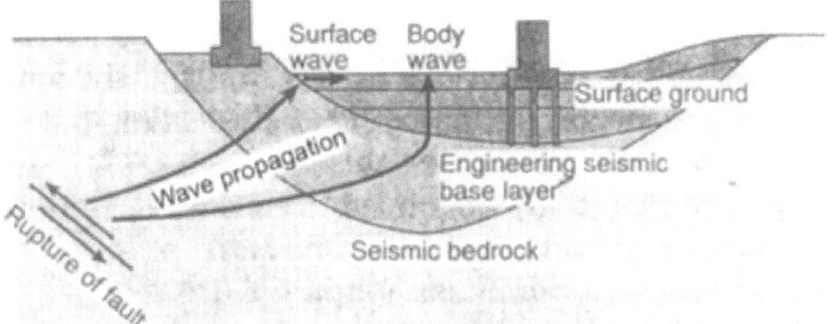

Fig. 5.1. Schematic illustration of the wave propagation from fault to ground surface (Yoshida and Iai, 1998; Reproduced with permission from the Swets & Zeitlinger Publ.)

Surface soil formations are the product of the long-lasting process of erosion, weathering and deposition; they are responsible for significant amplification and spatial variation of surface ground motion. Surface topography, in its simplest form, consists of convex (ridges, mountains, hills …) or concave surfaces (valleys, basins, canyons …) with different behaviour during an earthquake. In case of convex topographies, significant amplification is observed at the crest compared to that at the foot, while in the concave ones, the amplification varies at the lateral parts than at the base.

The effect of local geology on ground motion also depends on other parameters such as the intensity, the frequency and the incidence angle of the incoming wavefield (for strong or weak earthquakes) which in combination with the local site conditions might introduce non-linear phenomena. Generally, it could be stated that there is a large variety of parameters according to which, someone could categorize site effects, a fact that confirm the complexity and the need to understand the physical background of this phenomenon.

A general description of "site effects" could be defined as follows: "Surface soil formations and surface topography modify the characteristics (amplitude, frequency content and duration) of the incoming wavefield resulting to the amplification or deamplification of ground motion". A simple qualitative and quantitative estimation of site effects is often expressed by the amplification factor A_{max} and resonant – fundamental and higher ones - frequencies f_{res}.

5.2.1. SITE EFFECTS DUE TO LOW STIFFNESS SURFACE SOIL LAYERS

It has been long recognized that the amplitude of earthquake ground motion is affected by both the properties and configuration of the near surface material through which seismic waves propagate. These properties are impedance - resistance to particle motion - (Aki and Richards, 1980) and damping (attenuation).

Influence of Impedance and Damping in frequency and time domain

For horizontally polarized shear waves (SH), impedance can be defined (Equation 5.3) as the product of the density (ρ), the shear wave velocity (Vs) and the cosine of the angle of incidence (Figure 5.2).

$$I = \rho \cdot Vs \cdot cos\theta, \quad cos\theta \cong 1 \quad thus \quad I = \rho \cdot Vs \qquad (5.3)$$

Incidence angle, θ, is usually small near the surface of the earth and its cosine can be assumed to be equal to unity. As a seismic wave passes through a region of decreasing impedance, the resistance to motion decreases and, to preserve energy, the amplitude of the seismic wave increases. When there are sharp changes (decrease) in impedance below the earth's surface (such as sediments/rock interfaces), an increase in amplitude of the upwardly seismic wave is observed due to resonance, as some of the seismic waves transmitted into the upper layer get trapped in this layer and begin to reverberate.

Damping or inelastic attenuation is substantially greater in soft soils than in hard rocks and mitigates the increase in amplitude of seismic motion due to resonance. The fundamental phenomenon responsible for the amplification of motion in soil sediments is the trapping of seismic waves due to the impedance contrast between sediments and the underlying bedrock. For the simplest case of a soil layer with density ρ_1 and shear wave velocity V_{s1} overlying a stiffer layer with density ρ_2 and shear wave velocity V_{s2} (Figure 5.2a), the impedance contrast is expressed by the formula:

$$C = \frac{\rho_2 \cdot Vs_2}{\rho_1 \cdot Vs_1} \qquad (5.4)$$

To understand the basic concept of site effects, the simplification of the physical complex phenomena is instructive. Thus, when the structure is horizontally layered (1-dimensional structures), this trapping affects only body waves travelling up and down in the surface layers (Figure 5.2). When the sediments form a 2- or 3- dimensional structure due to soil thickness variations, this trapping also affects surface waves which develop on the sediments/bedrock interfaces and thus reverberate back and forth. In all cases, this effect is maximum when the reverberating waves are in phase with each other. The interference between these trapped waves leads to resonance.

Resonance therefore, is a frequency-dependent phenomenon related to the geometrical and mechanical (density, P-wave and S-wave velocities, damping) characteristics of the soil structure. While these resonance patterns are very simple in the case of a 1D structure (vertical resonance of body waves), they become more complex in 2D and 3D structures. The fundamental resonant frequency may vary between 0.2 Hz (for very thick deposits or for extremely soft materials) and 10 Hz or more (for very thin layers of deposits or weathered rocks).

The amplitude of fundamental resonant peaks is mainly related to the impedance contrast between surface soil layers and underlying bedrock, to the material damping of sediments and to a lesser degree with the characteristics of incident wavefield (type of waves, incidence angle, near or far field ...). For the simplest case discussed above, the amplification at the fundamental resonant frequency is given by the formula:

$$A_0 = \frac{2}{\dfrac{1}{C} + 0.5 \cdot \pi \cdot \zeta_1} \tag{5.5}$$

where C is the impedance contrast and ζ_1 the material damping of the sediments. For the case of very small damping ($\zeta_1 = 0$), the maximum amplification is simply double the impedance contrast. Another interesting observation is that when the wavelength, λ,

$$\lambda = Vs_1 \cdot T \tag{5.6}$$

is much longer than the thickness of the layer (meaning that $\omega H/Vs_1 \cong 0$), the amplitude of surface displacements is doubled. This is called the free surface effect and is caused by upgoing seismic waves being reflected off the free surface of the earth. At the surface, both upgoing and downgoing reflected waves are exactly in phase and the resultant amplitude at that location is doubled.

Figure 5.2 provides an illustration of the effect of resonance in the frequency domain, particularly a low resistance sedimentary layer overlying hard rock (impedance contrast c=5). Without taking into account the free surface effect (where the amplification would be doubled as mentioned previously), a 100 m thick layer produces peaks of amplification at about 0.5, 1.5, 2.5 Hz and higher. On the other hand, a 50 m thick layer produces peaks at 1.0, 3.0 Hz and higher. It can be stated therefore that, the amplification of higher peaks decreases with increasing frequency, due to the consideration of inelastic attenuation or damping, which in this specific case takes a relatively large value. It has been shown, both experimentally and theoretically that this amplitude very often reaches values between 6 and 10, while in the extreme case, exceeds 20 (high impedance contrast and small damping).

In case of 2D and 3D structures, fundamental frequency depends also on the geometry of the soil structures. The lateral geometry of these structures is affecting the amplification level at resonant frequencies especially when the material damping is small. Complex effects that are introduced due to the consideration of the finite lateral extent are due to the locally generated at the discontinuities (edges, faults, etc) and laterally propagated surface waves. The effect of these surface waves is manifested in two ways:

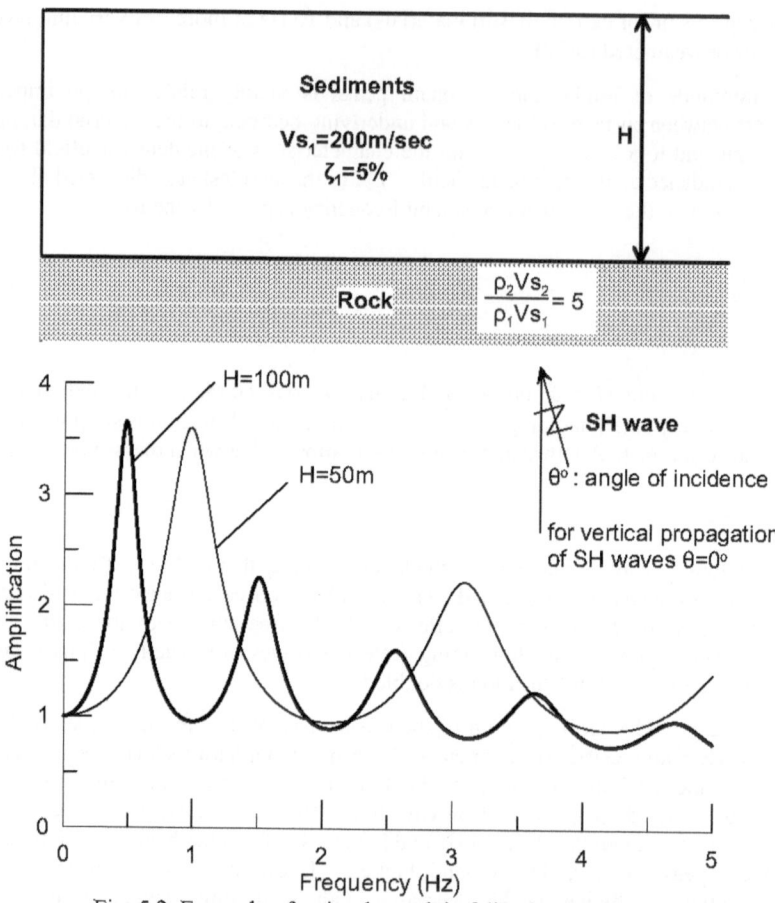

Fig. 5.2. Example of a simple model of 1D site amplification

When the semi-length of the soil structure is much larger than its maximum thickness (shallow basins), the waves have the same frequency characteristics as 1D resonance, thus increasing the 1D amplification level. When the semi-length of the soil structure is comparable to its thickness (deep basins), and the rebervarating back and forth surface waves are in phase, the waves interfere with each other leading to 2D resonance patterns. The same resonance effects are involved in the seismic wave modulation due to 3D soil structures. The consideration of the second and third lateral dimension in the wave propagation phenomena, in case of 2D and 3D resonance, leads to an increase in ground motion amplification and a shift towards higher values of the peak frequencies. An interesting comparison between 1D, 2D and 3D resonance, spectral peaks of amplification is presented in Figure 5.3. The differences between 1D and 2D resonance are much more pronounced than between 2D and 3D cases. This means that the consideration of the third dimension in the simulation of ground motion leads to quantitative differences relative to 2D analysis (much larger amplification and a small shift in resonant frequencies).

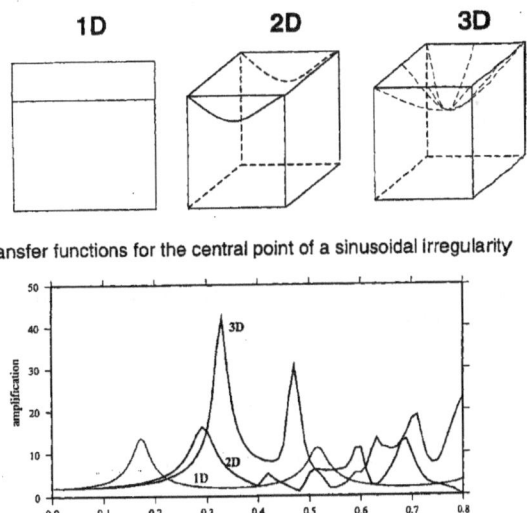

transfer functions for the central point of a sinusoidal irregularity

Fig. 5.3. Spectral responses computed at the basin center for 1D, 2D and 3D models of semi-shaped basin (Bard and Riepl 1999; Reproduced with permission of the publisher WIT Press)

In the time domain, these resonance patterns affect the peak amplitudes of ground motion (mainly peak ground acceleration, PGA and peak ground velocity, PGV), the waveforms and the motion duration, especially in 2D soil structures. Experimental evidence (records) from recent earthquakes (Mexico, Loma Prieta, Northridge etc) showed that PGA were up to 4 times larger at soil than at rock sites. Statistical analyses of records have shown that PGA is most likely to be amplified when the fundamental resonant frequency of a site exceeds 2-3Hz. On the other hand, it was also observed that liquefied sandy deposits induce important reduction of peak acceleration (i.e. Kobe case). Therefore, PGA values on sediments cannot be predicted straightforwardly from PGA values on rock and this issue is strongly related to the non-linear phenomena in soil behaviour. A general trend however do exists, for moderate accelerations levels (<0.2-0.3g), in the sense that amplification of PGA is expected at soil sites compared to rock sites.

This behaviour of PGA amplification may be attributed to a) the fact that in soils with low S wave velocity, the accumulated energy results in amplification and therefore, as the ground becomes "softer", amplification becomes larger (elastic range) and b) the fact that under strong dynamic loading the ground becomes "softer", (shear strength decreases) and hence, the peak acceleration becomes smaller and the predominant period of soil profiles is shifted to higher value (non-linear behaviour of soil materials). Consequently, amplification occurs under small ground shaking with decreasing absolute value as the ground shaking level is increased.

This observation has been already included in UBC97, UBC 2000, NEHRP and EC8-draft code previsions with the introduction of an amplification coefficient depending both on soil classification and input motion amplitude. Further discussion on these issues is given in subsequent part of this chapter.

Regarding the duration of ground motion, all recent studies report a significant increase of duration in sediments especially at longer periods when soil stratigraphy is complex. This fact is closely related to the geometry of the structure (2D or 3D) and the existence of strong lateral discontinuities; that will be discussed in the following sections.

5.3. Methods to Estimate Site Effects

There are various methods that may be used for site effect evaluation. The choice of the method usually related to the significance of the engineering project for which it is applied. Generally, the methods are classified in five main categories:

- Experimental-empirical techniques that utilise recordings of ground motion or ambient noise to estimate the basic characteristics of the expected ground motion - usually in the frequency domain.
- Empirical methods that evaluate parameters of earthquake motions such as acceleration, velocity and response spectra based on site classification, average S-wave velocity, topography, earthquake magnitude and existing amplification relationships; usually these methods are incorporated in seismic code provisions.
- Semi-empirical methods that compute time histories of earthquake motion by combining recorded earthquake motion of smaller earthquakes as element motions (i.e. Green's functions); these methods may account for the detailed fault rupture process and the effects of asperities.
- Theoretical methods where site effects are computed through an analytical and more often numerical 1D, 2D or 3D wave propagation model; different wave types with different incident angles may be used; the main advantage of these methods is the possibility to use complex constitutive relationships for describing soil behaviour under dynamic loading conditions and the ability to model accurately site stratigraphy inclusive of basin topography.
- Hybrid methods that compute time histories of earthquake motions by coupling a longer period component determined by a theoretical seismic fault model with a computational seismic wave propagation model having a shorter period component determined by a semi-empirical method.

The use of each method depends on many parameters and, in any case, requires an increased level of expertise. In the following paragraphs some aspects of the first four methods will be briefly discussed.

5.3.1. EXPERIMENTAL-EMPIRICAL

The majority of the experimental techniques that have been developed during the last decades analyze site effects in the frequency domain because this is a relatively easier way to handle earthquake recordings. It is reminded that earthquake recordings may be represented in the frequency domain as the product of Fourier spectra of the source effect, the path effect and the site effects. In order to estimate the influence of local geology (site effects), the removal of the influence of the first two terms (source and path effects) is necessary. For this purpose, several methods have been proposed which are classified in two major categories based on the criterion of the use of a "reference site"; the reference site can be generally defined as this particular control location that is free of all kinds of site effects and it is usually the nearby rock site. The most commonly applied experimental techniques are shortly presented in the following paragraphs.

Standard Spectral Ratio Technique (SSR)

The most popular and widely used technique to characterize site amplification has been the Standard Spectral Ratio, SSR, (Borcherdt, 1970), which is defined as the ratio of the Fourier amplitude spectra of a soil-site record to that of a nearby rock-site record from the same earthquake and component of motion (Figure 5.4). Source information is the same for this pair of records and when the two sites are closely located, the path effect is also considered the same. Hence, the ratio of the Fourier amplitude spectra expresses only the effect of the local soil conditions at the specific site. Theoretically speaking though, this technique is applicable only to cases that the data are derived from dense local arrays with at least one station on outcropping conditions defined as reference station.

A usual option for the selection of the reference station is a site of outcropping rock, while less frequently, a bedrock site having a downhole accelerometer installed in a borehole is the used for this purpose. The basic conditions for the application of this particular technique in the case of a surface reference station are: a) the existence of simultaneous recordings at a soil site and at the reference site, b) the reference site has to be free of any kind of site effects (sediments and topography) and c) the distance between the soil site and the reference one ought to be small (i.e. smaller than the epicentral distance), in order to consider that the effect of the propagating path of the seismic energy is the same for the two sites.

However, the condition that an outcrop rock reference site should be free of any kind of site effects often it is not valid. For this reason, a careful examination of the reference site is obligatory in order to correctly estimate amplification in sedimentary sites (Stiedl et al., 1996).

Generalized Inversion Scheme Technique (GIS)

Andrews (1986), having in mind that The Standard Spectral Ratio technique is reliably applicable only to data from dense, local arrays, proposed a generalized technique to look for all source, path and site effects in large data sets recorded in local or regional networks, by applying the solution of a large inverse problem. In this generalized inversion scheme, the path term is generally assumed to follow an a-priori known law. Thus, for a given data set, the unknown source and site effects Fourier amplitude spectra are simultaneously estimated from the whole data set generally through least square weighted inversion. The main advantage of this technique is the reliable estimation of the source and site effects terms from the whole data set especially in cases where there are not simultaneous records of all earthquakes at all sites of the network (Field and Jacob, 1995).

Coda wave technique

Phillips and Aki (1986) on the other hand, proposed an alternative method based on the use of coda waves. The estimation of the site effects term is exclusively based on the latest part of the recordings (coda waves), starting from that point, where time is double of that of the first S arrival. The spectral shape of these coda waves is independent of the source and of the recorded site because this part of the recordings is dominated by multi-scattered waves at the heterogeneities of the Earth's crust. More details could be found in Phillips and Aki (1986).

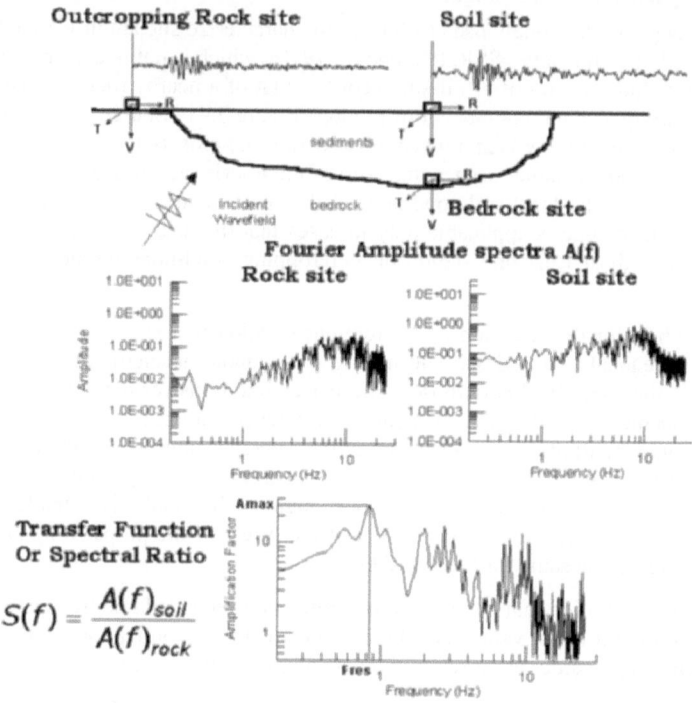

Fig. 5.4. General description of the Standard Spectral Ratio Technique (SSR)

Horizontal to Vertical Spectral Ratio Technique (HVSR)

All techniques mentioned above are using a reference site but in practice, appropriate reference sites are not always available. For this reason different methods that are not depending on reference sites have been developed. One of them is the Horizontal to Vertical Spectral Ratio. This extremely simple technique consists of using the spectral ratio of the horizontal to the vertical component of ground motion and estimates the Fourier amplitudes in different frequencies accordingly. The basic assumption of the method is that the vertical component of the ground motion in cases where the soil stratigraphy is flat and horizontal is supposed free of any kind of influence related to the soil conditions at the recording site. Figure 5.5 shows the general layout of the method which was first applied to the S wave portion of the earthquake recordings obtained at three sites in Mexico City by Lermo and Chavez-Garcia (1993). Generally, the Fourier spectra ratio exhibit similarities between SSR and HVSR technique, with a better fit in frequencies rather than amplitudes of the resonant peaks.

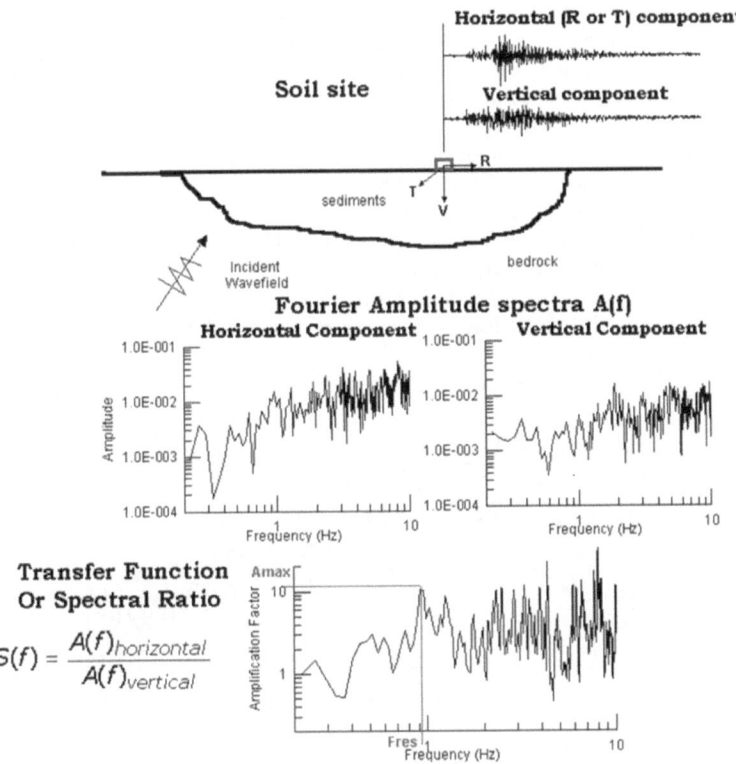

Fig. 5.5. Description of the Horizontal to Vertical Spectral Ratio Technique (HVSR)

Comments on SSR and HVSR

SSR and HVSR are the most commonly used experimental techniques for the estimation of site amplification due to local soil conditions; there are plenty of literature references on comparative results on their applicability and reliability. Herein, some of these works are briefly summarized and their main conclusions are highlighted.

Detailed comparisons between SSR technique and other reference station techniques (Field et al., 1992; Stiedl, 1993; Field and Jacob, 1995) led to few basic qualitative conclusions such as: a) the estimation of site effects with the use of SSR technique is relatively stable even if records are quite noisy, b) the process should be based on a significant number of earthquake recordings (the use of a limited number of records should be avoided) and c) the amplification level determined with SSR technique is quite similar with that determined from other techniques and especially with the GIS.

Other comparisons between results of SSR and HVSR techniques led to controversial conclusions. As it is already stated above, Lermo and Chavez-Garcia (1993) applied HVSR to the S wave portion of the earthquake recordings and found similarities between standard spectral ratios and these HVSR with a good fit in both frequencies and amplitudes of the resonant peaks. Some other researchers used HVSR technique on data sets from weak and strong motion records and concluded that the shape of the spectral ratios presents very good statistical stability with minor dependency on source and path effects and that it is quite well correlated with surface geology, while their amplification level seems to depend on the type of incident wave, a fact that does not affect the fundamental resonant frequency. Field and Jacob (1995) after systematic comparisons with other techniques concluded that the shape of the transfer function is satisfactorily reproduced by HVSR technique, although there is an underestimation of the amplification factor compared to SSR. On the same issue Raptakis et al. (1998, 2000), using a large and high quality date set from EUROSEISTEST experimental site, proved that the significant differences between SSR and HVSR amplitudes at the fundamental frequency are attributed to the considerable amplification of the vertical component due to diffracted Rayleigh waves at the lateral discontinuities of the basin (Figure 5.6).

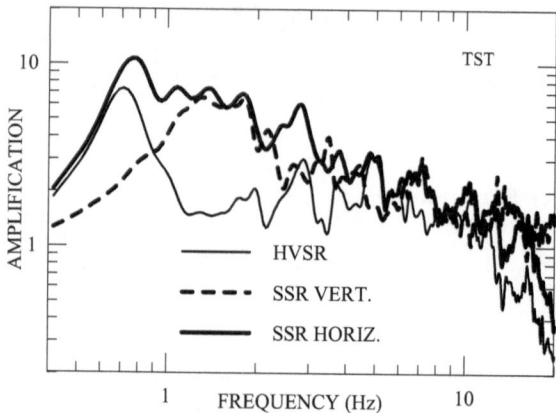

Fig. 5.6. Mean spectral ratios of HVSR technique compared to SSR horizontal and vertical components (after Raptakis et al., 1998; Reproduced with permission from D.Raptakis and the Earthquake Engineering Research Institute)

In conclusion, both SSR and HVSR techniques are reliable in estimating the fundamental frequency of the soil profile. However the amplification amplitude is comparable only when the soil layering is horizontal and there aren't lateral geometrical variations. In those cases, due the presence of in-ward propagating surface waves, it is expected that part of them will affect the vertical component and hence the amplitude of the HVSR method. For this reason, in cases where the stratigraphy is not flat and horizontal, which is pertinent in many real site conditions, the use of HVSR technique should be applied with caution at least for the derivation of the amplification factor at the fundamental frequency.

5.3.2. EMPIRICAL METHODS

Empirical methods are practically used either for preliminary analyses or in the frame of seismic code prescriptions with well specified amplification factors defined according to the soil classification and the earthquake intensity. Simple relationships giving the amplification factors for the peak acceleration or/and velocity with the average shear wave velocity of the soil profile are proposed in the literature (Joyner and Fumal 1984, Midorikwa 1987, Borcherdt et al., 1991). All these relationships should be used only for preliminary studies and with extreme caution.

The last part of the present chapter will be devoted to the discussion of code provisions regarding site effect estimates. In this paragraph a senior problem is discussed concerning the site characterization using exclusively the average S-wave velocity over the 30m from the surface, which was first introduced by Borcherdt (1994) and then adopted in most modern codes.

The major question is how accurate is the use of V_{s30} for soil and site characterization. Certainly, the main advantage is the simplicity in evaluating the site conditions by conventional geotechnical surveys which rarely exceed 30-40m. On the other hand, the question remains whether the simple knowledge of the s-wave velocity over the limited depth of 30m is an accurate parameter to estimate site amplification characteristics. It is interesting to notice two examples from recent down-hole recordings that prove the opposite (Figure 5.7).

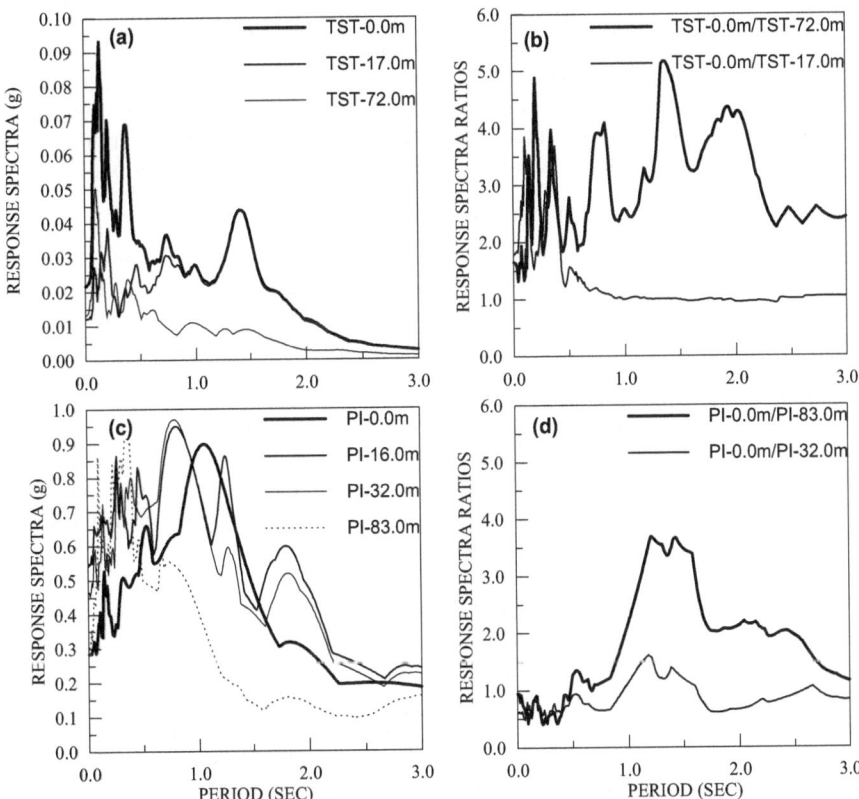

Fig. 5.7. Response spectra (left) and response spectra ratios (right) at the Euroseistest (up) and Port Island (down) vertical arrays. (Pitilakis et al.,1999; Reproduced with permission from the Swets & Zeitlinger Publ.)

In both sites the spectral ratios between the surface and down-hole records at various depths are considerably different at long periods (T>1sec). Large amplifications of the deep incident wave field are practically absent when we are computing the transfer ratio for shallower depths. Long period waves, mainly surface waves, generated at the lateral discontinuities disappear when only the uppermost layers are taken into account together with a 1D SH wave pattern (the case of Euroseistest valley). In the Port Island array in the U.S., due to liquefaction and strong inelastic behaviour of surface soils, the ground motion is de-amplified and the most severe response is observed at the fundamental period of the deposit (T=1sec). The recorded response spectral ratio between surface and -32m presents practically no amplification for T>0.5sec, while the amplification between surface and -83m is quite important.

In conclusion the use of $V_{s,30}$ as a basis for soil and site characterization is misleading in many cases. It should be used only when the actual site conditions are appropriate to that i.e. relatively shallow "seismic bedrock" or very firm soil conditions, flat stratigraphy.

In conclusion, empirical methods are mainly used for a quick simplified evaluation of the basic parameters of ground amplification: fundamental frequency of the soil profile and amplification ratio. They are useful (a) for microzonation studies and (b) with their special form of spectral amplification factors for different soil categories in seismic code prescriptions for the design of structures. In all cases they should be applied and used very carefully because their statistical background and the a-priori limited information required regarding site characterization may lead to serious errors.

5.3.3. SEMI-EMPIRICAL METHODS

The semi-empirical methods compute time histories of earthquake motions caused by large scenario earthquake by combining recorded earthquake motions by smaller seismic events. The Green's function technique is based on the idea that the total motion at a particular site is equal to the sum of the motions produced by a series of independent ruptures of many small parts on a causative fault. The method requires an approximate definition or estimation of certain parameters such as the geometry of the source, the slip functions describing the slip displacement vector with time for each elementary source, the velocity structure of the crustal materials between the source and the site and the Green's functions that describe the motion at the site due to an instantaneous unit slip at each elementary source. Normally the Green's function at each site, account implicitly the particular site specific ground behaviour in the linear elastic range.

The empirical Green's functions technique (EGF) (Hartzell 1978) bypasses these complicate computations by using the weak motions of small earthquakes as empirical Green's functions to simulate strong motion. Figure 5.8 illustrates the principles of the method. The method is essentially a deterministic one as it computes time histories for a defined earthquake scenario and other parametres. However it is possible to use statistical Green's functions which are computed as the statistical average of the recorded earthquake motions for different small seismic events. The EGF technique is particularly useful for generating near-field motions and when it is important to account the detailed fault rupture process and the effects of asperities. It is less accurate when strong non-linear behaviour is expected for local soils.

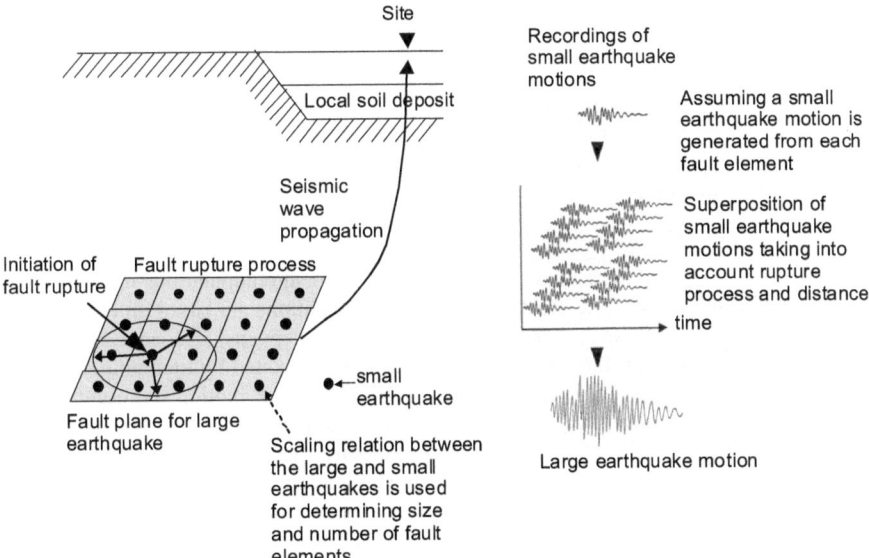

Fig. 5.8. Procedure for generating earthquake strong ground motions with the empirical Green's function technique (reproduced from ISO/WD 23469-draft).

5.3.4. THEORETICAL (NUMERICAL AND ANALYTICAL) METHODS

When the geological structure of an area and the geotechnical characteristics of the site are known, site effects can be estimated through theoretical analysis. The prerequisite of sufficient geotechnical knowledge of the soil structure including surface and deep topography is therefore obvious. Such an approach requires an in-depth understanding of the constitutive models describing the soil behaviour under dynamic solicitations and methods used to solve the wave propagation problem in 1, 2 or 3 dimensions. There are many models and methods which make the simple reference a rather difficult task and anyway beyond the task of the present chapter and book. Thus, in the present section the basis of the most conventionally used methods to account for site effects in ground response studies for microzonation purposes will be discussed.

Simple analytical models

As already mentioned, site amplification in soil sediments is related to resonance effects which are presented in the frequency domain in the form of spectral peaks in the Fourier transfer functions. A simple analytical approach which does not require any numerical computations, aims to estimate the fundamental period of the soil, τ_0, and the corresponding amplification factor A_0. A simple simultaneous estimation of these two parameters is possible only for sites that can be approximated as one layer over bedrock structure. This is a relatively easy way since only soil density, S-wave velocity, thickness, and damping of sediments as well as S-wave velocity of bedrock are required.

For multi layered sites, only the fundamental period could be satisfactorily estimated; Dobry et al. (1976) summarized the most significant methods. On the contrary there is no approximate and reliable formulae for the estimation of fundamental amplification factor A_0 in horizontally multi layered sites. Such formulae would imply many parameters, including damping, S-wave velocities and thickness of each layer. However, an upper bound of A_0 may be estimated using the impedance contrast between the lower stiffness surface layers and most rigid deep formations, together with the material damping of the surface soil deposits. The approximation is very crude and may lead to large overestimations and potential errors.

One dimensional response of "soil columns"

A number of analytical methods exist that allow numerical computations of the seismic response of a given site. The most widely used computations are based on the multiple reflection theory of S-waves in horizontally layered deposits (1D analysis of soil columns). According to this theory, "soil columns" are excited by incoming vertically incident plane S-waves that correspond to a surface bedrock motion representative of what is expected to occur in the area for a specific earthquake scenario. The parameters required for the analysis are the shear wave velocity, density, the material damping factor and the thickness of each layer. The above parameters may be obtained through in-situ geophysical and geotechnical surveys and appropriated dynamic laboratory tests. Alternatively, but with less accuracy, approximate correlations may be applied using conventional geotechnical parameters such as *SPT*, *CPT*, PI, D_r among others. These analyses may be performed considering either linear or non-linear behaviour for the soil. In the latter case, the non-linearity is usually approximated with an equivalent linear method that uses an iterative procedure to adapt soil parameters (i.e. stiffness and damping) to the strain level that each particular soil layer experiences during a specific earthquake motion. Specific curves expressing the degradation of shear modulus G and the respective increase of material damping, with the increasing shear strain level have been proposed by numerous researchers according. Figure 5.9 presents a typical set of G/Go-γ-$D\%$ curves. They have been estimated from resonant column tests on undisturbed specimens and they describe the dynamic behaviour of soil in the Euroseistest experimental site.

Average curves have been also proposed for different soil materials (clay with varying PI, sands, soil mixtures, etc). They must be used with caution because the actual behaviour for a given soil at a specific site may strongly vary from these average curves. This was the case in Mexico City where the lacustrine clayey deposits of extremely high plasticity index were found, through appropriate dynamic test (RC, CTX), to behave almost linearly despite the large strains experienced during the strong 1985 event and contrary to the previous belief that they should exhibit highly nonlinear behaviour because of their very low rigidity.

The last twenty years many interesting numerical codes have been developed with advanced non-linear and elastoplastic constitutive models that may also account for liquefaction phenomena. They certainly require additional parameters describing soil behaviour under complicated loading and drainage conditions which are not easily acquired even with sophisticated laboratory tests. Moreover the validation of these models with experimental results, mainly from actual seismic recording, is still a major unsolved problem. This fact combined with the need of complicate soil parameters is affecting seriously their wide use in practice.

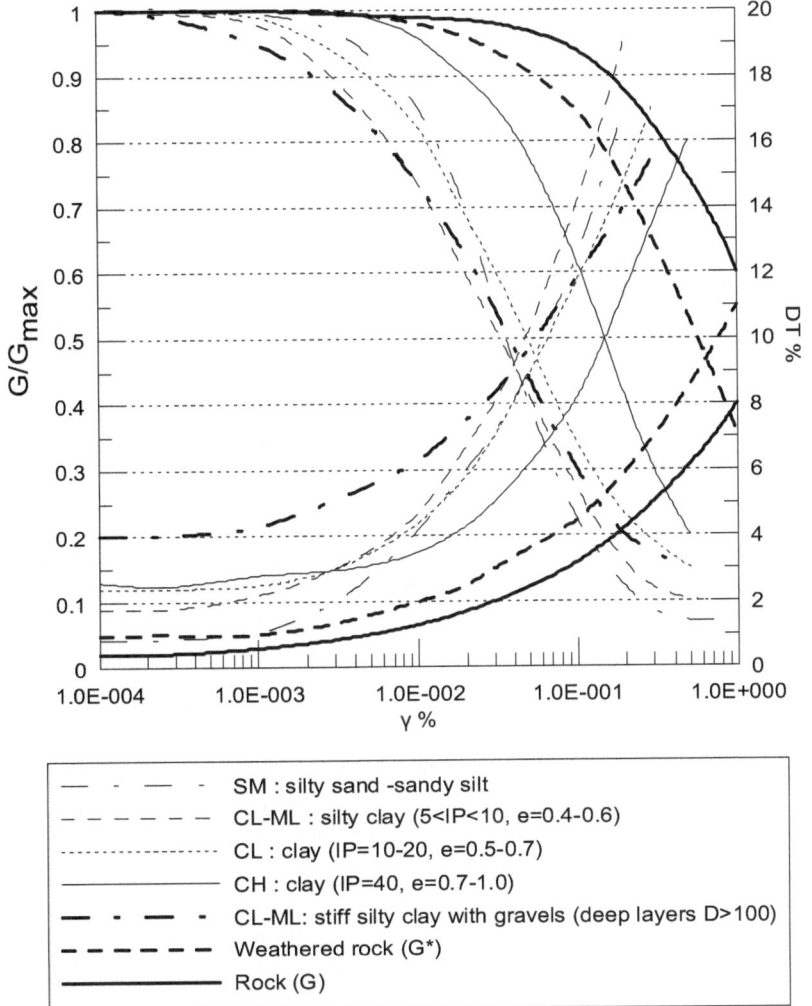

Fig. 5.9. Shear modulus and material damping dependency on shear strain for the soil formations at EUROSEISTEST site (after Pitilakis et al., 1999; Courtesy of the Journal of Earthquake Engineering)

Advanced 2D/3D models and methods

All numerical and analytical methods have the same theoretical basis (i.e. wave motion equations). However, many different models have been proposed to investigate several aspects of site effects which involve complex phenomena. For example, one has to consider for the various types of incident wave-field (near or far field, body and/or surface waves), the geometry of the structure (1D, 2D and 3D), the behaviour and the dynamic properties of soil materials (visco-elasticity, nonlinear behaviour, saturated media, etc). Typically, these advanced methods may be classified into four groups:

- Analytical methods which may be used for a limited number of simple geometries.
- Ray methods which are difficult to use when the wavelengths are comparable to the size of heterogeneities (usually the most interesting case).
- Boundary based techniques which are the most efficient when the site under consideration consists of a limited number of homogeneous geological units.
- Domain based models (finite difference and finite element methods) which allow accounting for very complex soil structures and constitutive models for the dynamic behaviour of soils but they are expensive from a computational point of view.

The development of these methods contributed significantly to the breakthrough in the understanding of site effects during the last three decades. They allow for parametric studies and more important the study of uncertainties of the seismic ground response at a site, considering the incomplete knowledge regarding the mechanical and geometrical characteristics of the site under consideration. However, there is still an important lack of reliable and detail validations.

5.3.5. CONCLUDING REMARKS

Theoretical models and computation tools have been developed during the last years. They allow detailed and advanced studies of wave propagation under complicated geometrical and material conditions. A complete comparative presentation and discussion on the advantages and the shortcomings of each model is beyond the scope of this chapter. Few examples of successful applications are therefore simply presented to highlight their capabilities.

More precisely the subsequent parts of this chapter are as follows: (a) 1D equivalent linear SH wave propagation computations performed in the frame of the detailed microzonation of Thessaloniki (Section 5.4), (b) 2D finite difference modelling of complex site effects in valleys (EUROSEISTEST and Thessaloniki, Section 5.5) and strong topographic irregularities (CORSEIS-Aegion Section 5.6), and 2D finite element equivalent linear computations in Thessaloniki (Section 5.5).

Future advances in these methods could be expected for the proper consideration of diffraction effects in complex surface or subsurface topography and realistic modelling of strong non-linear behaviour of soft soils, especially of sands (liquefaction phenomena). Nonetheless, the routinely use of these methods raises the following main concerns:

Theoretical models have been very rarely validated towards their ability to predict ground motion. Most of the comparisons between observations and theoretical results were made a-posteriori.

Numerical models are not panacea, but can be used only to some limited cases. The knowledge of their validity domain is a prerequisite in order to avoid erroneous estimations of site response attributes.

Their cost may sometimes be really high due to the detailed geotechnical and geophysical investigations required to provide a good knowledge of the constitutive soil properties needed as input parameters.

The deployment of instrumented experimental test sites in seismic regions with various local soil and site conditions that are deeply known in terms of their geological, geotechnical and seismological features, is a powerful tool to understand the nature of the complicate wave propagation phenomena, to study different aspects of the problem while it is a prerequisite condition for further development, validation and extension of all theoretical models and tools. Finally, experimental test sites and local arrays may be of great important to improve existing code prescriptions. To this extent, EUROSEISTEST (http://euroseis.civil.auth.gr) and CORSSA experimental sites (http://www.corinth-rift-lab-org) contribute significantly.

5.4. Site Effects in Horizontally Layered Soil Deposits

It has been shown previously that the dominant phenomenon governing the amplification of motion in sedimentary deposits overlying the rigid bedrock formations is the trapping of seismic waves due to the impedance contrast between sediments and rock. When the subsoil stratigraphy is almost horizontally layered, then the medium is practically 1-dimensional and the trapping phenomenon affects only body waves travelling up and down in the layered medium. For this reason, seismic ground response is very often calculated using only horizontally polarized shear waves (SH). When the soil stratigraphy is more complex forming a 2D or even 3D medium, then the trapping also affects surface waves, which are generated on the interface of sediments and bedrock exhibiting various inclinations. The trapping wave effect is maximized when the reverberating waves are in phase with each other. The interference between these trapped waves leads to resonance and hence to maximum ground motion amplitudes. The resonance which is a frequency depended phenomenon is quite simple in case of 1D medium (vertical resonance of body waves), and it becomes more complex for 2D/3D structures.

Material damping and generally, the non-linear soil behaviour particularly for seismic events of moderate and strong intensities, affect this trapping wave phenomenon by modifying amplitudes of ground motion in different frequencies and duration. Nevertheless, the fundamental phenomenon is still governed by the impedance contrast between sediments and bedrock. In turn, the impedance contrast, as expressed by the ratio of the product of shear wave velocities and densities between two layers, is the final and crucial result of the site characterization procedure in each particular site.

Some of the representative results are presented herein, involving a conventional 1D SH analysis, and implementing the well-known equivalent linear model for the soil behaviour, specifically for the city of Thessaloniki. The work is part of the microzonation study of the city (Pitilakis et al., 2003). In the following chapter the results of 1D analysis will be discussed in the light of 2D computations (finite difference and finite element method) as well as analyses of experimental data (recordings).

5.4.1. 1D SITE EFFECT COMPUTATIONS IN THE CITY OF THESSALONIKI

Thessaloniki, having almost one million inhabitants, is the second largest city of Greece (Figure 5.10). It has a long seismic history as it has been founded in the 4th century BC and it remained always through the centuries an important and big city. The more recent event was the large M=6.4, R=25 km event of June 20th 1978 that caused severe damage in the city buildings. Since then, considerable research has been undertaken leading to the microzonation study of the city, which has been concluded recently.

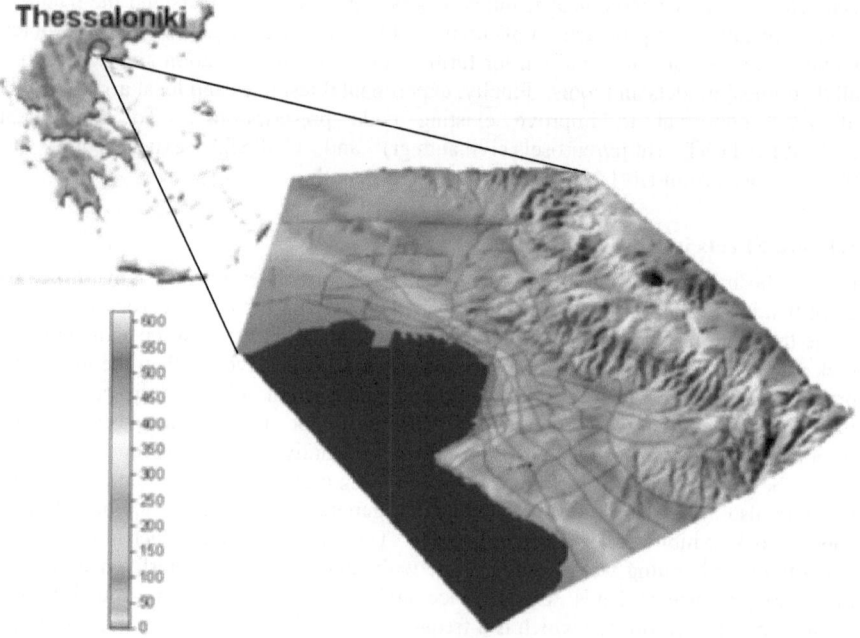

Fig. 5.10. Thessaloniki. Location and topographic relief

In the framework of this research project an extensive program of geotechnical and geophysical surveys and tests have been performed in order to construct an accurate geotechnical map appropriate for seismic response analysis of ground motion (Anastasiadis et al., 2002 and Pitilakis et al., 2003). Dynamic soil tests both laboratory and in situ were the most important part of all surveys. The detailed statistical analysis of all these data resulted in 9 main soil categories with full description of the dynamic soil properties (V_s, density, G/Go-γ-D) (Figure 5.11).

Various thematic maps have been also constructed; an example is illustrated in Figure 5.12 where four maps of the thickness corresponding to different soil categories are presented. These maps have been constructed based on a large number of cross sections for the whole area. Geological information and tectonics also played an important role. Based on these maps and cross-sections it can be concluded that the subsoil stratigraphy in Thessaloniki is not really flat and horizontally layered. Nevertheless, as the 1D analysis is always the basic reference study it was decided to perform a detailed "equivalent linear" analysis first in more than 300 representative "soil columns" estimated through the detailed geotechnical mapping of the city and then to examine the possible implications of 2D effects.

According to the seismic hazard analysis the design outcrop acceleration is PHGA=0.25g (i.e. 10% probability of exceedance in 50 years for design earthquake in many codes like NEHRP, EC8, EAK2000 for ordinary constructions). Five different input motions selected for outcrop conditions were applied, all scaled to 0.23g-026g to account for different epicentre distance from the focus. Figure 5.13 shows a typical example of the analysis in a specific location along the coastal area of the historical center of the city. Figures 5.14 and 5.15 illustrate in the form of GIS maps some of the most representative results of the detailed ground response analyses for the city of Thessaloniki.

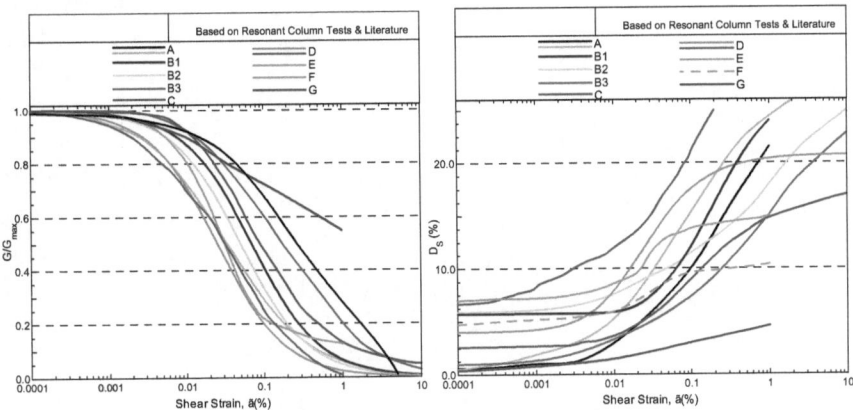

Forma tion		Description	V_S (m/s)	V_P (m/s)	Q_S
A		Artificial Fills, demolition materials and debris parts	200-350 (250)	400-1700	8-20 (15)
B1		Very Stiff sandy-silty clays to clayey sands, low plasticity	300-400 (350)	1900	15-20 (20)
B2		Soft sandy-silty clays to clayey sands, low to medium plasticity	200-300 (250)	1800	20-25 (20)
B3	Surficial	Stiff to hard high plasticity clays	300-400 (350)	1800	20-40 (30)
C		Very soft buy mud and silty sands	120-220 (180)	1800	20-25 (25)
D		Alluvium deposits, sandy-silty clays to clayey sands-silts, low strength and high compressibility	150-250 (200)	1800	15-25 (20)
E		Stiff to hard sandy-silty clays to clayey sands	350-700 (600)	2000	6-30 (30)
F	Subbase	Very stiff to hard low to medium plasticity clays to sandy clays Overconsolidated with rubbles and thin layers of gravels	700-850 (750)	3200	50-60 (60)
G		GreenSchists and Gneiss	1750-2200 (2000)	4500	180-200 (200)

Fig. 5.11. Dynamic properties, (mean "design" values), of the basic soil formations in Thessaloniki

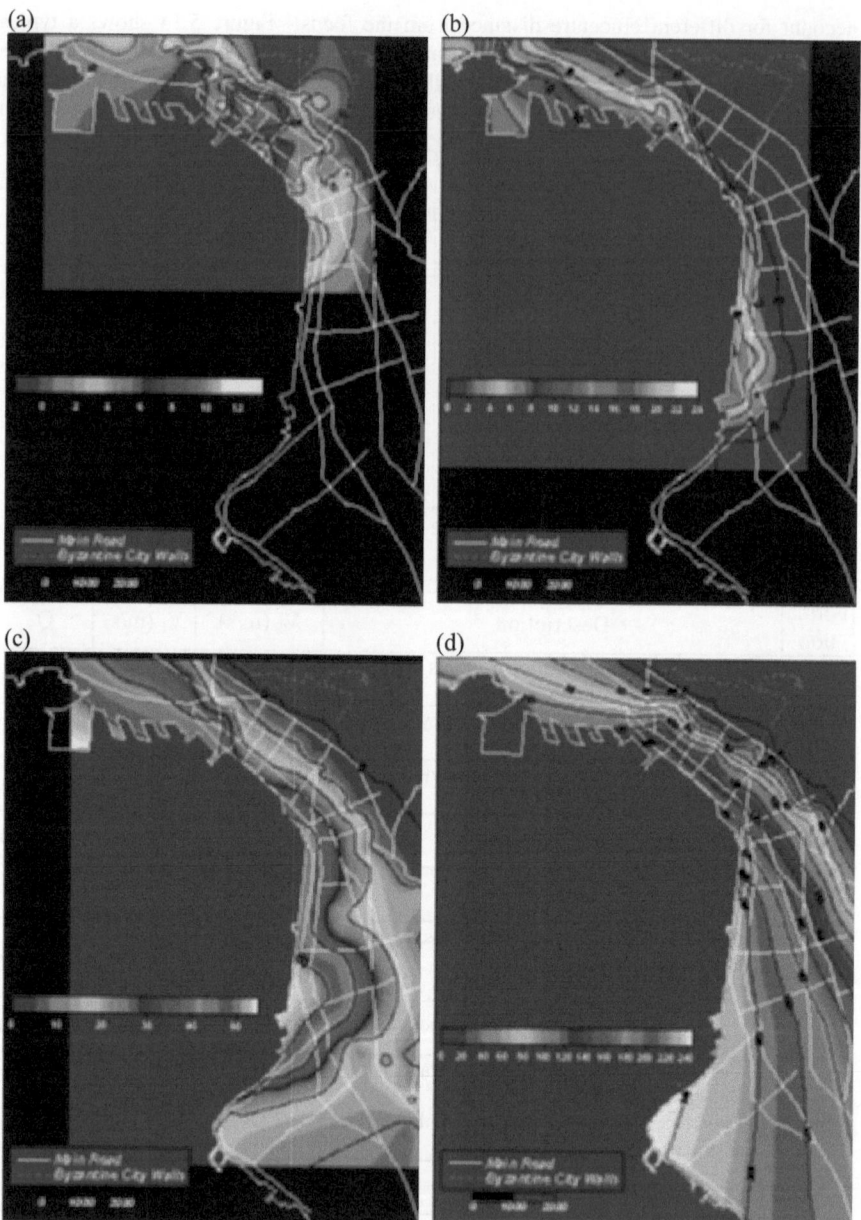

Fig. 5.12. Maps of different geotechnical formations, a) thickness of the historical center's debris and fills, b) thickness of the coastal loose silty sand and silty clay, c) depth of the upper surface of the basic stiff clay formation, and c) iso-depth of the "seismic" bedrock (Anastasiadis et al., 2001; Courtesy of the Birkhaeuser Publishing)

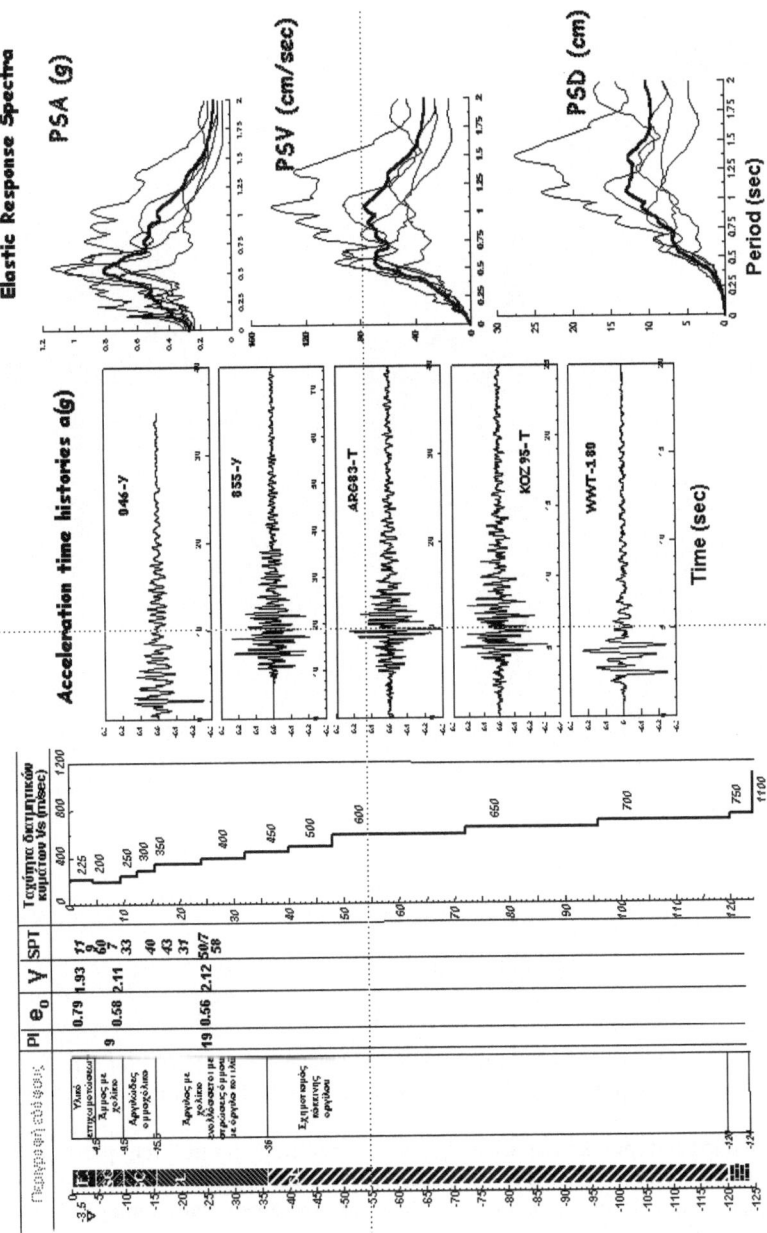

Fig. 5.13. Typical example of 1D analysis using 5 input motions, all scaled to the design outcrop acceleration (PHGA=0.25g)

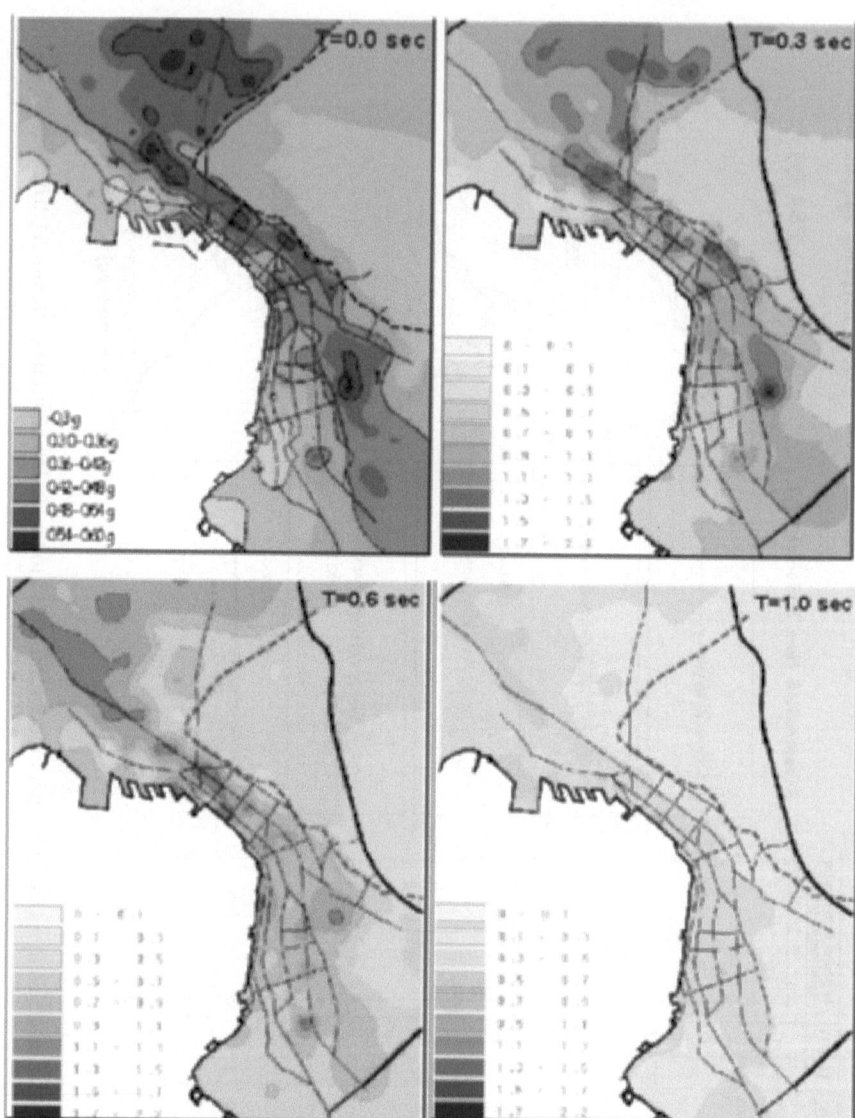

Fig. 5.14. Maps of the mean peak horizontal ground accelerations (m/s^2) for different spectral periods (T=0.0 sec, 0.3 sec, 0.6 sec and 1.0 sec)

The computed ground motion amplifications vary considerably in different parts of the city. This can be attributed to the local impedance contrast and the non-linear behaviour that exhibit some loose-soft soil materials, which are found especially along the shore line. The detailed site and soil characterization played a paramount role on the spatial variability and the amplitudes of the computed ground motion. An important practical conclusion drawn from the above observations is that the "effective values", (65% of the peak values), differ considerably from the "design value" proposed by the Greek Seismic Code (bearing in mind that according to EAK2000 Thessaloniki belongs in category II with a value of 0.16 g that is independent of the soil category). The question is therefore, whether this 1D body wave SH computations accurately describe the complex nature of wave propagation in case of non-horizontally layered media. In the following chapter an effort will be made to discuss this important issue.

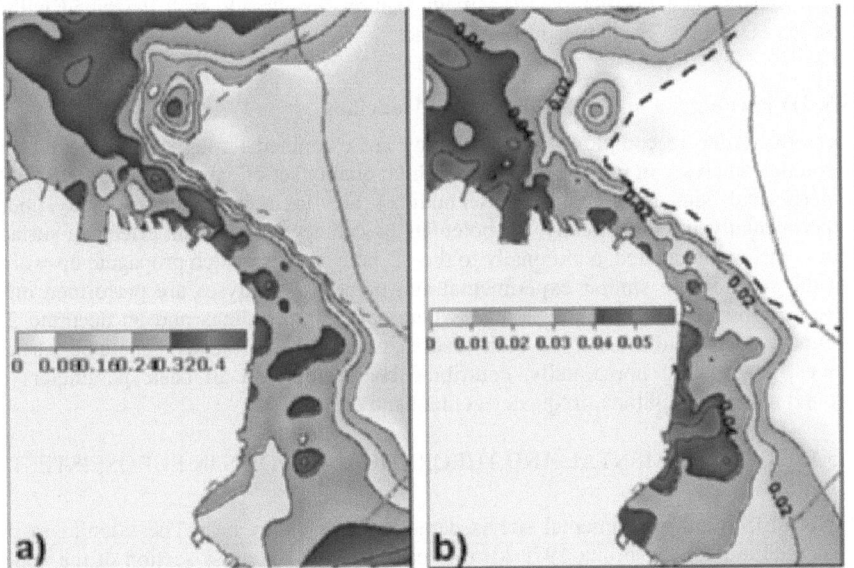

Fig. 5.15. Maps of the mean peak horizontal ground a) velocity PHGV (m/s) and b) displacement PHGD (m)

5.4.2. CONCLUSIVE REMARKS

One dimensional body wave propagation models are the basic tool for ground response analyses. In their simplest form (i.e. linear elastic or equivalent linear elastic soil behaviour) they are rather simple while they need for few parameters which are easily estimated even without performing specific dynamic field and laboratory tests, as there are many correlations with conventional geotechnical parameters (i.e. V_s-SPT, V_s-CPT, G/Go-γ-DT% with PI and D_R% for clays and sands etc). For fully non-linear and elastoplastic soil behaviour the evaluation of appropriate soil parameters and models is still a difficult task. Successful predictions of all parameters of ground response under strong seismic excitation (peak amplitudes, frequency content, spectral values and duration) for cases that soils exhibit highly non-linear behaviour (sometimes in the presence of liquefaction) are rare and often they are made a-posteriori.

Generally 1D models are reliable for nearly horizontally layered deposits and in cases when the impedance contrast between soil deposits and underlying rock is the controlling parameter of ground motion. The velocity of the bedrock and the incident wave field characteristics are playing an equally important role. With the 1D modelling the higher frequency parts of the expected ground motion can be captured quite accurately. Low frequency parts are less reliable and this is an important shortcoming for the case of deep basins (>300m).

In the case of Thessaloniki the 1D modelling of ground response was proven quite successful from an engineering point of view, both qualitatively and quantitatively. As it will be shown later on, the 2D ground response, in this particular case, is simply improving the general picture. The main features are already well described with a detailed 1D SH waves equivalent linear model.

5.5. 2D Phenomena in Ground Response Modelling

Site effects due to complex surface geology are examined through experimental and theoretical analyses in two cases. First the high quality set of data of EUROSEISTEST experimental site (http://euroseis.civil.auth.gr) are used in order to study, both experimentally and numerically, the potential appearance and relevant effects of surface waves that are generated, additionally to the 1D body waves, which propagate up-words. In the second case similar experimental and numerical analyses are performed on a specific cross-section of Thessaloniki, where seismic recordings and an accurate 2D ground model are disposed. In both cases it is proven that locally generated surface waves, propagating horizontally, contribute considerably in all basic parameters of ground motion (amplitude, frequency content and duration).

5.5.1. 2D EXPERIMENTAL AND THEORETICAL STUDIES IN EUROSEISTEST VALLEY

EUROSEISTEST experimental site is deployed in a valley near Thessaloniki in the epicentral area of the strong 1978 Ms=6.4 earthquake. The cross-section of the valley (Figure 5.16) is very well constrained through numerous geotechnical, geophysical and geological surveys (Pitilakis et al., 1999).

The database of ground motions recorded in surface and down-hole accelerographs comprises of many records from small and moderate earthquakes (M_w<5.8, 5<R<120km). An example of a far field event (Kozani M_w=6.5, R=100km earthquake) is given in Figure 5.17 together with the computed 1D SH wave synthetics that are derived using as input motion the record of the same transversal component at the reference site PRO. It is clear that 1D model is inadequate to simulate accurately the recorded motion, especially in the central part of the valley.

The reasons of the observed discrepancies have been discussed in detail in many studies (Raptakis et al., 2000; Chavez-Garcia et al., 2000; Makra, 2000; Makra et al., 2001; Pitilakis et al., 2001). In the frequency domain we observe through the transfer functions of P, S and SW waves composing the recorded signal (Figure 5.18.), large amplitudes of SW waves for frequencies up to 2Hz while the S window part has much lower amplitudes. This general trend is also observed in the downhole records, (TST-17m, TST-72m), which shows that surface waves affect the whole volume of the subsoil, at least in the central part of the valley and all three components. As surface waves are not observed at the reference site (PRO) or at the edges of the valley, it is concluded that the observed strong surface waves traces are locally generated on the lateral geological discontinuities of the valley, which is actually a graben (see Figure 5.16). Moreover the fact that observed maxima in both S and SW parts appear at the same frequency (0.7Hz), implies that both body S waves propagating upwards and locally generated surface SW waves contribute to the spectral amplification of the fundamental peak.

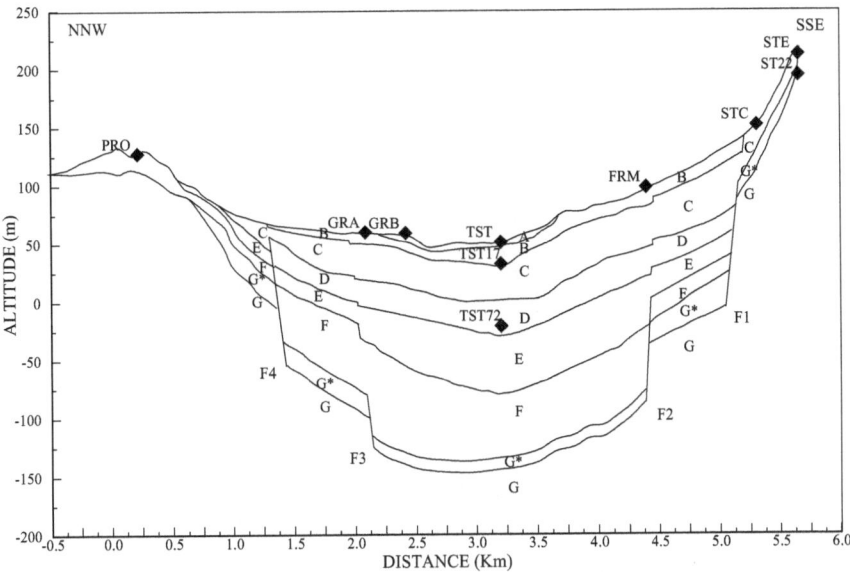

Layer	Description	S-wave	Density	Q_S
A	Silty, clayey sand	130	2.05	15
B	Silty sand and sandy clay	200	2.15	25
C	Marly silt and silty sand	300	2.075	30
D	Marly, sandy clay and clay silt	450	2.100	40
E	Alternating sublayers of clayey, silty sand and sandy clay with stones and gravels	650	2.155	60
F	Alternating sublayers of clayey, silty sand and sandy clay with stones and gravels	800	2.20	80
G*	Weathered schist bedrock	1250	2.50	100
G	Gneiss basement	2600	2.60	200

Fig. 5.16. EUROSEISTEST valley cross-section and strong ground motion instrumentation array layout amplification (Chavez-Garcia et al., 2000; Reproduced with permission from Elsevier)

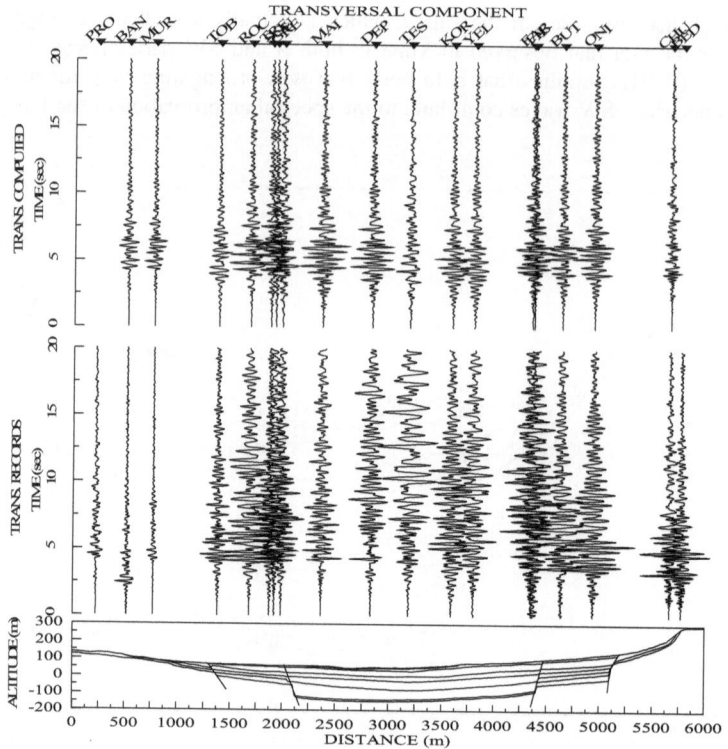

Fig. 5.17. Transversal seismograms from a far field event and 1D synthetics (Raptakis et al., 2000 Reproduced with permission from Elsevier)

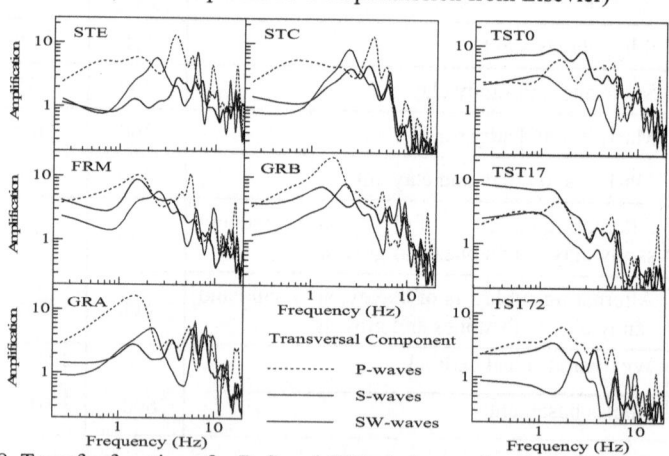

Fig. 5.18. Transfer functions for P, S and SW windows of surface (STE, STC, FRM, GRB, GRA, TST0) and downhole (TST-17, TST-72) transversal accelerograms (Raptakis et al., 2000 Reproduced with permission from Elsevier)

2D numerical analysis on the other hand, clearly indicates that the largest amplitudes of motion are not related to the vertically propagating SH waves, since the synthetic seismograms (Figure 5.19) are dominated by locally generated Love waves that converge to the center of the valley and thus, result to large amplitudes and a consequent increase of the duration; phenomena which are not seen outside the central part of valley. As a result, the synthetic ground motions verified the experimentally observed phenomenon of the existence and the importance of surface waves, which are locally generated at the edges of the valley and propagate to the center of the valley. A comparison between recorded and computed ground motion (Figure 5.20) proves that while the 2D model is reproducing successfully the recorded time histories, 1D modelling significantly fails to reproduce the observed long period surface waves.

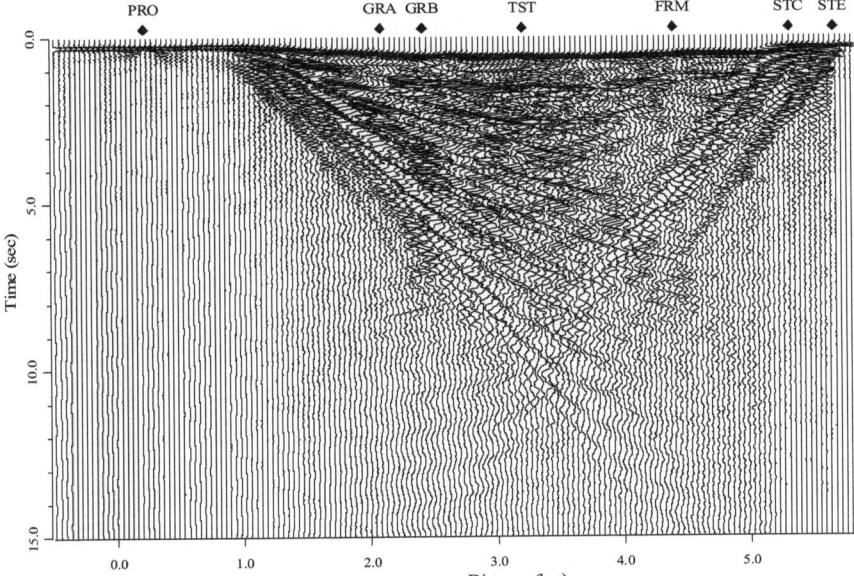

Fig. 5.19. 2D synthetics (f < 10 Hz) (Anastasiadis et al., 2001; Courtesy of the Birkhaeuser Publishing)

The important 2D effects described above, are further examined by introducing the notion of "aggravation factor" which is the ratio of the acceleration response between 2D and 1D ground response analyses (Makra 2000, Makra et al., 2001). After proper experimental verification this ratio along the whole cross-section of the valley (Figure 5.21) was calculated. 2D response spectra are significantly larger than 1D in a large band of frequencies and almost along the whole valley. This effect may have serious implications on design seismic motions since the vast majority of codes are based on the simple concept of 1D SH upward propagation. The case of Euroseistest valley is certainly a good and representative example; but in order to quantify the 2D and possibly the 3D effects for design input motion purposes, a number of additional cases should be examined. Thessaloniki is probably a second interesting case.

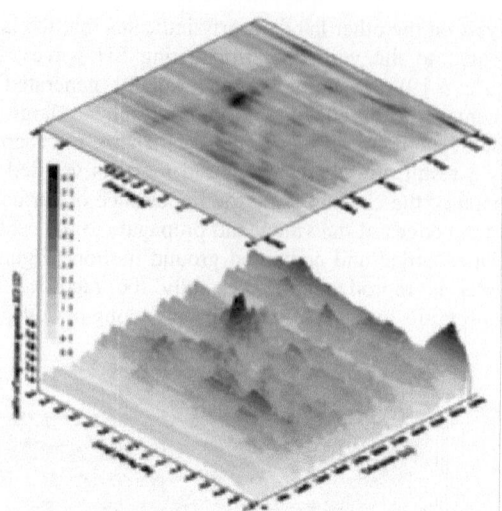

Fig. 5.20. Comparison between recordings (REC) of the transversal component and 1D, 2D synthetics at TST (centre), FRM and STC (south) stations all filtered at fc=3.5Hz.

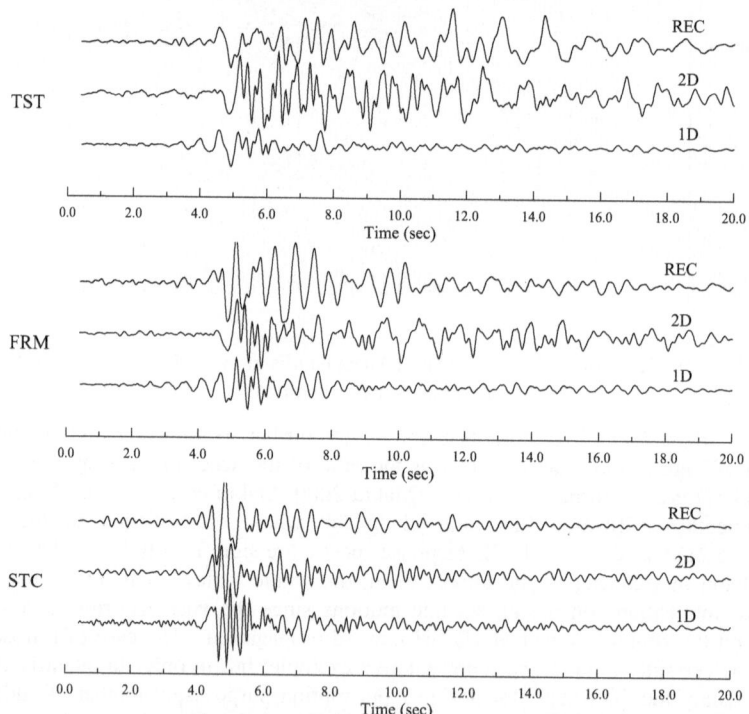

Fig. 5.21. Variation of the 2D/1D aggravation factor (ratio of acceleration response spectra) along the EUROSEISTEST cross-section for different periods

5.5.2. 2D EXPERIMENTAL AND THEORETICAL STUDIES IN THESSALONIKI

The complex site effects in Thessaloniki are examined in a characteristic cross-section (Figure 5.22) where there are three simultaneous recordings in broadband seismometers of a M_w=4.8 event occurred on 16.12.1993 (Figure 5.23). One of the stations (OBS) is on the outcrop and can be considered as a reference site. From the experimental transfer functions computed for stations LEP and ROT, it is shown that the transversal component, in both stations, correspond to higher amplifications and fundamental peaks at lower frequencies than the radial one (Figure 5.24).

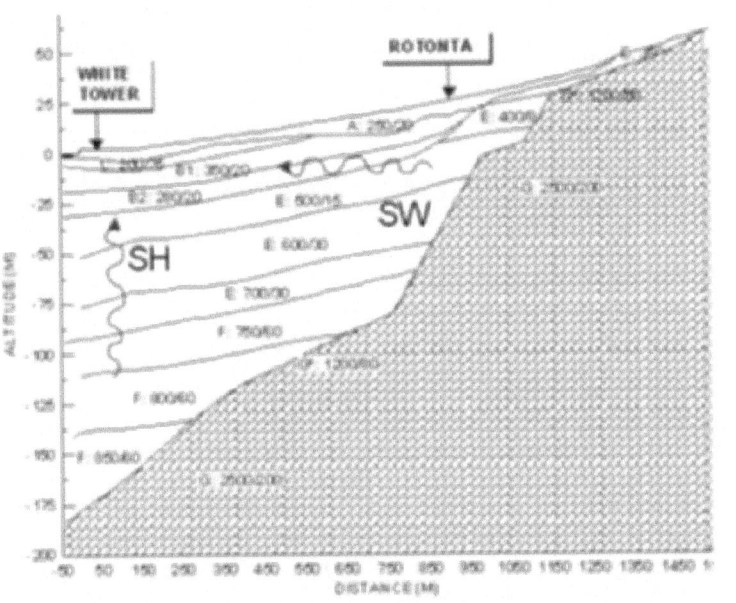

Fig. 5.22. Typical geotechnical cross-section in the historical centre of Thessaloniki (Anastasiadis et al., 2001; Courtesy of the Birkhaeuser Publishing)

The fact that the fundamental peak amplification of the transversal component is larger than the radial one and moreover shifted at lower frequencies is a strong indication of the appearance of diffracted Love waves and of their pronounced role in the recorded ground motions. Furthermore, the higher amplification of the vertical component at the most distant station (LEP) compared to ROT which is closer to the edges, is an indication that Rayleigh waves are also generated at the edges affecting the vertical component. Examining separately the S and SW parts of the recorded motions at the most distant station LEP (Figure 5.25), it was observed that S and SW waves appear having the same frequencies along the entire time history, and the major portion of the amplitude amplification is due to the SW part. The same phenomenon has been observed at Euroseistest; locally generated surface waves at the edges of the valley induce additional amplification to the 1D body wave propagation particularly at the fundamental 1D frequency. All three components were amplified (the vertical due to the Rayleigh waves), while a clear increase of the duration of the shaking was also observed

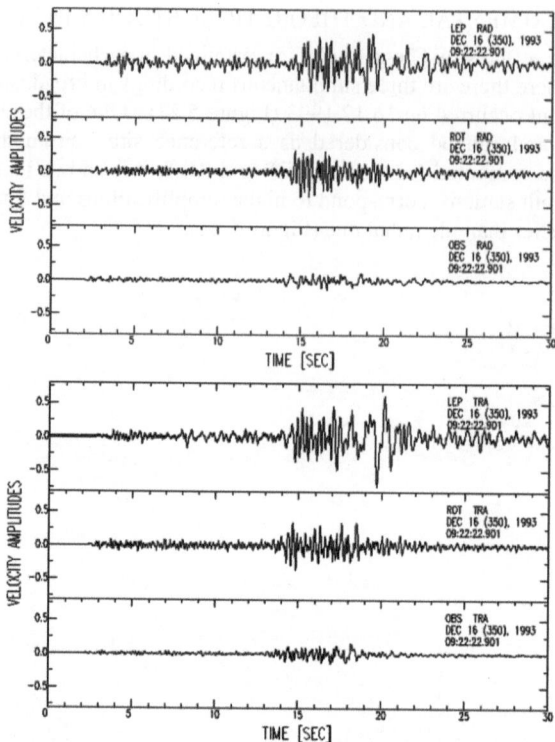

Fig. 5.23. Radial (left) and transversal (right) component of an M_w=4.8, R>100km event (16.12.93) recorded at 3 stations (seismometers) along the cross-section (Raptakis et al., 2003a)

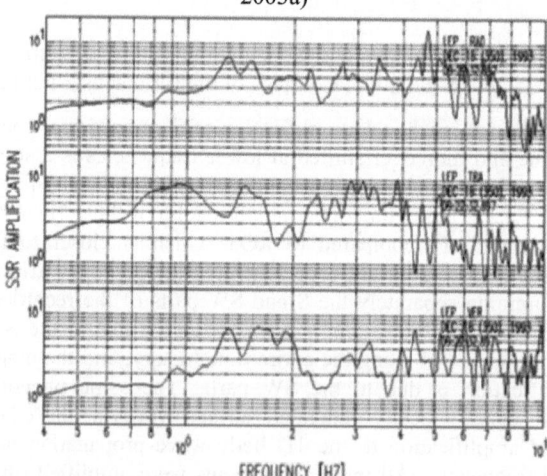

Fig. 5.24. Experimental transfer functions (reference station at OBS), and time windows of S and SW parts of signal recorded at station LEP (on the shore line) (Raptakis et al., 2003a)

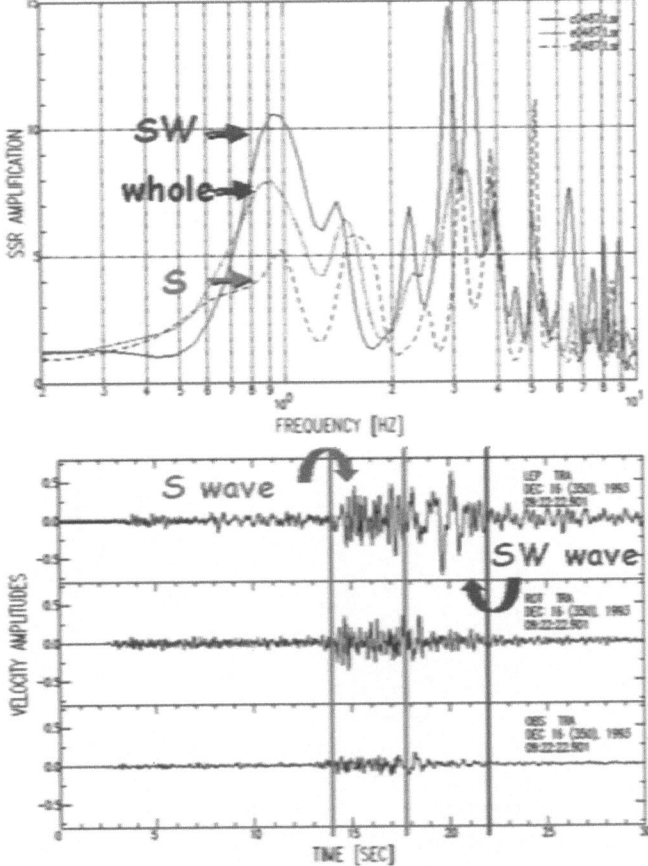

Fig. 5.25. Experimental transfer functions of S, SW and S+SW parts of the complete signal (Raptakis et al., 2003a)

The experimental observations of the complex site effects, mainly in the frequency domain, are followed by a numerical study of a 2D model of the cross-section (Figure 5.26) The synthetic ground motion is clearly dominated by two locally generated Love wave trains; the first appearing at the fundamental mode (latest part of the seismograms), while the second one represents a higher mode of Love waves. It is shown that the topmost layers with small V_s values dominate the fundamental mode, while higher modes are guided by the deeper soil layers with higher Vs propagation velocities.

The presence of the surface waves modifies considerably the ground motion characteristics and as it is illustrated in Figure 5.27, the 2D simulations, using either a finite difference or a finite element model, are much better simulating the actual recordings. 1D model is proven rather inadequate to accurately describe the complicated site effect pattern at the most distant station (LEP), which is dominated by the presence of surface waves.

Kyriazis Pitilakis

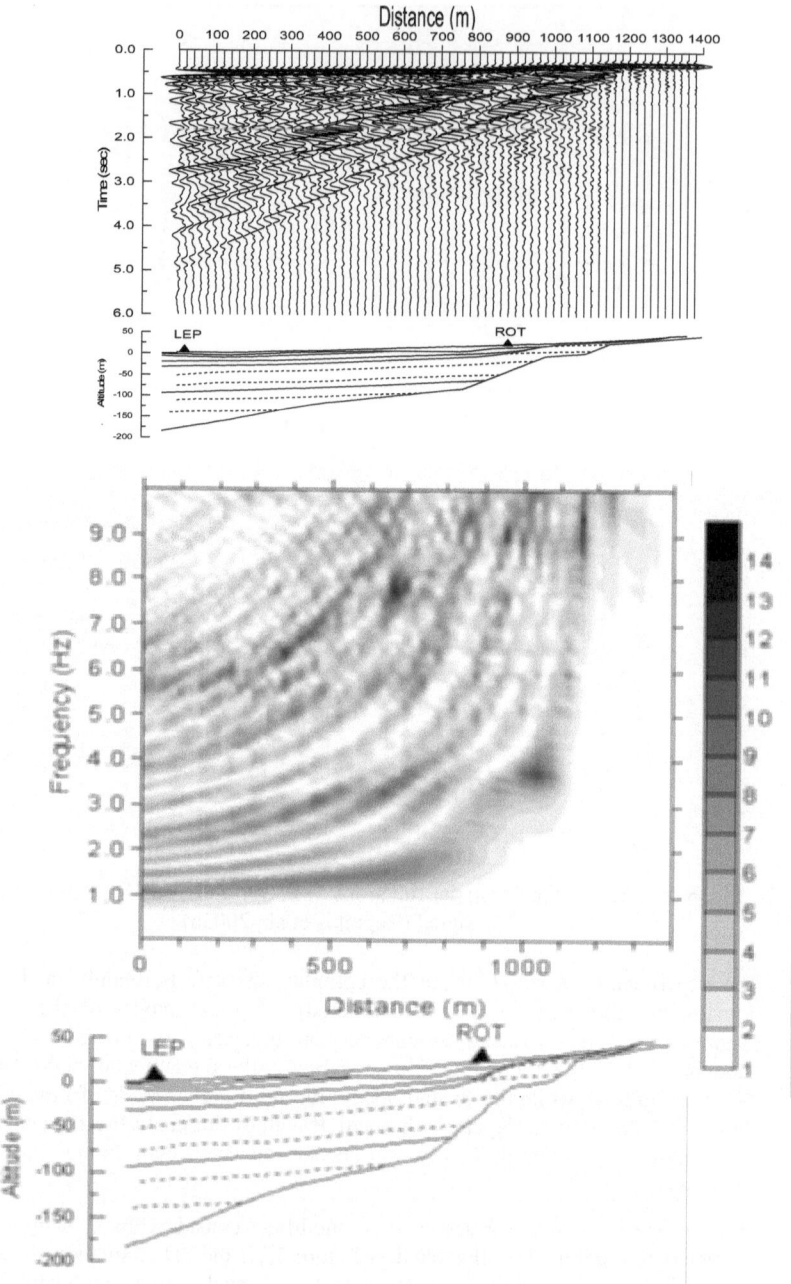

Fig. 5.26. a)2D model. Synthetic seismograms, b)2D Finite difference method (Raptakis et al., 2003b)

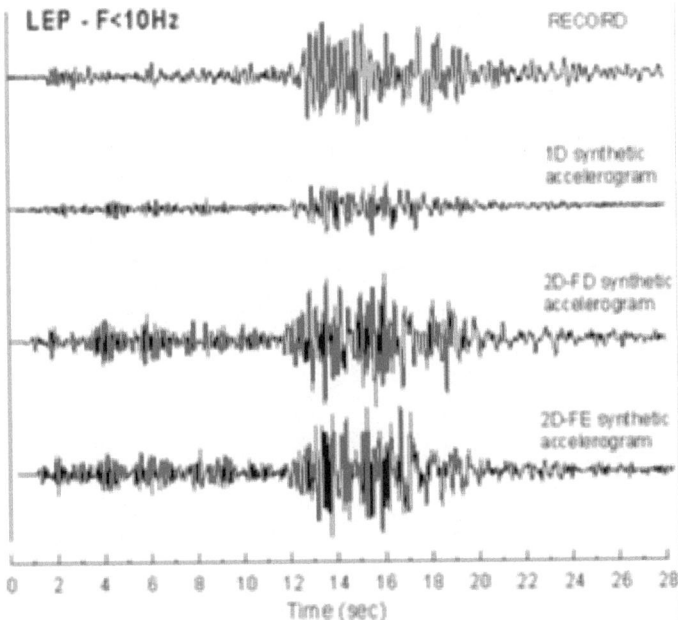

Fig. 5.27. Comparison between 1D, 2D synthetics (Finite Difference and Finite Element models) and recorded acceleration time histories at station LEP for the same event as in Figures 5.23 and 5.24 (Raptakis et al., 2003b)

To further examine the importance of the 2D effects compared to conventional 1D analyses the PGA values were computed at T=0.0sec along the cross-section for two potential strong seismic events. The input motion records correspond to outcrop ground accelerations from the Thessaloniki 1978 major aftershock and have been recorded on outcrop conditions (station OBS); they are scaled at 0.1g and 0.3g. The 1D analysis has been performed on vertical 1D profiles extracted from the 2D model using an equivalent linear model. The 2D computations are using a finite element model, which allows for equivalent linear simulation of the soil behaviour as well. The amplification ratio is higher for the 2D model almost all along the sedimentary part of the cross-section. The differences between 2D and 1D modelling are increased for the stronger excitation while the 1D model in certain areas presents no amplification at all.

As PGA (T=0.0sec) is not always the best indicator of the site amplification severity, especially for engineering applications, the amplification ratio was computed for the whole range of periods (Figure 5.29). It is shown that the 1D spectral amplification at the fundamental frequency (around 1Hz) is about 6.25 for a region between 100m and 500m from the coastal section for two events (input motion 0.1g and 0.3g) (Raptakis et al., 2003b) area. On the contrary, the highest 2D spectral amplification factor, reaching the value of 10, is found practically close to the shoreline of the cross-section. This is due to the predominant effect of surface waves, as discussed earlier, which is certainly more pronounced at longer distances from the edges where the surface waves are generated. No important differences have been observed for 0.1g and 0.3g input motion with respect to the above remarks.

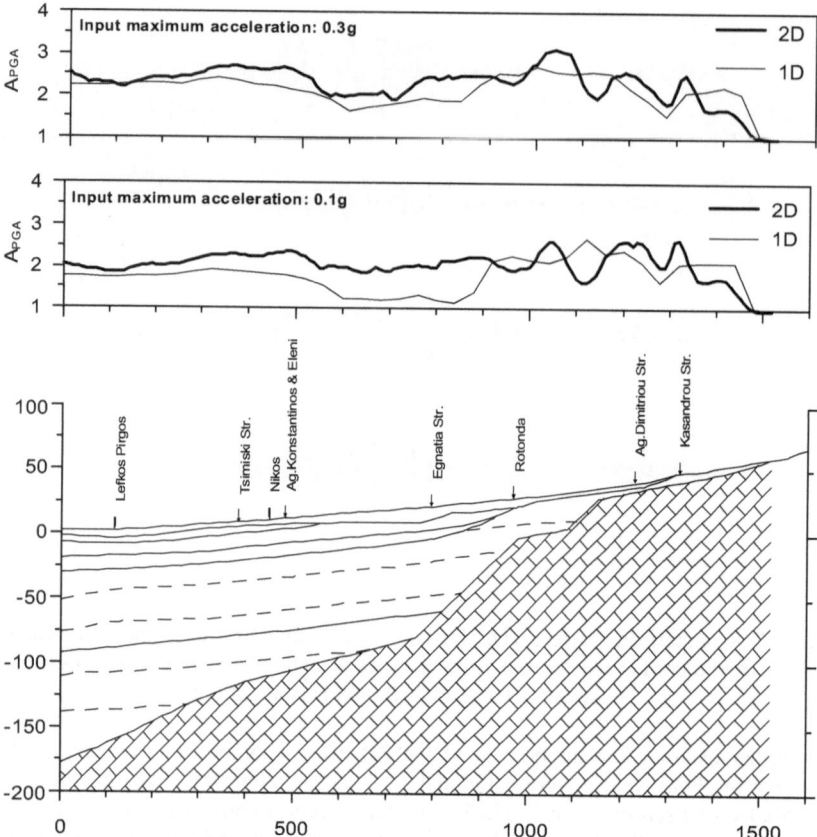

Fig. 5.28. Computed peak ground acceleration amplification factor along the cross-section at T=0.0sec (PHGA-soil/PHGA-rock) (Raptakis et al., 2003b)

5.5.3. CONCLUSIVE REMARKS

It is readily shown both experimentally and theoretically that in case of rather shallow sedimentary valleys, such as in EUROSEISTEST, that the wave field and the ground response is strongly dominated by surface waves locally generated at the lateral geological discontinuities. These waves contribute significantly to the 1D fundamental amplification and to the elongation of the strong motion duration. 1D models cannot capture all these characteristics of the complex wave-filed in geologically complex sites contrary to the 2D modelling.

Similar phenomena are observed in the case of Thessaloniki which is presenting similar geological and topographical features. Again surface waves were generated at the northern lateral geological discontinuities and propagating horizontally "in-wards" modify the spectral amplification pattern all along the city. The phenomenon is more pronounced at longer distances from the "generating" discontinuities, i.e. along the shore-line of the city. The 2D ground response is certainly better portraying the "reality". However for practical engineering applications the computed 1D SH-wave "averaged and smoothed" spectral amplitudes, especially at the fundamental mode, are not very much different. When there is an accurate knowledge of site conditions and soil characteristics, 1D modelling is still offering a good description of the average ground response. It is perfectly adequate for microzonation studies where a large number of computations is needed using many different input motions (at least five for each seismic scenario) to get average and smoothed design motion response spectra.

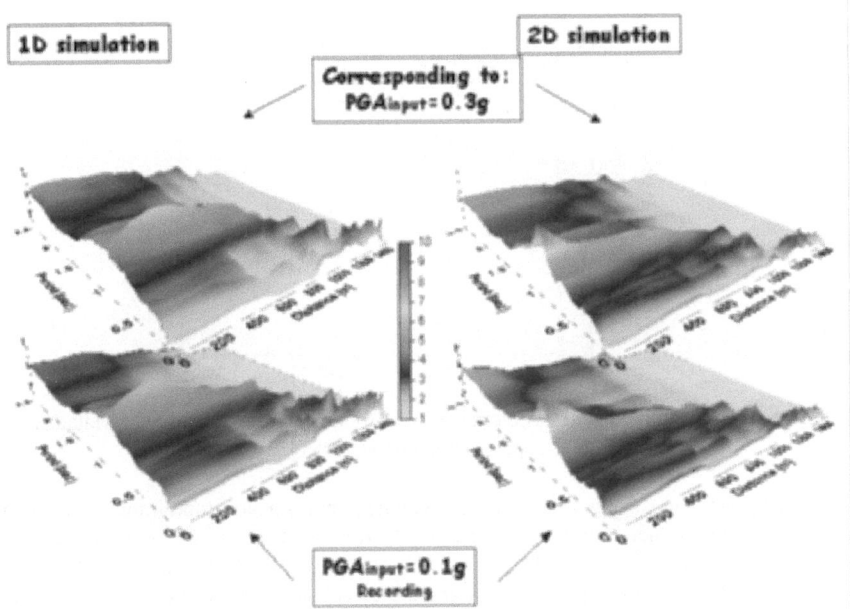

Fig. 5.29. Spectral Amplification factor SA(T) soil / SA(T) rock along the cross

There is no doubt that 2D analyses offer a much more detailed description of the complicated ground response phenomena in geologically complex sites. For the moment most of 2D or even 3D models are linear elastic with constant damping and moreover their applications and results are "site and case depended". Numerical 2D/3D models with more advanced soil models are very "heavy" from computational point of view and they cannot be applied easily, especially for parametric studies. For this reason, so far, 2D models are mainly used to understand the physics of complex wave propagation phenomena, less for design purposes and even less for detailing code prescriptions. Moreover a full and reliable validation of each model, including the simplest 1D ones, is always a major requirement. To this respect EUROSEISTEST 3D accelerometric array and all other combined surveys and studies programmed in the frame of this experiment , together with other similar experiments in Japan and USA and the analyses of data coming from many accelerometric arrays installed all over the world, are expecting to contribute seriously in better understanding and modelling of complex site effects.

5.6. Site Effects Due to Surface Topography

5.6.1. BRIEF LITERATURE REVIEW

It has been reported after destructive earthquakes (Friuli, Italy 1976, Irpinia, Italy 1980, Chile 1985, Whittier Narrows 1987, Kozani, Greece 1995, Aegion, Greece, 1995 and Athens, Greece 1999) that buildings located at hill tops suffer more intensive damage than those located at the base (Brambati et al., 1980; Siro, 1982; Celebi, 1987; Kawase and Aki, 1990).

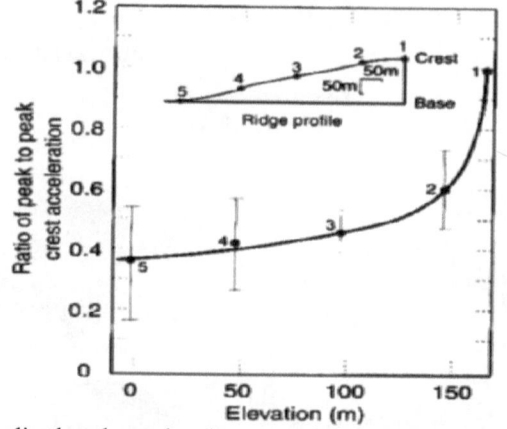

Fig. 5.30. Normalized peak accelerations recorded on mountain ridge at Matsuzaki, Japan (after Jibson, 1987)

There are few but strong instrumental evidence that surface topography affects the amplitude and frequency content of ground motion (Geli et al., 1988; Faccioli, 1991; Finn, 1991; Chavez-Garcia et al., 1996; 1997, LeBrun et al., 1999; Jibson, 1987). Two well known examples of apparent topographic effects are observed on the abutment of Pacoima Dam in San Fernando 1971 Earthquake (Boore, 1972), where an impressively high acceleration of 1.25g was recorded at the crest of a narrow ridge adjacent to the dam in Tarzana station during the Northridge 1994 earthquake for which the amplification was of the order of 5 in the vicinity of 3Hz reaching comparable peak acceleration values. In Europe weak motion measurements (Bard and Meneroud 1987 and Nechtschein et al., 1995; Chavez-Garcia et al., 1996; LeBrun et al., 1999) reported similar observations of large amplifications, almost with a ratio of ten, in a narrow frequency band around 5Hz.

Theoretical models predict a systematic amplification of seismic motion at convex topographies while de-amplification phenomena are observed over concave geometries such as valleys. According to Bard (1999) these effects are related mainly to three physical phenomena: (a) the sensitivity of surface motion to the incidence angle around the critical especially for SV waves, (b) the focusing and d-focusing of seismic waves along the topographic relief and (c) the diffraction of body and surface waves which propagate downwards and outwards from the topographic features and lead to interference patterns between direct and diffracted waves. Different researchers have contributed to define the importance of various parameters.

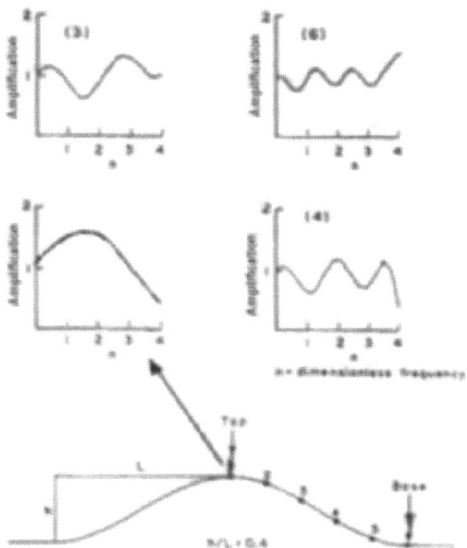

Fig. 5.31. SH Fourier transfer functions to homogeneous half-space outcrop motions (after Geli et al., 1988; Courtesy of the Seismological Society of America)

- The nature of the incident waves is studied by Ashford and Sitar (1997) who reported that the amplification of incident SV is higher because reflection and diffraction of SH waves does not generate other wave types.
- The incident angle has been studied also by Ashford and Sitar (1997) who found that in generally the greatest absolute acceleration at the crest is observed for the case of vertically propagating waves.
- Boore et al. (1981) and Ashford and Sitar (1997) studied the effect of slope inclination and observed that the amplification becomes higher when the slope becomes steeper.
- The spatial variation of ground motion along the bank of canyons with different geometry (triangular, semi-cylindrical, semi-elliptical etc) have been studied by Sanchez-Sesma et al. (1982), Trifunac (1973) and Wong and Trifunac (1974). Both outlined the importance of the ratio of the canyon width to the wavelength of the incident SH waves and of the incident angle.
- The frequency content of the input motion is also a key parameter. It is observed that for long wavelengths, i.e. very low frequencies, topographic effects are negligible, while the effects become significant for wavelengths comparable to the geometric characteristics of the relief. (Ashford and Sitar, 1997, Ohtsuki and Harumi, 1983).
- Other parameters like soil stratigraphy and dynamic soil properties (G_o, material damping etc) have also an important effect on the qualitative and quantitative modulation of the ground motion in topographic irregularities.

The main conclusions of the instrumental and theoretical results can be summarized in the following:

- Amplification is generally larger for the horizontal components that for the vertical.
- The steeper the slope, the higher the crest amplification
- Qualitatively the maximum effects correspond to wave lengths comparable to the horizontal dimension of the topographic feature.
- The absolute value of the amplification ratio cannot be estimated or computed among other reasons because the amplification of the displacement at the crest is generally combined with a de-amplification at the base of the topographic irregularity and their respective absolute values are not easily estimated a-priori.

5.6.2. SEISMIC CODES

It may be seen from the preceding paragraph that quantification of topographic effects on seismic ground motion is a very difficult task where many parameters of different nature are involved. The lack of high quality experimental data to better understand the physics of topographic effects and to validate the numerous theoretical and numerical analyses and results makes ambiguous the incorporation of the acquired experience in seismic codes.

However the ongoing version of the European Seismic Code EC8 is an exception. Based on earlier French Code AFPS-1990 it is proposing a correction factor, called aggravation factor F, for both ridge and cliff type topographies as a function of the height H and the slope angle i (Figure 5.32). The aggravation factor is defined as (a) $2D = F_{topo}$ (b)1D and takes values according to Figure 5.32; an extra 20% increase is anticipated when a surface "soft" layer with thickness more than 5m is present.

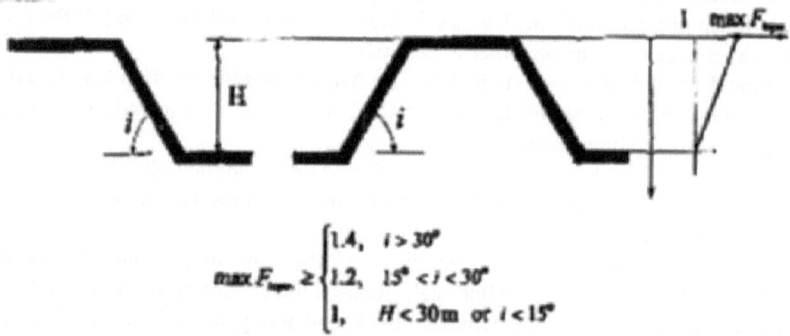

$$max\,F_{topo} \geq \begin{cases} 1.4, & i > 30° \\ 1.2, & 15° < i < 30° \\ 1, & H < 30m \ or \ i < 15° \end{cases}$$

Fig. 5.32. European Seismic Code provisions (EC8-2000) for topographic effects

5.6.3. THEORETICAL STUDIES IN AN EXPERIMENTAL SITE IN GREECE

The array

The Gulf of Corinth is one of the most seismically active regions in Greece. The most recent destructive earthquake occurred in 1995 (Ms=6.5) and caused serious damages in the city of Aegion (Figure 5.33) which is built in a sort of terrace right on the foot wall of the Aegion fault. Part of the city, mainly harbour facilities and small old masonry factory buildings are on the hanging wall part of the fault along the coastal area. The fault of Aegion itself has not been activated during the recent Aegion earthquake sequence. Most of the damages are reported up-hill. The earthquake has been recorded in one station also up-hill (see Figure 5.33 site: OTE).

Important spectral accelerations are recorded (Figure 5.34) especially in the transversal component which are not fully justified by the magnitude and the epicentral distance of this particular event. The simplified cross section of the area is given in Figure 5.35. It is based on an extensive geotechnical and geophysical survey contacted in the framework of European research program (CORSEIS http://www.corinth-rift-lab.org and Pitilakis et al 2003). The main part of the city terrace is dominated by hard-stiff soils (CL-GC with $N_{spt}>40$ and average $V_s=400m/s$); the "seismic" bedrock may be considered at the level of conglomerates found at -20m to -25m. On the contrary at the coastal area the surface soils are mainly very loose saturated silty- sands (ML-SM with $N_{spt}<8$ and $V_s<200$ m/s) presenting high liquefaction susceptibility and exhibiting clear non linear behaviour. The equivalent to seismic bedrock conglomerate layer is found at -180m from the level of the sea (approximately 250m from the terrace).

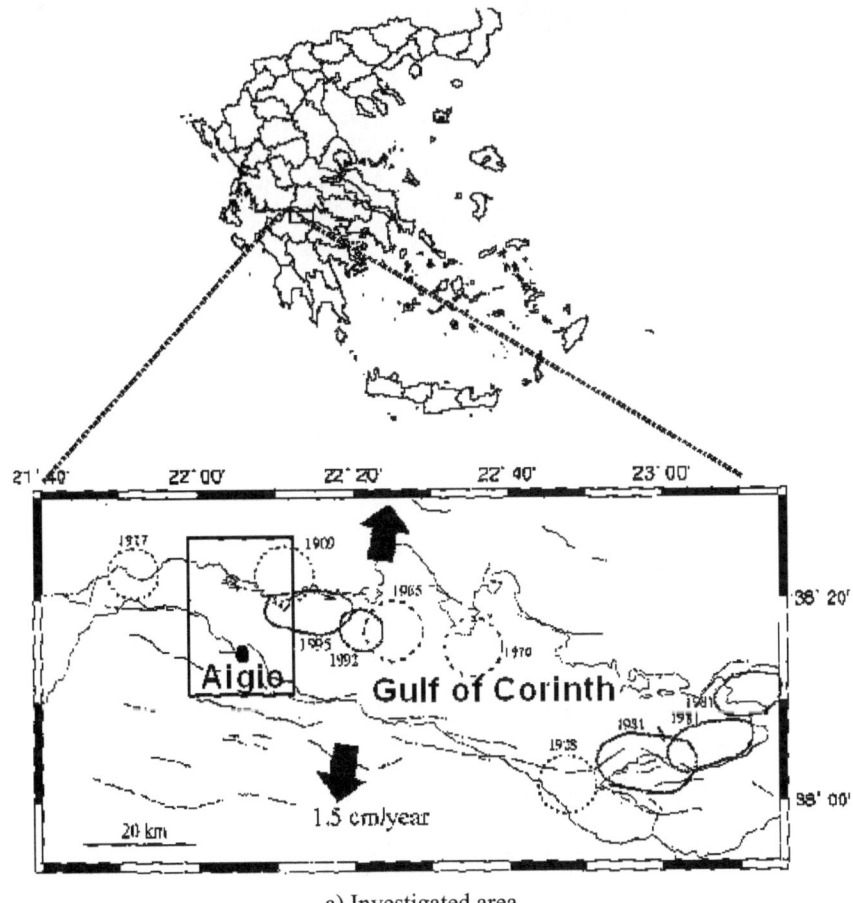

a) Investigated area

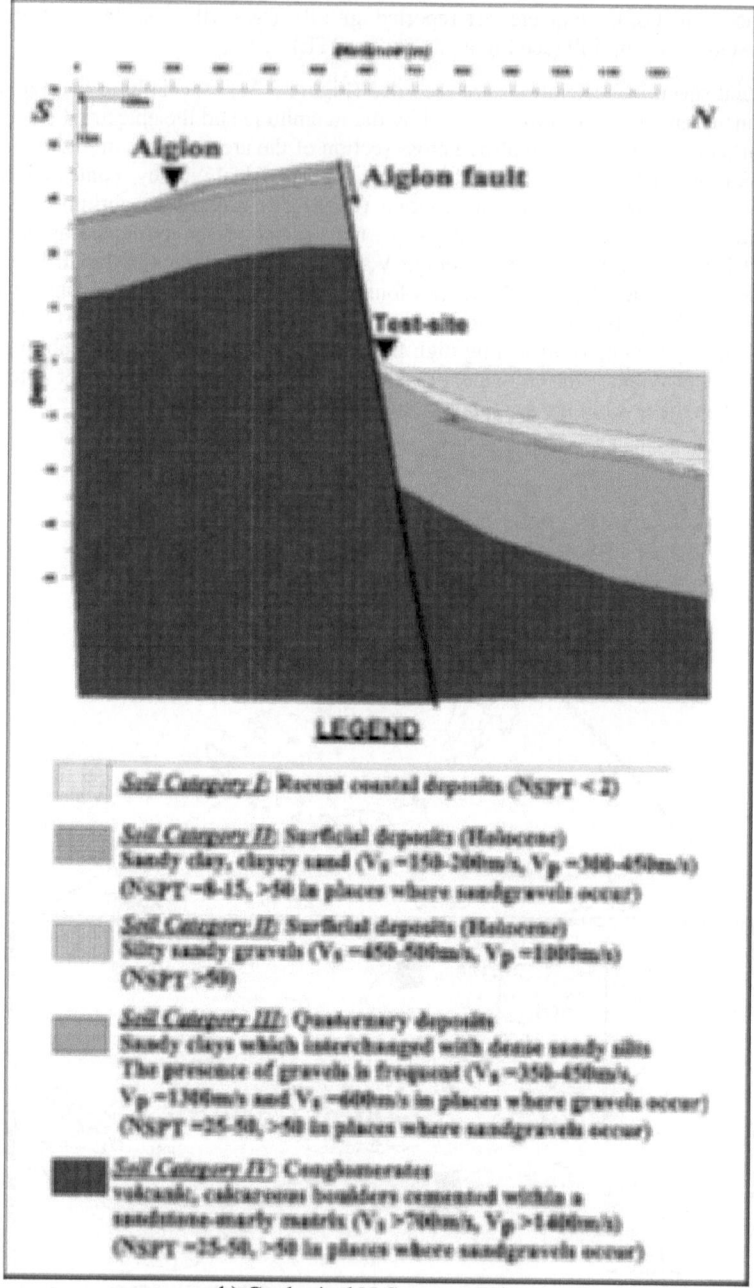

b) Geological N-S cross section

Fig. 5.33. (a) Location and (b) Geological cross section of the investigated area

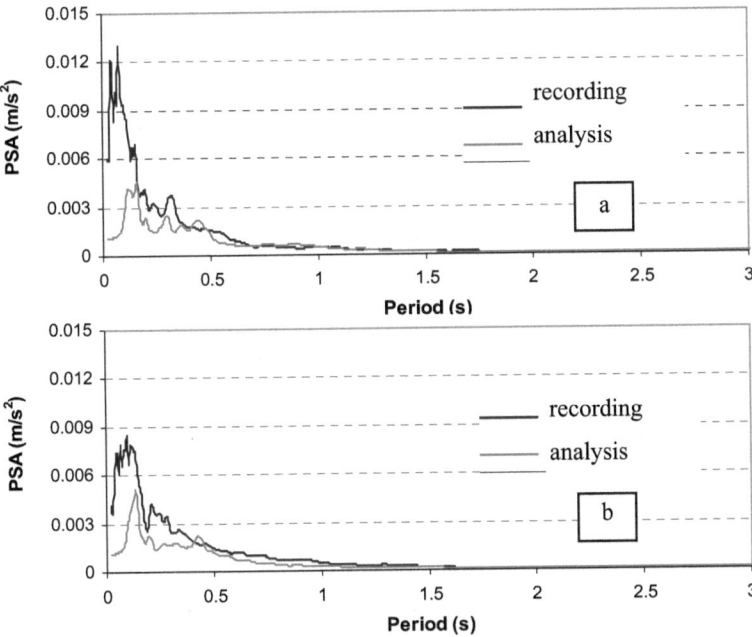

Fig. 5.34. Computed (1D SH waves) acceleration response spectra at the free surface resulting from the recording at depth z=-178m; a) x-direction, b) y-direction, Comparison with recordings

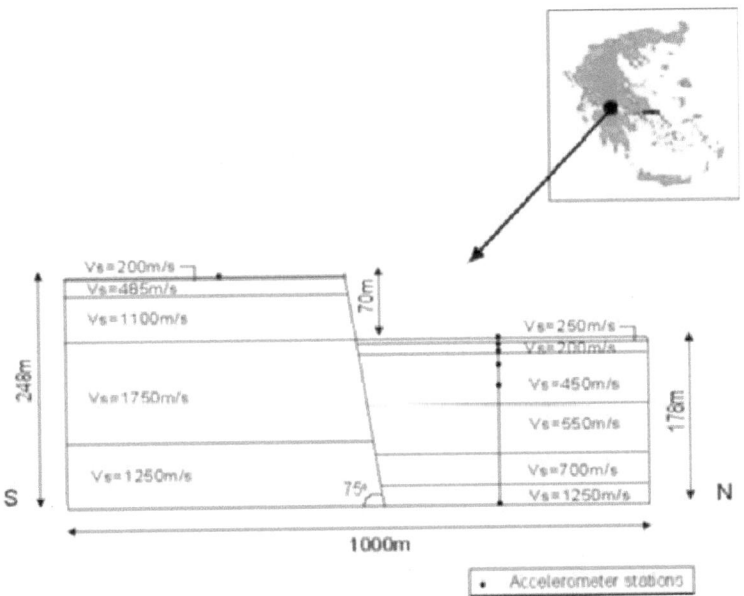

Fig. 5.35. Simplified cross section of the area

In the frame of CORSEIS project a vertical down hole array has been installed at the coastal area of Aegion consisting of five 3D broad band accelerometrers and two dynamic pore pressure transducers. The deepest station is installed in the conglomerate layer at -178m and the other four stations at -60m -30m -14m and at the free surface. The pore pressure transducers are installed in the saturated loose marine deposits at -6m and -14m depth where liquefaction phenomena are very prominent. The vertical array (CORSSA, after CORinth Soft Soil Array) combined with two more surface accelerometric stations operating on the terrace, form a unique array to study site effects in the presence of soil liquefaction, soil nonlinearities, near or far field conditions and topographic irregularity.

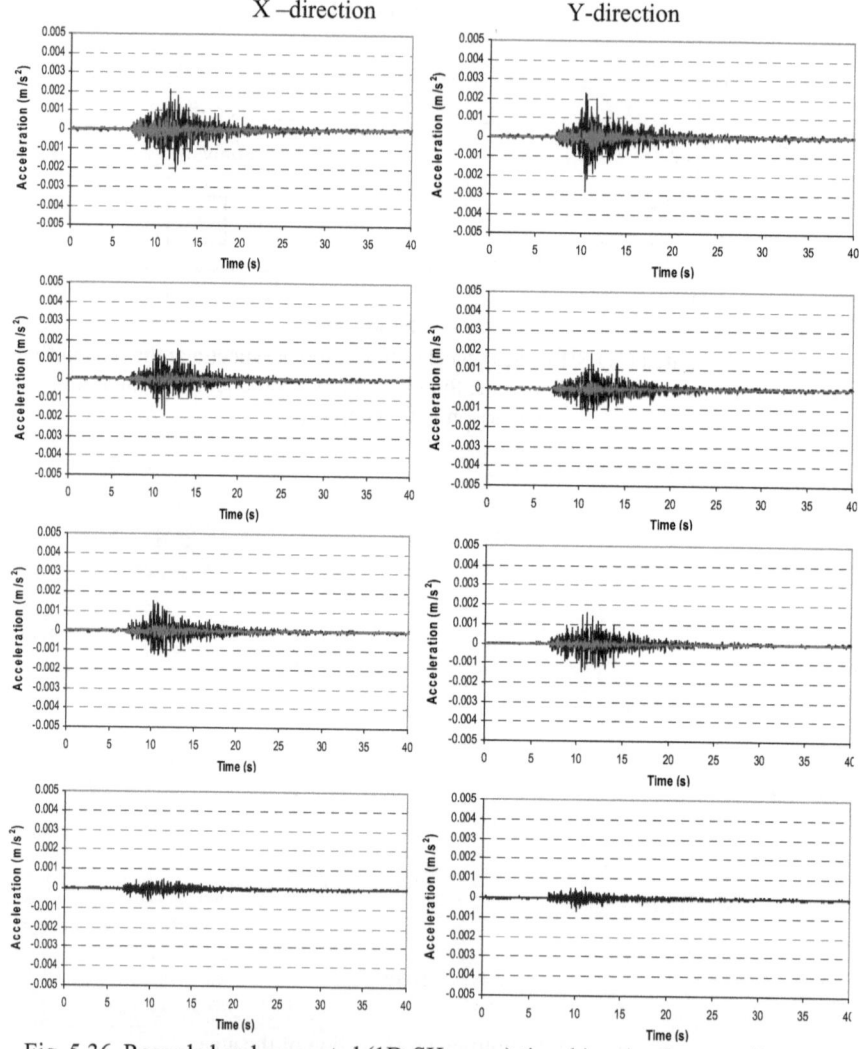

Fig. 5.36. Recorded and computed (1D-SHwaves) time histories for a small event

Preliminary 1D linear elastic computations of small event recorded during the validation period of the instrumentation, (Pitilakis, D., 2002), proved that 1D modelling is enable to simulate successfully the wave propagation pattern in the foot-hill of Aegion, most probably due to strong 2D effects which could not be captured by the SH waves propagating up-wards with input motion the recorded signal at -178 m (Figure 5.36). Hence the 2D wave propagation is necessary even for weak motions.

Two-Dimensional modelling

A two dimensional wave propagation analysis is then performed using the finite difference code FLAC (Ktenidou, 2003). The motivation of the present analysis is to a-priori study the effects of topography using simple input motions of Ricker type with different frequencies in the spatial variation and the intensity of ground shaking along the irregularity. Apart from the understanding of the physics of wave propagation in the presence of Aegion irregularity, we are expecting to check the theoretical results with actual small-moderate and maybe strong ground motion recordings when available in the whole surface and downhole array.

The finite difference discretization of the two dimensional model is shown in Figure 5.37 together with the studied cases for the case of SV waves. The frequency f_o=0.88 Hz of the selected Ricker wavelet is comparable to the fundamental frequency of the foot-hill side of the mesh; f_o= 1.43 Hz is the fundamental 1D frequency of the left up-hill side of the mesh and the highest (f_o=3 Hz) is selected for a more realistic incident motion. All analyses are linear elastic.

MODEL	INPUT (RICKER WAVELET)
Linear elastic model with global damping ξ=2%	f = 0.88, 1.43 & 3 Hz
Linear elastic model with global damping ξ=1%	f = 3 Hz
Equivalent linear model with damping according to soil layer	f = 3 Hz

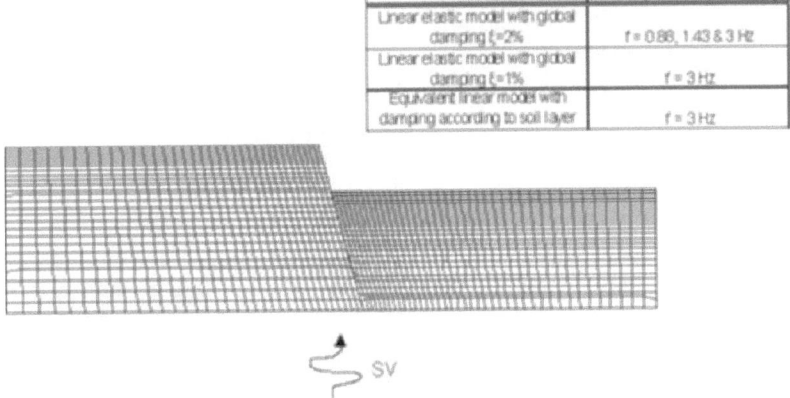

SV

Fig. 5.37. Finite difference discretization of the analysed configuration

The vectorial representation of the computed velocities, having arbitrarily selected scales, for the three incident wavelets is very interesting (Figure 5.38). After 1-2 sec the ground motion is dominated by surface waves with relative high amplitudes. The absolute amplitudes are decreased considerably after the first 2 sec but the motion continues. The frequency of the incident monochromatic motion modifies considerably the wave pattern. The upper part of the mesh seems to be less sensible comparatively to the hanging wall part when higher frequencies than its fundamental one are present in the signal.

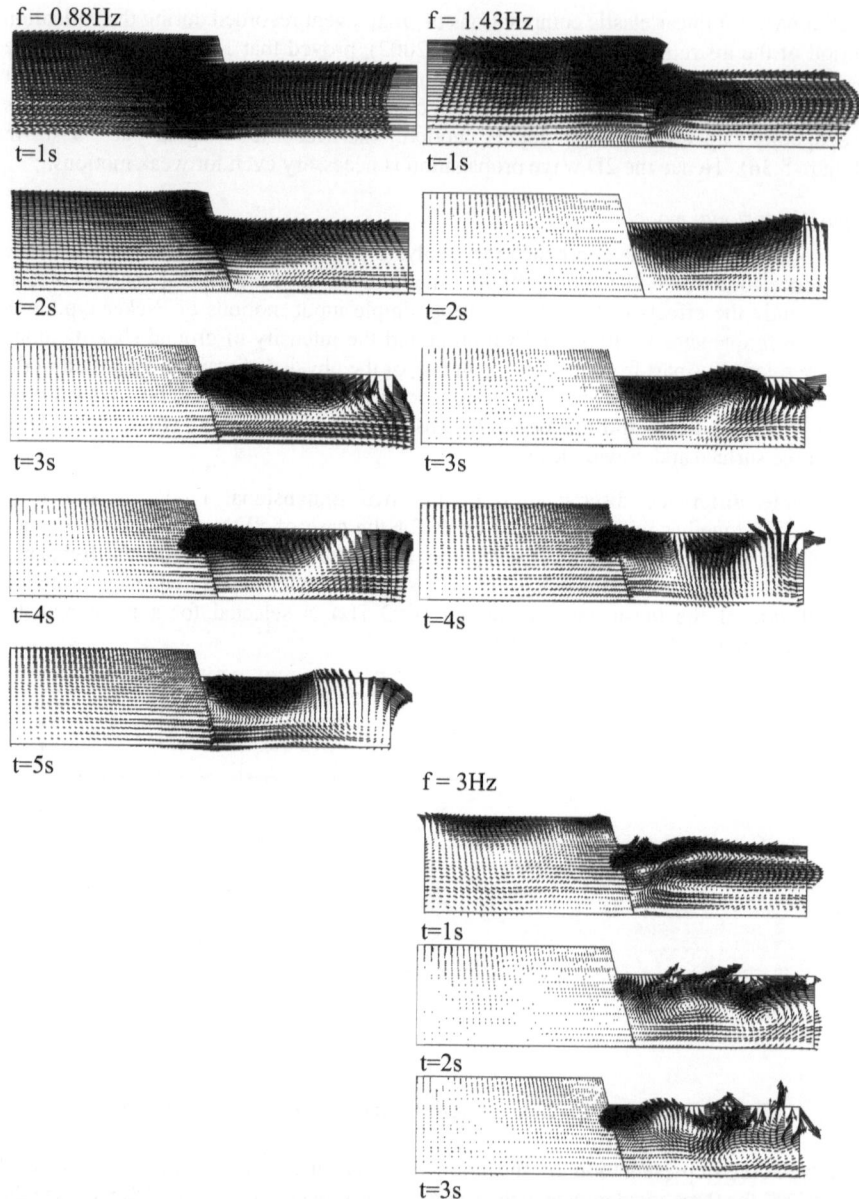

Fig. 5.38. Velocity vectors for three Ricker wavelets (fo=0.88Hz, 1.43Hz and 3Hz); linear elastic analysis with constant damping ξ=2%. Scale is magnified for late spots

The spatial distribution of the computed horizontal and vertical accelerations along the surface is shown in Figures 5.39 and 5.40.

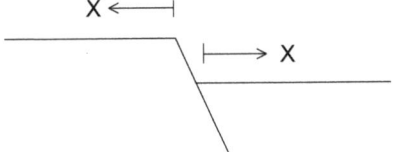

(a) ABSOLUTE MAXIMUM HORIZONTAL ACCELERATION: $a^H max(x)$

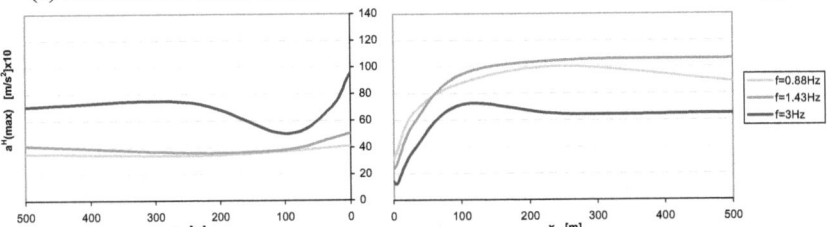

(b) NORMALIZED MAXIMUM HORIZONTAL ACCELERATION: $a^H max(x)/a^H max(500)$

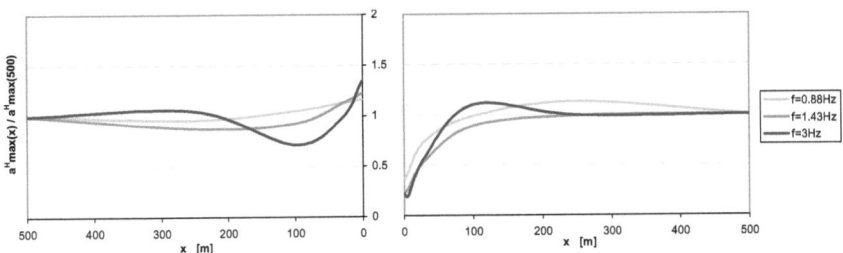

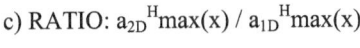

c) RATIO: $a_{2D}^H max(x) / a_{1D}^H max(x)$

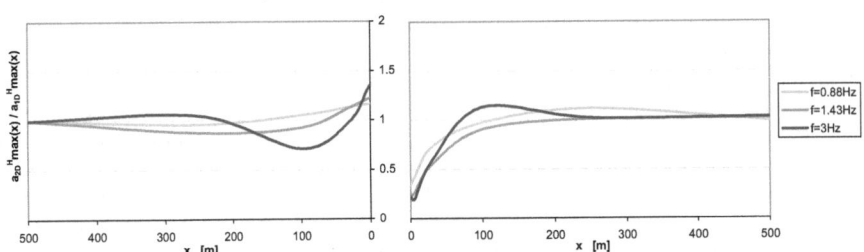

Fig. 5.39. Computed horizontal accelerations at the free surface (a) maximum accelerations for the 2D model and different frequencies of input motion; (b) effect of the frequency content of the input motion on the normalized peak horizontal accelerations; (c) 2D peak ground accelerations normalized to 1D peak accelerations for Ricker f_o=3Hz

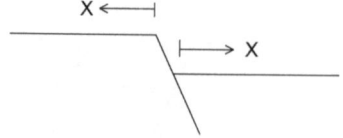

(a) ABSOLUTE MAXIMUM VETRICAL ACCELERATION αVmax(x)

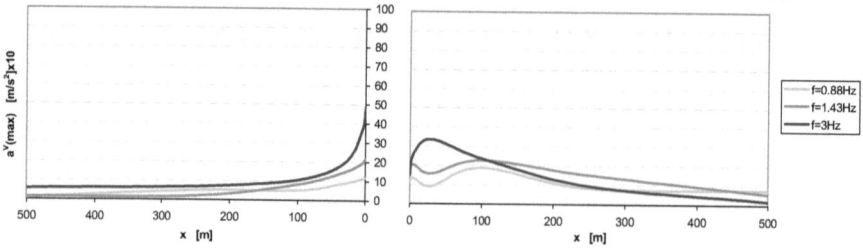

(b) NORMALIZED MAXIMUM VETRICAL ACCELERATION
αVmax(x) / αVmax(500)

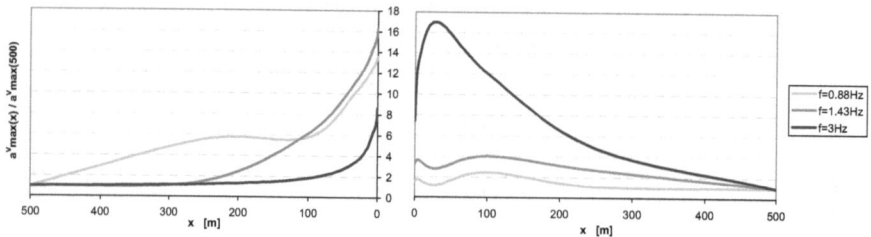

Fig. 5.40. Computed vertical accelerations at the free surface (a) maximum accelerations for the 2D model at different frequencies of input motion; (b) effect of the frequency content of the input motion on the normalized peak vertical accelerations

From the velocity vectors it is evident the generation of surface waves, mainly Rayleigh which propagate inwards modifying the motion. The cales have been magnified after 2 sec because the motion has decreased considerably. These surface waves produces complicated wave patterns which involves, among other issues, the development of strong vertical acceleration components at the ground surface on both sides of the topographic irregularity. The amplitudes of these vertical accelerations, which are not related to the incident wave field, are not negligible compared to the peak horizontal acceleration (Figures 5.39a and 5.40a). Bearing in mind that real excitations are containing vertical acceleration components as well, and based on these result, it should be expected that in reality the vertical acceleration will be very strong in the vicinity of topographic irregularities.

Another important observation is that the peak horizontal accelerations are increased (about 30%) near the crest; on the contrary they are considerably decreased near the foot of the cliff (almost 60%in average). Comparing the 2D horizontal accelerations with the "free topography field" (i.e. at least 500m far from the crest) accelerations (Figure 5.39b), it is observed that the influence of the topography is concentrated at a short distance from the crest (about 200m) and the important variations in amplitudes even more close to that. The influence of the frequency content of the incident SV wave is affecting seriously the response, especially the vertical component. Higher frequencies amplify more the "parasitic" vertical acceleration of the computed ground motion in both sides of the irregularity.

In Figure 5.41 the so called "Topographic Aggravation Factor" (denoted as TAF) is plotted. It is defined as the ratio of the Fourier amplitude spectra at certain location (i.e. at the crest) and at long distance from it (x=500m) which may be considered as "free-field". It can be readily seen that the topographic effect is more important at high frequencies. In the range of frequencies of interest for ordinary buildings (1-4 Hz), the Fourier amplification factor due to the topography (TAF) is varying between 1.2 and 1.5 which may have an important effect. This amplification is well compared with the amplification factor proposed in EC8-Draft (Figure 5.32).

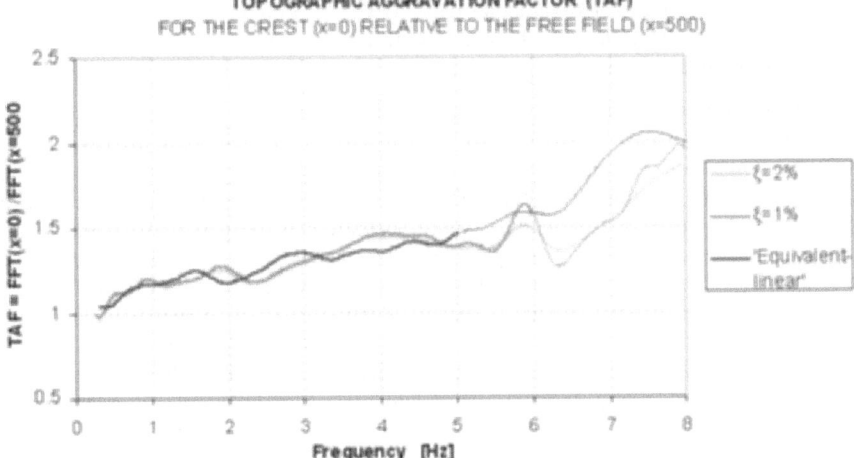

Fig. 5.41. Spectrum of Topographic Aggravation Factor at the crest for a Ricker wavelet f_0=3 Hz excitation and different values of material damping.

5.6.4. CONCLUSIONS

The analyses performed have proven that there is a strong diffracted wave field at the topographic irregularity which generates important modifications on the ground motion at both sides of the cliff. Additional amplifications in the horizontal and vertical component of the motion are observed near the crest. At the foot side we noted the existence of very strong "parasitic" vertical accelerations while the horizontal component is seriously decreased.

The frequency content of the incident wave field has a serious influence at the amplitudes of the motion. The Topographic Aggravation Factor computed at the crest has been proven practically invariant for different frequencies of the incident wave. It has been found in particular that a 25%-50% of extra amplification due to topographic effects should be expected at least for the case considered herein. In general it is concluded that site topographic effects maybe quite important. There are very few experimental observations and records to better understand the complex phenomena related to the topography and more important to quantify them. It is expected that the CORSSA array will play a very positive role to this issue.

5.7. Site Effects and Seismic Codes

Seismic ground response characteristics, defined generally as "site effects", are inevitably reflected in seismic code provisions. The selection of appropriate elastic response spectra according to soil categories and seismic intensity is the simplest way to account for site effects both for engineering projects and for a general purposes microzonation study.

Modern seismic codes (IBC2000, UBC97, NEHRP, EC8) all introduced in the last few years, after the recent strong earthquakes in America, Europe, Japan and worldwide, which produced numerous valuable data, have incorporated the most important experimental and theoretical results with the necessary adjustments and simplifications for purely practical reasons.

The main improvement is that amplification factors of spectral values are varying with the seismic intensity; lower shaking intensity earthquakes introduce higher amplification factors due to the more linear elastic soil behaviour, contrary to higher intensities where soils are exhibiting non-linear behaviour resulting to a decrease of peak spectral values. Additionally, a more accurate soil categorization is introduced based on a better description of soil profiles using standard geotechnical parameters (i.e. plasticity index PI, undrained shear strength Su) and average Vs values. In IBC2000 and other codes of the same family, a special attention is given to near field conditions introducing higher amplification factors for the same earthquake magnitude. Also for soil layers of small thickness presenting high impedance contrast, the new version of codes attribute higher amplification factors which is compatible with observations and theory.

In general the parameters describing site effects in seismic codes are expressed through (a) soil categorization and (b) spectral amplification factors and shapes. The pioneering work of Seed et al. (1976) is still the basic reference. Since then all improvements are basically resulting from a more detailed soil categorization and a large number of numerical analyses from numerous researchers all over the world. 1D site effect computations using the equivalent linear model is the main and almost universal tool for all improvement and modifications introduced so far (i.e. Dickenson and Seed 1996, Pitilakis et al., 2003). The increasing number of records during the last two decades gave the minimum necessary validation background to these theoretical efforts and simulations. Full non-linear and elasto-plastic models are not really used mainly because of the difficulty in defining soil parameters for all soil categories, while it is known that probably for strong earthquakes, the use of elastoplastic models shall lead to an even more important decrease of ground amplification, especially in low resistance soil layers (Pitilakis et al., 1999).

Modern codes are certainly a serious step forward for a better evaluation of design input motions at least for ordinary buildings. Future improvements should be addressed to the following issues: Azimuthal effects (i.e. different spectral values for the two horizontal components), basin edge and deep basin effects, evaluation of ground motion for large and very large shear strains, vertical component, topographic effects, velocity and displacement spectra, spatial variability of ground motion etc.

The complete description of the real ground motion at a specific site under a hazardous in time and space seismic event is still a utopia. Seismic codes should always reflect the basic knowledge and technology of the present time, keeping in mind that they must be simple and realistic, having an acceptable level of accuracy, compatible among others, with the tools used for the seismic design of the structures.

In the following paragraphs the two more recent seismic codes (IBC2000 and EC8-Draft) are presenred, together with an improved site categorization proposal and the corresponding spectral amplification values and shapes which is the result of a long research effort in Aristotle University in Greece.

5.7.1. THE CONCEPT OF EUROCODES

The purpose of the Eurocodes is to achieve harmonization between structural and geotechnical design in Europe. The Commission of the European Communities (CEC) initiated a work in 1975 of establishing a set of harmonised technical rules for the structural and geotechnical design of buildings and civil engineers works based on article 95 of the Treaty. In a first stage it was planned to serve as alternative to the national rules applied in the various Member States and finally when reaching an acceptable level of harmonization to replace them and serve as a unique document.

Since 1989, 'EN 1990: Basis of Structural Design' is considered the primary document in the Eurocode suite and establishes the principles and requirements for safety and serviceability of structures. EN 1990 must be applied whenever the Eurocodes 1 to 9 are used. The Eurocode 8 (EC8) "Design of Structures for Earthquake Resistant" deals with the design and construction of buildings and civil engineering works in seismic regions.

5.7.2. INTERNATIONAL BUILDING CODE 2000

The need and motivation for a 'uniform' design code has always been very strong in the Unites States. Due the different geographic regions of the U.S., and the corresponding significantly different level of probabilistic seismic risk, seismic codes have been issued primarily for the Western coast by the Californian authirities. At the end of 1994, the three existing model code groups together formed the InternationalCodeCouncil (ICC) with express purpose of developing a single set of construction codes for the entire United States, leading to the combination of UBC (Uniform Building Code) and NEHRP (National Seismic Hazard Reduction Program) into the IBC2000 (International Building Code). Although IBC2000 is not a 'real' international code, but rather, a common code for all States of America, it is indeed a significant step forward towards harmozation and a major scientific breakthrough.

5.7.3. SOIL AND SITE CLASSIFICATION

Soil and site characterization is provided in Eurocode 8 and not in Eurocode 7 on account of being directly related to the design spectra proposed for considering the seismic force that is statistically expected to act on a structure. In its latest revision, the soil is classified in five major categories (Table 5.1) and two spesific sub-categories that correspond to very loose or liquefiable material respectively. The advantage of such a classification is that the three parameters that are used for soil identification (shear wave velocity, Nspt values and undrained strength) is relatively easy to me measured, but on the other hand the soil stiffness is determined by the V_s values of the 30 uppermost layers only. The average shear wave velocity $V_{s,30}$ is computed according to the following expression:

$$V_{s,30} = \frac{30}{\sum\limits_{i=1,N} \dfrac{h_i}{V_i}} \tag{5.7}$$

where h_i and V_i denote the thickness and shear wave velocity of the N formations of layers existing in the top 30 meters. In case that the value of $V_{s,30}$ is unknown, the value of N_{SPT} will be used. Nevertheless, its has been shown and discussed in previous paragraphs that the above approach may be a particular simplification which can potentially lead to eroneous results especially in cases of deep soil formations or abrupt stiffness change between the soil layer at -30m and the bedrock laying deeper.

Table 5.1. Soil classification according to Eurocode 8 (prEN1998-1, Draft 4, 2001)

	Description	N_{SPT}	s_u (kPa)	$Vs_{,30}$ (m/sec)
A	Rock or other rock-like geological formation, including at most 5m of weaker material at the surface	-	-	> 800
B	Deposits of very dense sand, gravel or very stiff clay at least several tens of m in thickness characterized by a gradual increase of mechanical properties with depth.	>50	>250	360-800
C	Deep deposits of dense or medium-dense sand, gravel or stiff clay with thickness from several tens to many hundreds of m	15-50	70-250	180-360
D	Deposits of loose-to-medium cohesionless soil (with or without some soft cohesive layers) or of predominantly soft-to-firm cohesive soil	<15	<70	<180
E	A soil profile consisting of a surface alluvium layer with $Vs_{,30}$ values of class C or D and thickness varying between about 5 and 20 m, underlain by stiffer materials with $Vs_{,30}$ >800 m/sec			
S_1	Deposits consisting – or containing a layer at least 10m thick – of soft clays/silts with high plasticity index (PI>40) and high water content	-	10-20	<100 (indicative)
S_2	Deposits of liquefiable soils, of sensitive clays, or any other soil profile not included in classes A-E or S_1			

For sites with ground conditions matching the two special subsoil classes S1 and S2 special studies for the definition of the seismic action are required. For these classes and particularly for S2 the possibility of soil failure under the seismic action must be considered. Further sub-division of this classification is permitted to better conform to special soil conditions. The seismic action defined for any sub-class shall not be less than those corresponding to the main class specified in Table 5.3, unless this is supported by special site-classification studies foreseen in the National Annex.

The classification according to IBC2000 as seen in Table 5.2 is identical to the provisions of Uniform Building Code 1997, and practically distinguishes soil profiles in five main categories while a special condition F case is also prescribed. Conceptually, the soil categorization is similar to that of EC8, both in terms of the $V_{s,30}$ assumption and the V_s threshold values that separate the subsoil classes. The latter is illustrated in the comparative Table 5.3 where most of the present seismic code soil classifications are compared.

Table 5.2. Site Classes – Specifications according to Uniform Building Code 1997 and International Building Code 2000.

	Description	$V_{s,30}$ (m/sec)
A	HARD ROCK-Eastern United States sites only	≥ 1500
B	ROCK	760-1500
C	VERY DENSE SOIL AND SOFT ROCK Undrained shear strength $s_u > 100$ kPa or $N_{SPT} > 50$	360-760
D	STIFF SOILS Stiff soil with undrained shear strength 50 kPa $< s_u <$ 100 kPa or $15 < N_{SPT} < 50$	180-360
E	SOFT SOILS Profile with more than 3 m of soft clay defined as soil with PI > 20, moisture content w>40%, undrained shear strength $s_u < 50$ kPa and $N_{SPT} < 15$	≤ 180
F	SOILS REQUIRING SITE SPECIFIC EVALUATIONS 1. Soils vulnerable to potential failure or collapse under seismic loading: e.g. liquefiable soils, quick and highly sensitive clays, collapsible weakly cemented soils. 2. Peats and/or highly organic clays: 3 m or thicker layer 3. Very high plasticity clays: 8 m or thicker layer with PI>75 4. Very thick soft/medium stiff clays: 36 m or thicker layer	

Table 5.3. Comparison of soil classification in modern seismic codes worldwide

$V_{s,30}$ *(m/sec)*	180	360	760	1500	
UBC/97 **IBC/2000**	S_E	S_D	S_C	S_B	S_A
GREEK SEISMIC CODE EAK2000	D – C	C B A		A	
EC8 (ENV1998)	C	C B A		A	
EC8 (prEN1998) (Draft4, 2001)	D	C	B	A	
New Zealand, 2000 (Draft)	D (T>0.6s =>$V_{s,30}$<200)	C (T<0.6s =>$V_{s,30}$>200)	B	A	
Japan, 1998 (Highway Bridges)	III (T>0.6s =>$V_{s,30}$<200)	II (I) (T=0.2-0.6 s => $V_{s,30}$=200-600)		I (T<0.2s =>$V_{s,30}$>600)	
Turkey/98	$Z_4 - Z_3$	$Z_3 - Z_2$	$Z_3 - Z_2 - Z_1$	Z_1	
AFPS/90	$S_3 - S_2$	$S_3 - S_2 - S_1$	$S_1 - S_0$	S_0	

Table 5.4. Soil amplification parameter and corner periods as a function of subsoil class for earthquake Type 1 (Eurocode 8).

Ground Type	S	T_B	T_c	T_D
A	1.0	0.15	0.40	2.00
B	1.20	0.15	0.50	2.00
C	1.15	0.20	0.60	2.00
D	1.35	0.20	0.80	2.00
E	1.40	0.15	0.50	2.00

Table 5.5. Soil amplification parameter and corner periods as a function of subsoil class for earthquake Type 2 (Eurocode 8)

Ground Type	S	T_B	T_c	T_D
A	1.0	0.05	0.25	1.20
B	1.35	0.05	0.25	1.20
C	1.50	0.10	0.25	1.20
D	1.80	0.10	0.30	1.20
E	1.60	0.05	0.25	1.20

A more refined soil and site characterization (Table 5.6) has been performed by the Lab. of Soil Mechanics and Foundation Engineering of the Aristotle University Thessaloniki, based on a more comprehensive approach using a large date base of high quality records in perfectly known soil conditions and over 800 1D computations of ground response with different soil profiles in terms of impedance contrast, soil type and relative thickness, depth of the rigid or non-rigid bed-rock etc and of course input motion characteristics (Pitilakis et al., 2003).

5.7.4. COMPATIBILITY OF DESIGN FORCES

Although technically feasible, designing a structure to respond elastically the design earthquake input leads to disproportionally increased construction cost. Moreover, the actual demand is related to the seismic energy dissipation ability and not a particular performance under given seismic forces. For these reasons, seismic design codes prescribe and allow inelastic response under certain hierarchy, while ensuring collapse prevention and structural integrity. The correlation of the elastic with the design spectra therefore, is inevitably related to the level of energy absorption which is anticipated to take place through cycles of inelastic behaviour and damage of the structural members. The latter is usually described as the behaviour factor q (in Eurocodes, the Greek Seismic Code and a number of other codes) and the response modification factor R in the U.S. codes. It is therefore necessary, before proceeding to the critical and comparable evaluation of the code-defined spectra to account for the different assumptions on the reduction of forces prescribed in different codes and proceed to the appropriate calibrating calculations.

5.7.5. SPECTRAL AMPLIFICATION

Typically, the shape of the spectrum is a function of the soil category that the foundation soil can be classified into, according to the soil and site characterization procedure described previously: for softer materials, the spectrum is generally shifting rightwards representing a higher level of bedrock ground motion amplification that is expected to occur when higher period pulses propagate through the (relatively) loose soil profile. Moreover though, it has been widely recognized in recent versions of seismic codes (i.e. UBC since 1997) that the epicentral distance also has a crucial role on the spectral amplification at particular period ranges, i.e. a near field earthquake is expected to be rich in high frequencies with respect to a distant event for which the amplification at long periods is expected to be relatively lower. Additionally, both IBC2000 and the latest version of Eurocode 8 (prEN1998-1, Draft 4, 2001) account for the effect of earthquake magnitude; in Eurocode 8 by distinguishing to Type 1 ($M_s>5.5$) and Type 2 ($M_s<5.5$) spectra as seen in Figure 5.42. Figure 5.43 and the accompanying Tables 5.8 and 5.9 are illustrating the proposed acceleration response spectra in IBC2000.

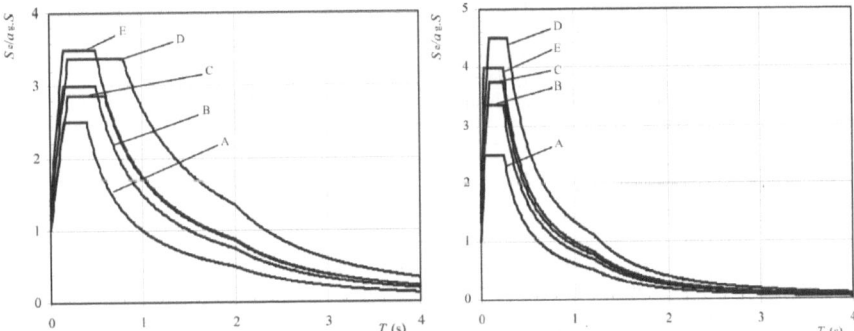

Fig. 5.42. Recommended Type 1 (left) and Type 2 (right) elastic response spectrum for soils A to E according to Eurocode 8 (prEN1998-1, Draft 4, 2001).

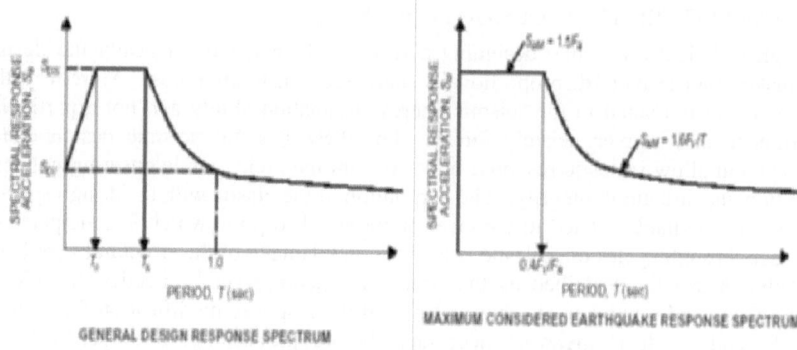

GENERAL DESIGN RESPONSE SPECTRUM

MAXIMUM CONSIDERED EARTHQUAKE RESPONSE SPECTRUM

Fig. 5.43. Design spectra according to IBC2000.

The new normalized acceleration response spectra suggested by Aristotle University for the soil classification presented in Table 5.6 are given in Figure 5.44. The parameter S is expressing, as in EC8, the spectral amplification at T=0.0sec while the parameter β is the maximum spectral acceleration coefficient describing with the three corner periods T_B, T_C and T_D, the shape of the acceleration response spectrum for each soil category.

Table 5.6. Soil and Site Classification (Pitilakis et al., 2003)

CATEGORY		DESCRIPTION	To (sec)	REMARKS
A	A_1	Healthy rock formations		$V_s \geq 1500$ m/sec
	A_2	Slightly weathered / segmented rock formations, provided that the weak, highly weathered surficial layer has a thickness of less than 5.0m	≤ 0.2	Surface weathered layer: $\overline{V}_s \geq 300$ /sec. Rock formations: $V_s \geq 800$ m/sec
		Geologic formations which resemble rock formations in their mechanical properties and their composition (e.g. conglomerates)		$V_s \geq 800$ m/sec
B	B_1	Highly weathered rock formations whose weathered layer has a considerable thickness of 5.0 - 30.0m	≤ 0.4	Weathered layer: $\overline{V}_s$ [2] ≥ 300 m/sec
		Soft rock formations of great thickness or formations which resemble these in their mechanical properties (e.g. stiff marls)		$\overline{V}_s = 400$-800 m/sec N_{SPT} [3] >50 S_u [4] >200 kPa
		Soil formations of very dense sand – sand gravel and/or very stiff clay, of homogenous nature and small thickness (up to 30.0m)		$\overline{V}_s = 400 - 800$ m/sec $N_{SPT} > 50$ $S_u > 200$ kPa
	B_2	Soil formations of very dense sand – sand gravel and/or very stiff clay, of homogenous nature and medium thickness (30.0 - 60.0m), whose mechanical properties increase with depth	≤ 0.8	$\overline{V}_s = 400 - 800$ m/sec $N_{SPT} > 50$ $S_u > 200$ kPa
C	C_1	Soil formations of dense to very dense sand – sand gravel and/or stiff to very stiff clay, of great thickness (> 60.0m), whose mechanical properties and strength are constant and/or increase with depth	≤ 1.2	$\overline{V}_s = 400 - 800$ m/sec $N_{SPT} > 50$ $S_u > 200$ kPa
	C_2	Soil formations of medium dense sand – sand gravel and/or medium stiffness clay (PI > 15, fines percentage > 30%) of medium thickness (20.0–60.0m)	≤ 1.2	$\overline{V}_s = 200 - 400$ m/sec $N_{SPT} > 20$ $S_u > 70$kPa
	C_3	Category C_2 soil formations of great thickness (>60.0 m), homogenous or stratified that are not interrupted by any other soil formation with a thickness of more than 5.0m and of lowerr strength and V_s velocity	≤ 1.4	$\overline{V}_s = 200$-400 m/sec $N_{SPT} > 20$ $S_u > 70$ kPa

D	D₁	Recent soil deposits of substantial thickness (up to 60m), with the prevailing formations being soft clays of a high plasticity index (PI>40), with a high water content and low values of strength parameters	≤ 2.0	$\overline{V}_s \leq 200$ m/sec $N_{SPT} < 20$ $S_u < 70$KPa
	D₂	Recent soil deposits of substantial thickness (up to 60m), with prevailing fairly loose sandy to sandy-silty formations with a substantial fines percentage (so as not to be considered susceptible to liquefaction)	≤ 2.0	$\overline{V}_s \leq 200$ m/sec $N_{SPT} < 20$
	D₃	Soil formations of great overall thickness (> 60.0m), interrupted by layers of category D₁ or D₂ soils of a small thickness (5 – 15m), up to the depth of ~40m, within soils (sandy and/or clayey, category C) of evidently greater strength, with	≤ 1.2	$\overline{V}_s \geq 300$ m/sec
E		Surface soil formations of small thickness (5 - 20m), small strength and stiffness, likely to be classified as category C and D according to geotechnical properties, which overlie category A formations ($V_s \geq 800$ m/sec)	≤ 0.5	Surface soil layers: $\overline{V}_s = 150 - 300$ m/sec
X		Loose fine sandy-silty soils beneath the water table, susceptible to liquefaction (unless a special study proves no such danger, or if the soil's mechanical properties are improved) Soils near obvious tectonic faults Steep slopes covered with loose lateral deposits Loose granular or soft silty-clayey soils, provided they have been proven to be hazardeous in terms of dynamic compaction or loss of strength Recent loose landfills Soils with a very high percentage in organic material		

Table 5.7. Values of Fa as a function of site class and shaking intensity (IBC2000).

SITE CLASS	SHAKING INTENSITY				
	$A_s \leq 0.1$	$A_s = 0.2$	$A_s = 0.3$	$A_s = 0.4$	$A_s \geq 0.5$
A	0.8	0.8	0.8	0.8	0.8
B	1.0	1.0	1.0	1.0	1.0
C	1.2	1.2	1.1	1.0	1.0
D	1.6	1.4	1.2	1.1	1.0
E	2.5	1.7	1.2	0.9	*
F	*	*	*	*	*

Table 5.8. Values of Fa as a function of site class and shaking intensity (IBC2000)

SITE CLASS	SHAKING INTENSITY				
	$A_s \leq 0.1$	$A_s = 0.2$	$A_s = 0.3$	$A_s = 0.4$	$A_s \geq 0.5$
A	0.8	0.8	0.8	0.8	0.8
B	1.0	1.0	1.0	1.0	1.0
C	1.7	1.6	1.5	1.4	1.3
D	2.4	2.0	1.8	1.6	1.5
E	3.5	3.2	2.8	2.4	*
F	*	*	*	*	*

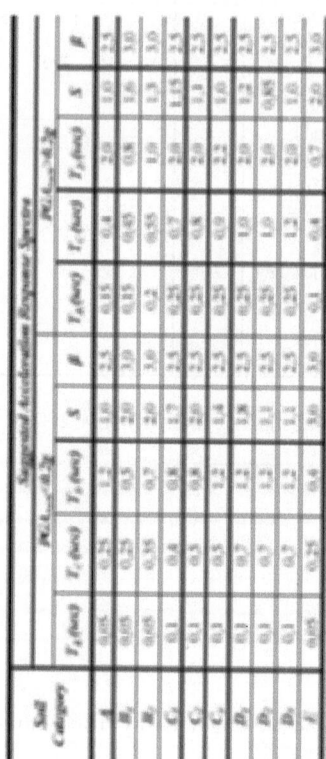

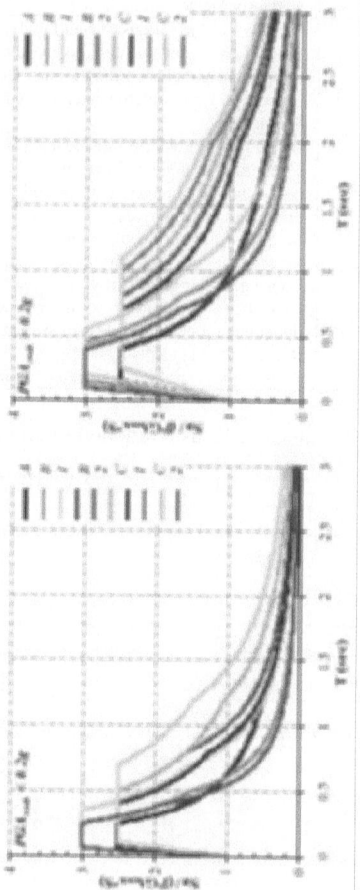

Fig. 5.44. Suggested acceleration response spectra for each soil category normalized to the maximum acceleration value ($PGA_{rock} * S$)

In conclusion, it can be stated that, the spectral values are a function of local soil properties, distance from the source (only for IBC2000) and intensity of the seismic event. On the other hand, there is a number of inherent uncertainties in the definition of the appropriate local soil profile determination due to the inadequate process to evaluate the overall dynamic response of the soil formations from the Vs profile of the uppermost 30 meters. This is the reason that in the new soil categories suggested by Aristotle University instead of $V_{s,30}$ we are using the average Vs over the whole soil column under consideration. Moreover, it has been extensively shown that the geometrical characteristics of soil profiles (considered as 1-Dimensional soil columns) are rather inadequate to describe the complex nature and physics of the multilayered, non-linear, 3-Dimentional and possibly inclined soil layers that lay over a non-rigid bedrock and for which the actual amplification of the amplitude of incoming waves in particular frequencies is a function of the overall soil complexity, topography and geology, hence a very complex and multi-parametric phenomenon codified as 'site effects'. It is clear therefore, that, although considerable steps have been taken with respect to the consideration of the role played by soil on the seismic input, there are a number of issues that are either scientifically unresolved or the progress made has not yet been reflected to modern seismic codes.

Acknowledgements

The work presented herein is reflecting the scientific research work performed the last 10 years in Aristotle University by many graduate and post graduate students and mainly by my students and presently collaborators Dr. D. Raptakis, Dr. K. Makra and Dr. A. Anastasiadis. Text notes prepared by Dr. D. Raptakis and Dr. K. Makra as well as by Dr. A. Sextos were most valuable. Dr A. Sextos and Ms. V. Terzi, MSc-AUTH, kindly took care of the final version of the text and Figures. I would like to express my acknowledgements to all of them.

CHAPTER 6
EVALUATION OF LIQUEFACTION-INDUCED DEFORMATION OF STRUCTURES

Susumu Yasuda
Department of Civil and Environmental Engineering, Tokyo Denki University, Japan

6.1. Introduction

In the current design of countermeasures against liquefaction, about 200 gals or less of the maximum surface acceleration has been applied to estimate cyclic shear stress. However, far greater shaking such as 400 to 600 gals of the maximum surface acceleration caused severe liquefaction-induced damages during the 1995 Hyogoken-Nambu (Kobe) and 1999 Kocaeli earthquakes. Liquefaction-induced flow also occurred in Kobe. After these earthquakes, new design concepts and methods for evaluating liquefaction, which can be applied for strong ground shaking, have been studied by conducting model tests, laboratory tests and analyses. This chapter focuses on these recent studies and introduces the new design methods. The liquefaction-induced damage due to the strong ground shaking by the two earthquakes is introduced at first. Then, the behaviour of structures in liquefied dense or silty grounds is discussed. Design methods for liquefaction-induced deformation of structures and countermeasures are introduced. Finally design methods and countermeasures for liquefaction-induced flow are shown.

6.2. Design Procedures for Liquefaction

6.2.1. CURRENT DESIGN PROCEDURES

In 1964 the Niigata and Alaska earthquakes inflicted huge damage to buildings, bridges and other structures by liquefying loose-sandy soils. After these earthquakes many studies on the liquefaction of sandy soils have been conducted by laboratory cyclic shear tests, shaking table tests, site investigations and analyses. Then many kinds of methods for the prediction of the occurrence of liquefaction have been developed. Countermeasure methods against liquefaction have proposed and applied also. These prediction methods were introduced by TC4 (1999), JGS (1998), Yasuda (1999), etc. Current countermeasures were summarized by JGS (1998). Based on these methods, the conventional approach shown in Figure 6.1 has been used for the design of liquefaction. In this approach, the assessment of liquefaction potential is done first. Then the acceptability of the likely degree of damage is roughly judged and, if necessary, appropriate countermeasures are selected. However, in general, the degree of damage expected from liquefaction is not evaluated because it is difficult to evaluate.

In the current design, around 200 gals or less of the maximum surface acceleration has been used as design input motion. It is not always necessary to judge the degree of damage because it is easy to improve the ground not to liquefy under this level of shaking. However, liquefaction cannot be prevented by current countermeasures under strong ground shaking, such as 600 gals, which was observed during the 1995

A. Ansal (ed.), Recent Advances in Earthquake Geotechnical Engineering and Microzonation, 199–230.
© 2004 *Kluwer Academic Publishers.*

Hyogoken-Nambu (Kobe) and Kocaeli earthquakes. Therefore, it is necessary to introduce a new design concept based not on the occurrence of liquefaction but on the likely degree of damage to structures. This new design concept is rational and will be used for not only for strong ground shaking but also normal ground shaking.

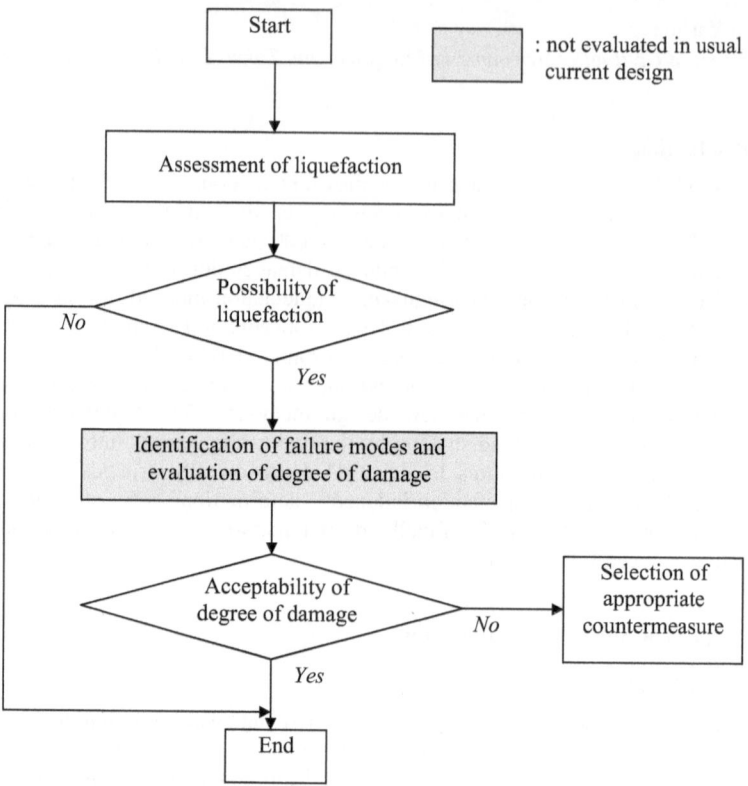

Fig. 6.1. Current design procedure

6.2.2. EFFECT OF THE 1995 KOBE EARTHQUAKE

The 1995 Hyogoken-Nambu (Kobe) earthquake, with a magnitude of 7.2, caused liquefaction of many artificially reclaimed land areas and alluvial plain deposits along Osaka Bay, as shown in Figure 6.2. In Kobe City, several waterfront areas and two large islands had been constructed by reclamation along Osaka Bay. Liquefaction occurred in almost all of the artificially reclaimed lands and islands because the reclaimed soil was loose and the ground shaking was very strong, i.e., more than 600 gals of maximum surface acceleration.

A large number of structures, such as bridges, tanks, quay walls, buildings, houses, earth dams, river dikes and buried pipes, were seriously damaged by liquefaction, as shown in Photo 6.1. Moreover, many quay walls and revetments moved toward the sea and brought about extensive flow of the ground behind them as shown in Photo 6.2.

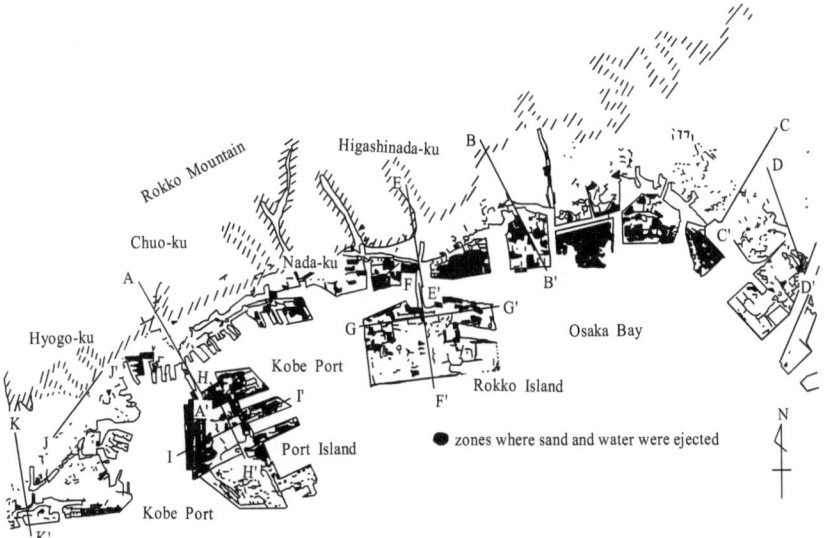

Fig. 6.2. Sites on Kobe and surrounding cities liquefied during the 1995 Hyogoken-Nambu earthquake

Photo 6.1. Settled school houses in Kobe

Photo 6.2. Liquefaction-induce flow behind a quay wall in Kobe

For example, the average horizontal and vertical displacements of quay walls on Port Island were 2.7m and 1.3m, respectively. The ground behind quay walls liquefied and flowed toward the sea due to the movement of the quay walls, as shown schematically in Figure 6.3. The ground flow brought severe damage to many structures, such as bridges, buildings, houses and pipelines. Piles of bridges and buildings were deformed, continuous footings of wooden houses were ruptured, and buried pipes were bent. The displacement of the ground was measured by surveying the opening of the ground cracks (Ishihara et al., 1996; Ishihara, 1997) and by aerial photogrammetry. Figure 6.4 shows the distribution of ground surface displacement behind quay walls, measured by Ishihara et al. (1996). Displacements just behind quay walls reached about 2 to 3.5m and decreased with distance from the wall. Flow extended back almost 100 to 150m behind the quay walls.

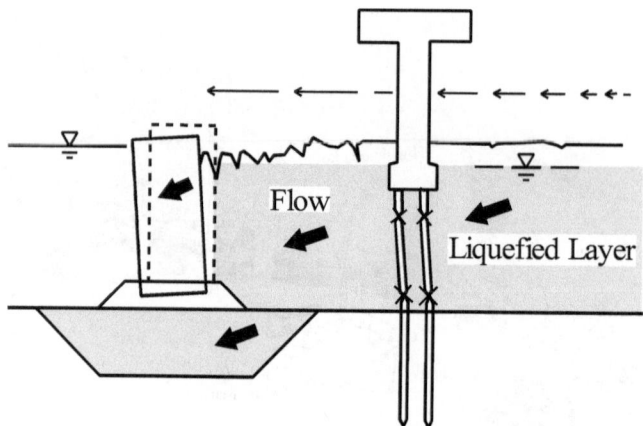

Fig. 6.3. Outline of the ground flow behind quay wall

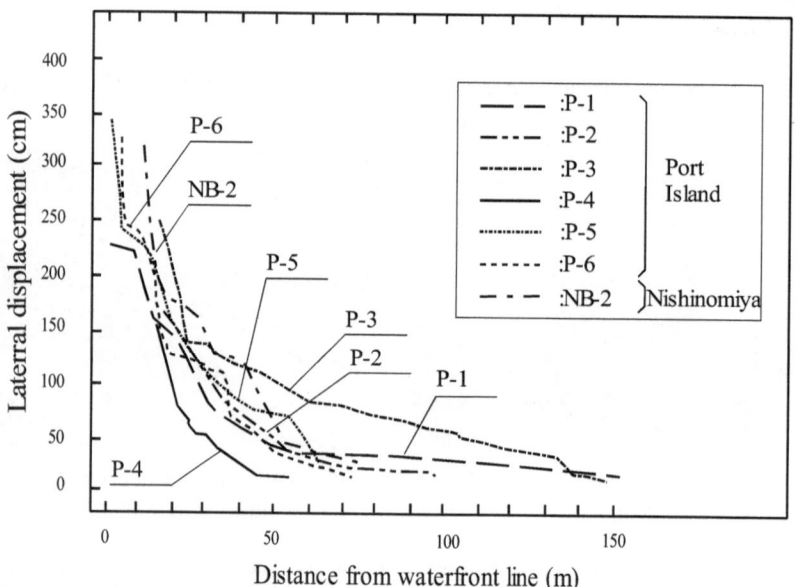

Fig. 6.4. Distribution of horizontal displacement behind quay wall (Ishihara et al., 1996; Courtesy of the JSSMFE)

This earthquake pointed out the following two problems in the current design procedure for liquefaction:

1. In the current design procedure, as shown in Figure 6.1, the design acceleration on the surface is about 200 gal or less in general. However, very intense seismic motion, such as 600 to 800 gals of the maximum surface acceleration, hit Kobe and surrounding cities and caused severe liquefaction-induced damage.

If the input acceleration is very strong, even medium dense and dense sand would liquefy. However, it is not clear that severe damage would occur or not in the medium dense or dense sandy ground.

2. The damage of structures due to liquefaction-induced flow must be considered in the design.

6.2.3. LIQUEFACTION-INDUCED SETTLEMENT DURING THE 1999 KOCAELI EARTHQUAKE

During the 1999 Kocaeli, Turkey earthquake, many RC buildings of 4 to 6 stories settled and tilted in the zones shown in Figure 6.5, which were recognized by the reconnaissance team of the Japanese Geotechnical Society (2000). Typical damage to buildings is shown in Photos 6.3 and 6.4. The building in Photo 6.3 tilted about 30° just after the earthquake without damage to glass windows then gradually tilted to 60°. Cracks in the ground occurred due to a slide at the site shown in Photo 6.4. The building on the crack was torn, as shown in the photo. Boiled sands were observed at several sites. However, the number of sites where boiled sands were observed and the volume of the boiled sands were not much, compared with the liquefied sites during past earthquakes, such as Niigata and Dagupan cities during the 1964 Niigata and the 1990 Luzon Philippines earthquakes, respectively.

Photo 6.3. A settled building at Site 16 and   Photo 6.4. A settled building at Site SK4 in
S4 in Adapazarı   Adapazarı

The reconnaissance team of the Japanese Geotechnical Society tried to measure the settlement and angle of inclination for about 200 buildings (JGS, 2000). Among them the data for settled buildings only were selected and the relationship between average settlement and angle of inclination was drawn as shown in Figure 6.6 (Yasuda et al., 2001a). The average settlement of each building was calculated from the maximum settlement, inclination and the width of the building. The largest settlement was about 50cm. Typical settlements of the buildings and inclination were 20cm to 40cm and 1 degree to 3 degrees, respectively. These values were smaller than those induced in Niigata and Dagupan cities.

After the earthquake, Swedish weight sounding tests, borings and undisturbed soil samplings were carried out by a research team (Yasuda et al., 2001b, Kiku et al., 2001, Tsukamoto et al., 2001). Figure 6.7 shows the data investigated at site SK4 shown in Figure 6.5. Buildings around site SK4 were severely damaged, as shown on Photo 6.4. Weak soil layers which are silt layers and a sandy silt layer lay to a depth of about 6m.

The upper two silt layers were especially quite soft. Underlying sandy silt layer is relatively weak. According to the Swedish weight sounding tests, the N_{sw} values of the weak soft silt layers were mostly below zero. Undistributed samples were taken at several depths. Undrained cyclic triaxial tests and sieve analyses were carried out on samples from depths of 2.40 to 4.75m. The fines contents, F_c of the samples from GL.-2.5m and GL.-4.5m were 97.5% and 65.2 to 99.7%, respectively. The Plasticity Index, I_p was 38.4 and NP to 25.1, respectively.

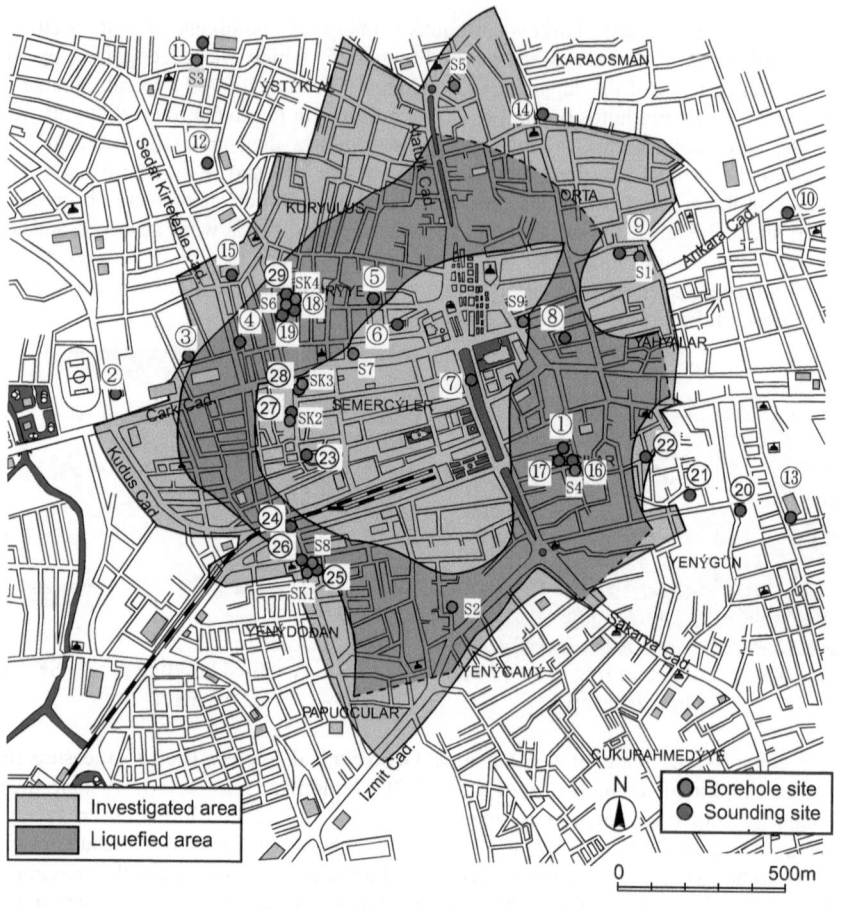

Location of site (Swedish Sounding)

Fig. 6.5. Liquefied zone and investigated sites in Adapazarı City (Yasuda et al., 2001b)

The liquefied soils at Niigata and Dagupan were clean sands. Therefore the reason why there were fewer sand boils and settlement of buildings were few at Adapazarı than at Niigata or Dagupan must be the differences between liquefied soils. The liquefied soil in Adapazarı was silt and had some cohesion, which decreased the settlement.

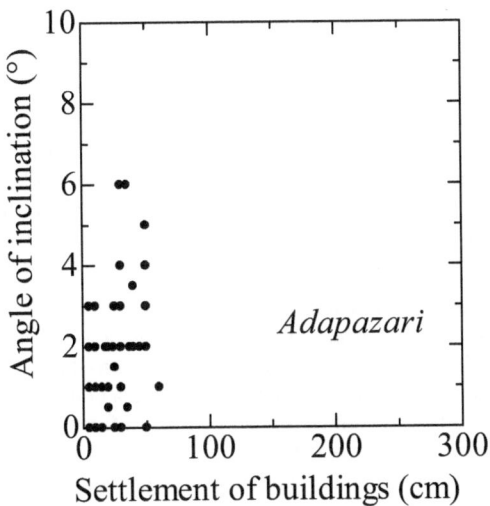

Fig. 6.6. Relationship between settlement of building and angle of inclination (Yasuda et al., 2001a)

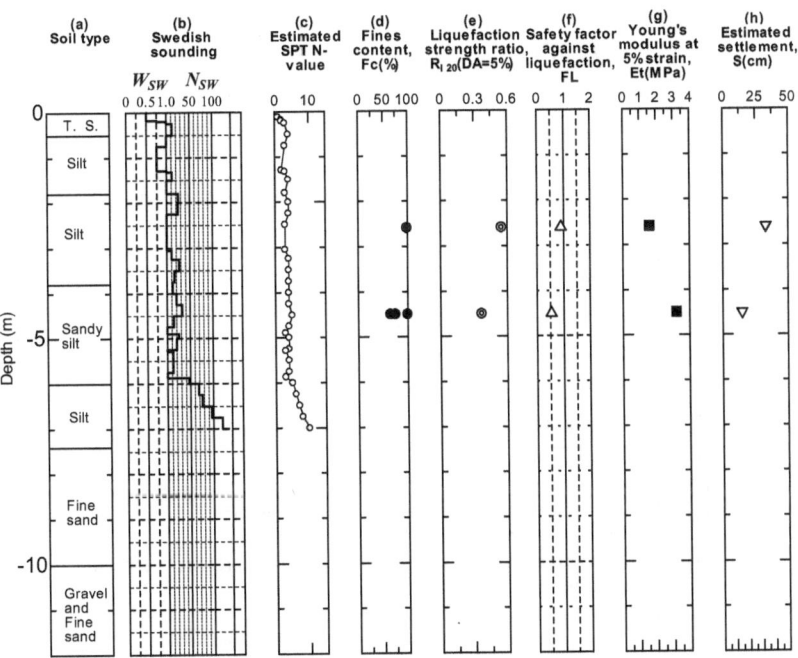

Fig. 6.7. Results of in-situ and laboratory tests at Site SK4 (Yasuda et al., 2001b)

6.3. Studies on Liquefaction-induced Deformation of Structures in Dense Sand or Silty Sand Grounds

6.3.1. NEW METHODS FOR THE PREDICTION OF THE OCCURRENCE OF LIQUEFACTION UNDER STRONG SHAKING

The shaking in Kobe during the 1995 Hyogoken-Nambu (Kobe) earthquake was very strong, as the maximum acceleration was 600 to 800 gals on the ground surface. In Japan, about 200 gals of the maximum surface acceleration had been considered in the estimation of the liquefaction potential before the Kobe earthquake. After that earthquake, it became necessary to develop a new design concept to consider the strong shaking which is called as "Level 2 shaking".

Some studies have been conducted to develop a new method for the estimation of liquefaction potential under strong shaking. In the new specification for highway bridges (The Japanese Road Association, 1997), the formula for evaluating undrained cyclic strength was revised because the previous formula could not be applied to "Level 2 shaking". Several cyclic triaxial tests were carried on frozen samples and case studies, and a new formula, shown in Table 6.1 was proposed. Figure 6.8 shows the relationship between N_I and R_L for clean sand. The big difference compared with the relationship introduced in the previous specification is that R_L increases rapidly with N_I in the range of $N_I > 20$.

Table 6.1. (a) Method to evaluate liquefaction potential (JRA, 1997)

$$R = c_w R_L$$

$$R_L = 0.0882\sqrt{N_a/1.7} \qquad\qquad (N_a < 14)$$

$$R_L = 0.0882\sqrt{N_a/1.7} + 1.6 \times 10^{-6} \cdot (N_a - 14)^{4.5} \qquad (N_a \geqq 14)$$

<Sandy soil case>

$$N_a = c_1 N_I + c_2$$

$$N_I = 1.7 N/(\sigma'_v + 0.7)$$

$$\begin{cases} c_1 = 1 & (0\% \leqq FC < 10\%) \\ c_1 = (FC + 40)/50 & (10\% \leqq FC < 60\%) \\ c_1 = FC/20 - 1 & (60\% \leqq FC) \end{cases}$$

$$\begin{cases} c_2 = 0 & (0\% \leqq FC < 10\%) \\ c_2 = (FC - 10)/18 & (10\% \leqq FC) \end{cases}$$

<Gravelly soil case>

$$N_a = \{1 - 0.36\log_{10}(D_{50}/2)\}N_I$$

R : Dynamic shear strength ratio
R_L : Cyclic triaxial strength ratio
c_w : Modification factor based on earthquake motion properties
N_a : Corrected N-value accounting for the effects of grain size
N_I : N-value convertd to correspond to effective overburden pressure of 1kgf/cm²
N : N-value obtained from standard penetration testing
σ'_v : Effective overburden pressure (kgf/cm²)
c_1, c_2 : Modification factor of the N-value based on the fine-grained fraction
FC : Fine-grained fraction (%)
D_{50} : Mean grain diameter (mm)

One more important revision is the correction factor C_w. Two types of ground motion: ℵ generated by interplate fault in the ocean (named Type 1), and ℑ generated by inland fault (named Type 2) are introduced in the specification. The maximum surface acceleration for the two types of ground motions are 0.3g to 0.4g and 0.6g to 0.8g in high seismic zones, respectively. The factor to correct the irregularity of seismic shear stress, C_w is defined as follows:

[Type 1] $C_w=1.0$ (6.1)

[Type 2]$C_w=1.0$ ($0 < R_L \leq 0.1$), $C_w=3.3\ R_L+0.67$ ($0.1 < R_L \leq 0.4$)

$C_w=2.0$ ($0.4 < R_L$)   (6.2)

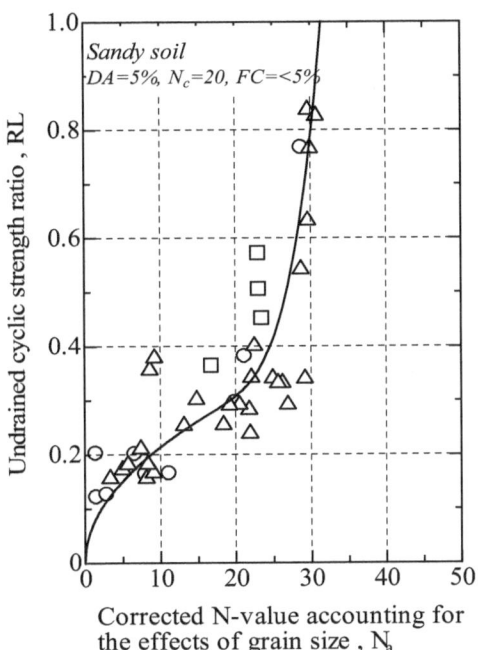

Fig. 6.8. N_a vs. R_L in the new design code

6.3.2. SOIL DENSITY AND SPT N-VALUE WHICH CAUSE LIQUEFACTION UNDER STRONG SHAKING

As the maximum surface acceleration in the "Level 2 shaking" is very high, even dense sand ground is judged to induce liquefaction. This means that a low degree of compaction cannot prevent liquefaction under a strong earthquake. If a clean sand ground such as that shown in Figure 6.9 is assumed, the maximum *SPT N*-value which causes liquefaction under different maximum surface accelerations can be calculated based on Table 6.1. Figure 6.10 shows the relationship between the maximum surface acceleration and the calculated critical *SPT N*-value at the depth of GL-8.1m, where effective overburden pressure $\sigma'_v=98$kPa. As shown in the figure, the critical *SPT N*-value increases with the maximum surface acceleration.

Susumu Yasuda

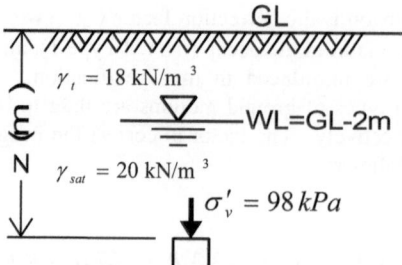

Fig. 6.9. Model ground to estimate critical condition of liquefaction

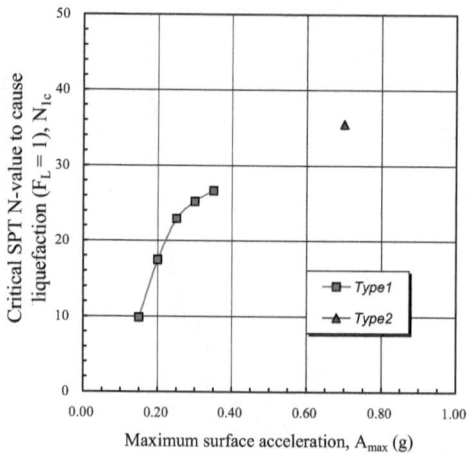

Fig. 6.10. Critical *SPT N*-value to cause liquefaction under different maximum surface acceleration

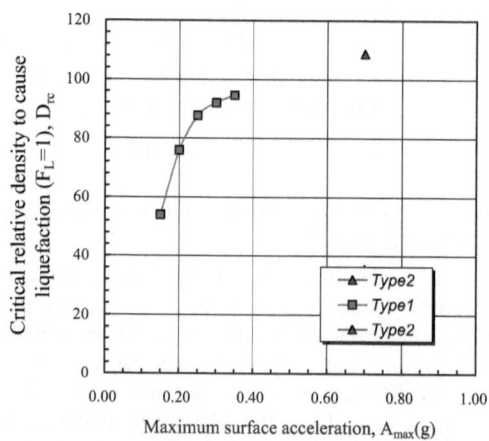

Fig. 6.11. Critical relative density to cause liquefaction under different maximum surface acceleration

The author proposed the following relationship between *SPT N*-value, D_r and mean confining pressure, p' for Toyoura Sand, which is a clean sand in Japan.

$$N=(-2.25\times10^{-6}p'+3.37\times10^{-6})\times(D_r+60)^{(0.477p'+2.95)} \tag{6.3}$$

By using this relationship, critical *SPT N*-value, shown in Figure 6.10 can be converted into relative density, as shown in Figure 6.11. If the maximum surface acceleration is 0.15g to 0.20g ("Level 1 shaking") the critical *SPT N*-value is about 10 to 15 and the critical relative density is 50% to 70%. On the contrary, under the strong shaking of 0.35g to 0.60g ("Level 2 shaking") the critical *SPT N*-value is about 25 to 30 and the critical relative density is about 90% to 100%. Therefore, it can be said that medium dense and dense sand with relative density of about 50% to 90% is the soil that liquefies under Level 2 shaking but not liquefies under the Level 1 shaking. However, it may be that structures are not severely damaged by the liquefaction of medium dense or dense ground. This basis for this conclusion is discussed in the following paragraph based on an experience during the 1995 Hyogoken-Nambu earthquake and several test results.

6.3.3. BEHAVIOUR OF STRUCTURES IN LIQUEFIED DENSE SANDY GROUND

Subsidence of Compacted Dense Ground during the 1995 Hyogoken-Nambu Earthquake

Kobe City is built on a narrow alluvial plain facing Osaka Bay. Coastal areas have been reclaimed for many years to enlarge the flat land areas. Liquefaction occurred in these reclaimed lands and two large man-made islands, Port and Rokko islands. In Port Island, some zones were improved by installing sand drains and preloading. In addition, some zones were compacted with sand compaction piles or rod (vibro) compaction. The purpose of the soil improvement was not the mitigation of liquefaction of reclaimed sand but the consolidation of the underlying soft clay layer. Figure 6.12 shows the zones of soil improvement on Port Island (Watanabe, 1981). The central areas, which are residential, were improved by installing sand drains, preloading and a combination of the two methods. High-rise apartments and office buildings are constructed in these areas. The ground for amusement park, tanks, some structures and tram depot was improved using sand compaction piles or rod (vibro) compaction. Most of this soil improvement work was applied to the bottom of the alluvial soft clay.

SPT N-values of uncompacted soils below the ground water table were mostly 10 or less. Figure 6.13 compares *SPT N*-values in untreated zones with those in zones treated with sand drains and sand compaction piles derived from other published data (Yasuda et al., 1996). The *SPT N*-value in zones compacted by sand compaction piles and rod compaction were 18 to 31. The *SPT N*-values in the zones treated by sand drains were 14 to 25. The sand drain method does not normally densify the subsoil. However, densification of the reclaimed soils must have been induced by the special construction condition in Port and Rokko islands. Since the reclaimed sandy soil is 15m to 20m thick and contains much gravel, strong vibration forces and a long time were necessary to advance the casing down to the alluvial clay layer.

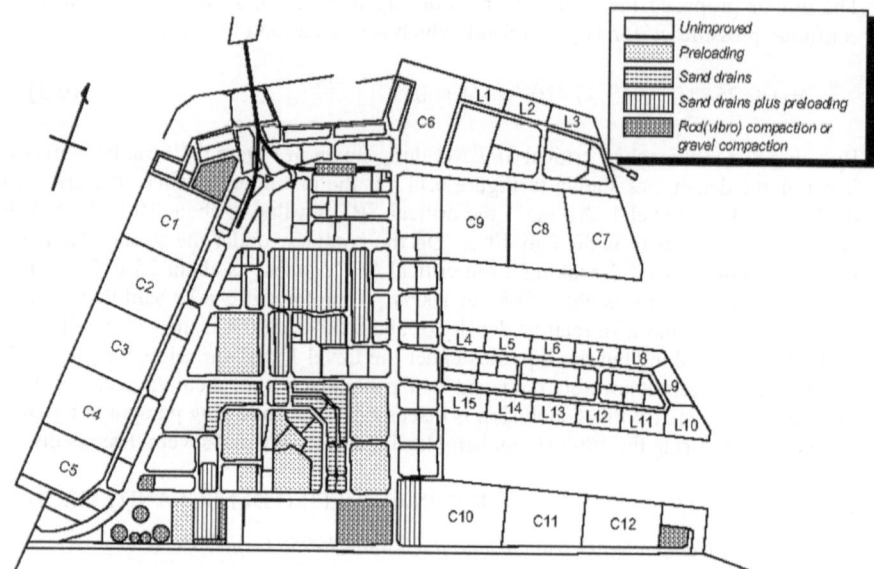

Fig. 6.12. Improvement zones in Port Island (quoted and modified from Watanabe, 1981)

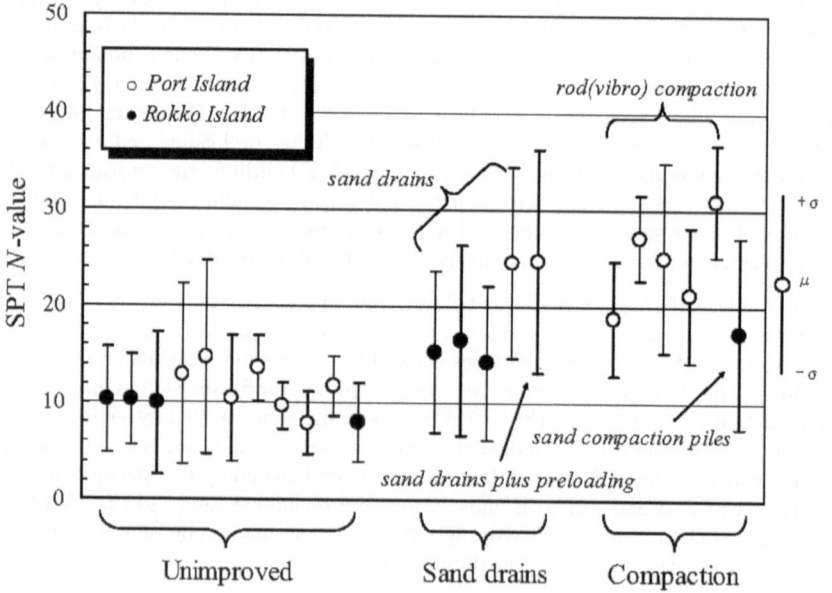

Fig. 6.13. Comparison of *SPT N*-value before and after improvement (Yasuda el at., 1996; Courtesy of the JGS)

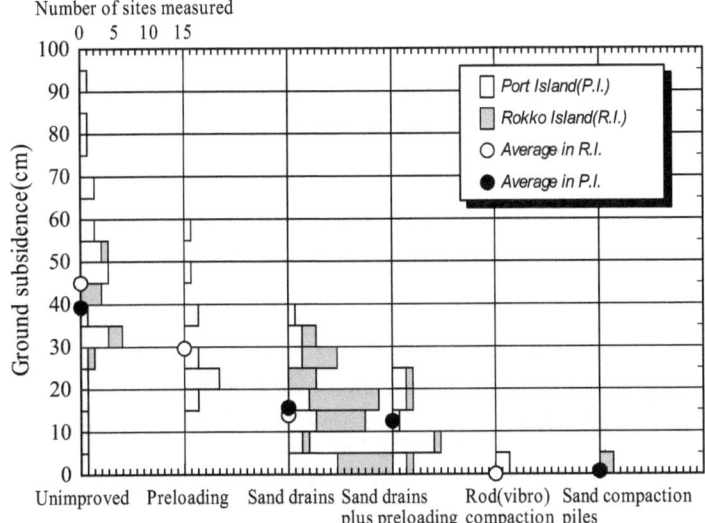

Fig. 6.14. Comparison of ground subsidence in zones treated with different methods
(Yasuda el at., 1996; Courtesy of the JGS)

By comparing Figure 6.12 with Figure 6.2, it was determined that no sand and water were ejected in zones treated with sand compaction piles, rod (vibro) compaction, sand drains plus preloading and sand drains. Sand and water were ejected in un-treated zones and in a few locations in zones treated by preloading. Large ground subsidence, of up to several tens of centimetres, was observed in the zones where sand and water were ejected. Many buildings settled and tilted in these zones. On the contrary, no subsidence and no damage to structures were observed in the zones densified with sand compaction piles and rod (vibro) compaction, and only slight subsidence was observed in the zones treated by other methods. The average subsidence in the untreated zones was about 40cm to 45cm as shown in Figure 6.14. Subsidence decreased with the degree of compaction. The average subsidence in the zones treated by preloading, sand drains, sand drains plus preloading, rod (vibro) compaction piles was about 30cm, 15cm, 12cm and 0cm, respectively. The order of decrease of subsidence is the same as the order of increase in N-values in *SPT*, mentioned before.

Stress-strain Curves of Dense Sand obtained by Laboratory Tests

The author has studied post-liquefaction behaviour by conducting torsional shear tests. Toyoura sand, which is a clean sand, was used at first to study the effects of density, confining pressure and severity of liquefaction. Then several sands were tested to study the effect of fines content (Yasuda et al., 1998). Specimens were saturated by applying back pressure and were consolidated. Then a prescribed number or prescribed amplitude of cyclic loadings was applied in undrained condition. The safety factor against liquefaction, F_L, which implies the severity of liquefaction, was controlled by the number of cycles or amplitude of the cyclic loadings. After that a monotonic loading was applied under undrained condition with a relatively high speed of $\gamma = 10\%$ per minute, as shown in Figure 6.15.

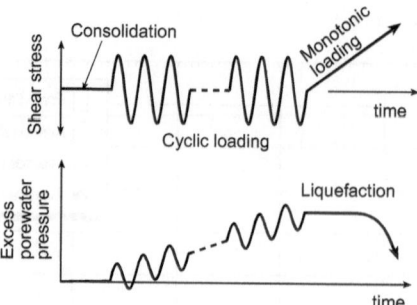

Fig. 6.15. Procedure of cyclic and monotonic loading (Yasuda el at., 1998)

Relationships among shear stress, τ, excess pore pressure, Δu, and shear strain, γ in the monotonic loading were measured. Figure 6.16 shows stress-strain curves and excess pore water pressure-strain curves in the case of Toyoura sand with different relative densities, D_r. Scales of axes in Figure 6.16(c) are enlarged scales of Figure 6.16(a). Shear strain increased with very low shear stress up to very large strain. Then, after a resistance transformation point, the shear stress increased comparatively rapidly with shear strain, following the decrease of pore water pressure.

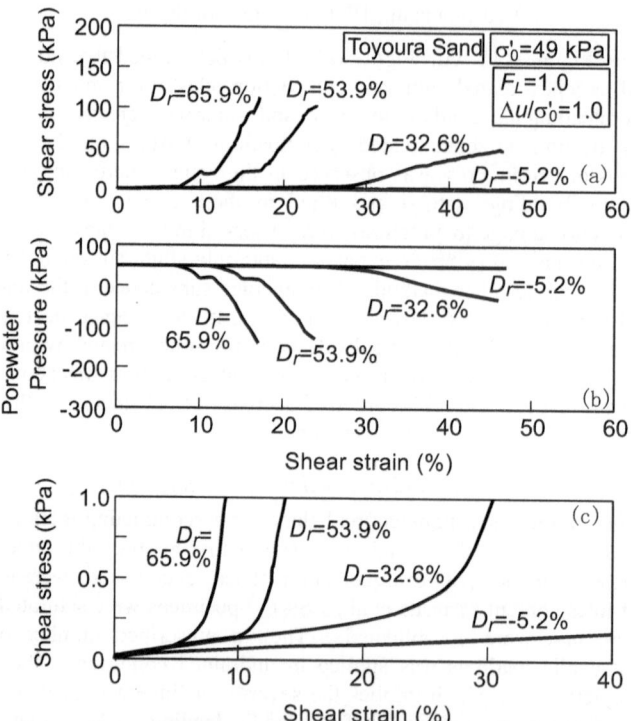

Fig. 6.16. Stress-strain and strain-pore water pressure curves for Toyoura sand (Yasuda el at., 1999b; Courtesy of the JSCE)

The amount of strain up to the resistance transformation point is called the "reference strain at resistance transformation, γ_L" as shown in Figure 6.17. The reference strain at resistance transformation increased with the decrease of $\Box_r$ as shown in Figure 7.16 (c). Stress-strain curves before and after the reference transformation point can be presented approximately by a bilinear model with G_1, G_2, and γ_L :

$$\tau = G_1\gamma \text{ for } \gamma < \gamma_L \tag{6.4}$$

$$\tau = G_1\gamma_L + G_2(\gamma\text{-}\gamma_L) \text{ for } \gamma \geq \gamma_L \tag{6.5}$$

where G_1 and G_2 are the shear moduli before and after the reference transformation point, respectively. The G_1 decreased with the decrease of D_r as shown in Figure 7.16(c).

To estimate the reduction rate of shear modulus due to liquefaction, the rate of shear modulus G_1/G_0, which is the ratio of shear modulus after and before liquefaction, was calculated. Two types of G_0 were selected: ℵ $G_{0,I}$: secant modulus of stress-strain curves at $\gamma=0.1\%$ in the case of no cyclic loading ($\Delta u/\sigma'_v=0$) and ℑ G_N : estimated from *SPT* N-value by the formula of $G_N(kPa)=2800N$. In the first shear modulus, 0.1% of shear strain was selected because the strain of soils which occurs in the ground due to overburden pressure is estimated as around $\gamma =0.1\%$ in the usual case. The second one is widely used for the design of foundations in Japan. Figure 6.18 shows relationships among the reduction rate of shear modulus, relative density and F_L. The reduction rate of shear modulus of $D_r=90\%$ is about 10 times the rate of $D_r=50\%$. This means that if liquefaction occurs in dense ground, shear modulus does not decrease so much.

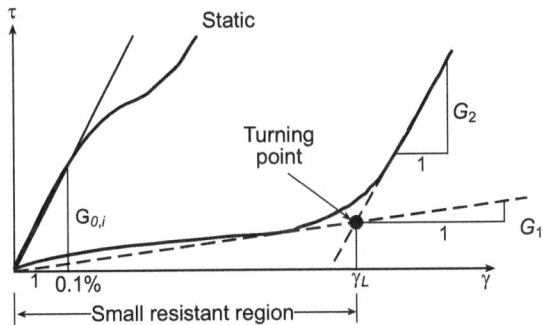

Fig. 6.17. Definition of $G_{0,i}$, G_1, G_2 and γ_L (Yasuda el at., 1998)

Behaviour of Structures in Liquefied Dense Ground obtained by Model Tests

The author has conducted several shaking table tests to demonstrate liquefaction-induced deformation of structures in grounds with different density. Figure 6.19 shows a schematic diagram of equipment for plate loading tests. In these tests, the soil container was shaken for 10 cycles with 3Hz to cause liquefaction. The severity of liquefaction F_L was controlled by the amplitude of shaking. Then a plate was pushed into the ground quickly. Figures 6.20 and 6.21 show relationships among load and settlement of the plate and F_L. As shown in these figures, the plate settled up to several centimetres with very low resistance. Then resistance recovered. Settlements up to the recover of the load for the grounds of $D_r =50\%$ and $D_r =90\%$ were 5 to 8cm and 1 to 2cm, respectively.

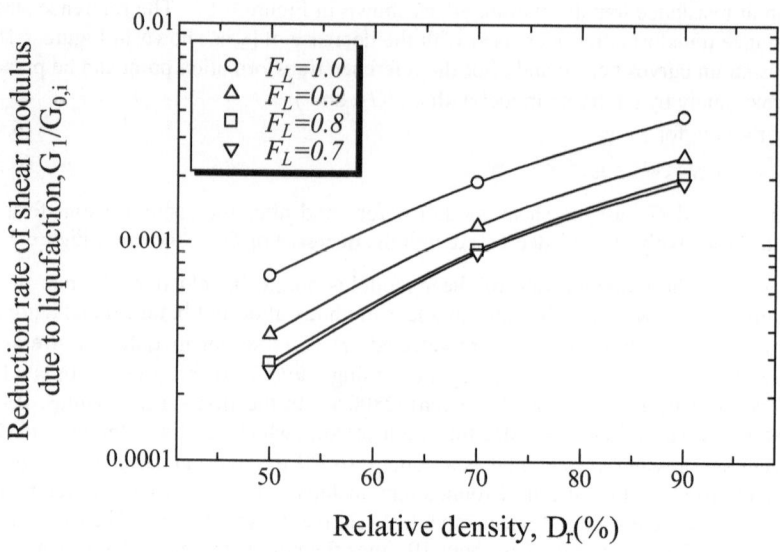

Fig. 6.18. Relationship among relative density, F_L and reduction rate of shear modulus

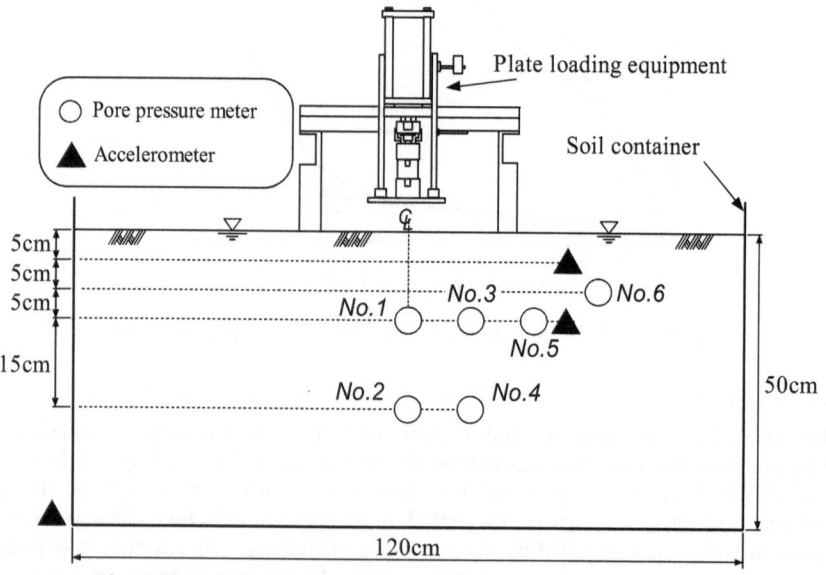

Fig. 6.19. Schematic diagram of equipment for plate loading test

Using the same shaking table, the floatation of a buried common utility duct was tested. Figure 6.22 shows schematic diagram of the test equipment. Shaking was applied for

10 cycles to cause liquefaction with different amplitude of acceleration. Figure 6.23 compares the floatation of the model duct for the ground with different relative densities. As shown in this figure, the floatation of the model duct in the ground of D_r =50% was about several times that in the ground of D_r =90%.

As shown in the two series of tests, the settlement or floatation of structures due to liquefaction is strongly affected by the density of the ground.

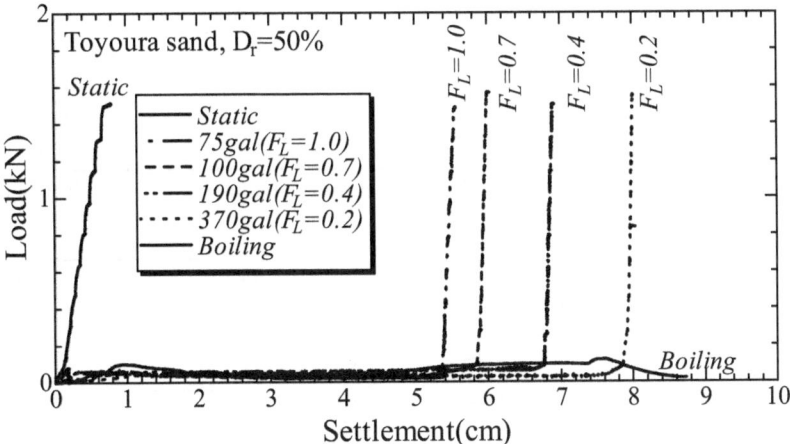

Fig. 6.20. Load-settlement curves in plate loading tests for liquefied loose sand

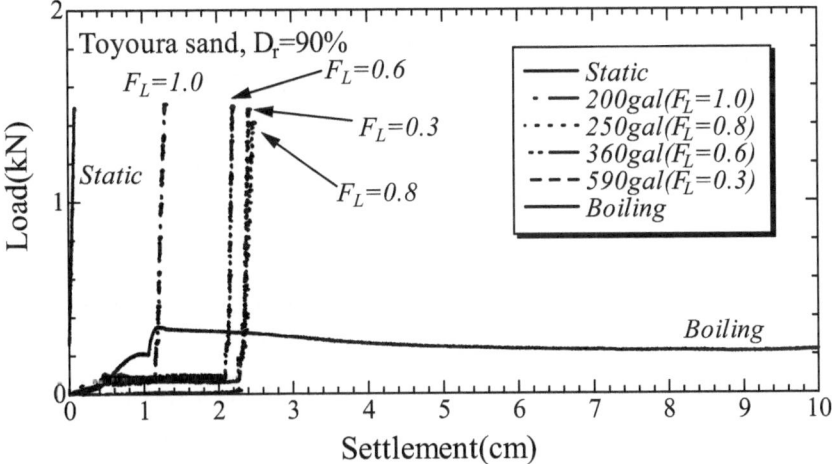

Fig. 6.21. Load-settlement curves in plate loading tests for liquefied dense sand

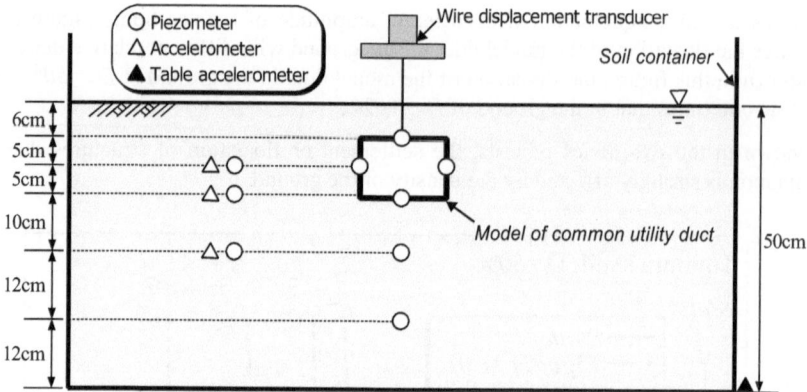

Fig. 6.22. Schematic diagram of model test for floatation of common utility duct

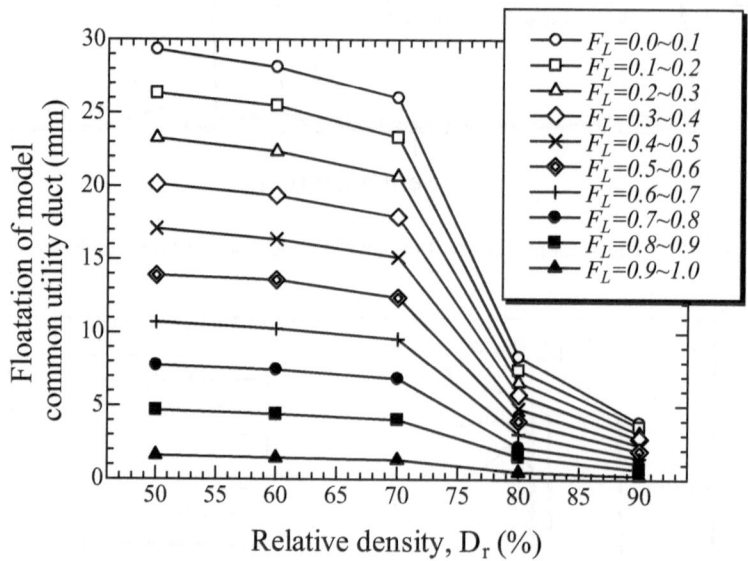

Fig. 6.23. Relationship among relative density, F_L and floatation of model common utility duct in shaking table tests

6.3.4. BEHAVIOUR OF STRUCTURES IN LIQUEFIED SILTY GROUND

Comparison of the Settlement of Buildings in Different Grain Size of Grounds

As mentioned in 6.2.3, many buildings settled and tilted due to soil liquefaction in Adapazarı during the 1999 Kocaeli, Turkey earthquake. The settlement of buildings was induced in Niigata, Dagupan, Kobe, and Yuanlin cities during the 1964 Niigata, Japan earthquake, the 1990 Luzon, Philippines earthquake, the 1995 Hyogoken-Nambu (Kobe), Japan earthquake and the 1999 Chichi, Taiwan earthquake. However the settlement of buildings differed by city. Many buildings settled more than 1m in Niigata and Dagupan. On the contrary, the settlement of buildings in Adapazarı, Kobe and Yuanlin was less than 0.5m. The author compared grain size distribution curves of liquefied sands in these cities as shown in Figure 6.24 (Yasuda et al., 2001a). The sand in Niigata has no fines. On the contrary, the sands in Adapazarı and Yuanlin have about 20% and 100% of fines (less than 0.075mm). Figure 6.25 compares the relationship between fines content of the liquefied soils and the maximum settlement of buildings in five cities. The settlement of buildings decreased with the fines content. It seemed that the grain size, especially fines content, of liquefied soil strongly affected the settlement of buildings.

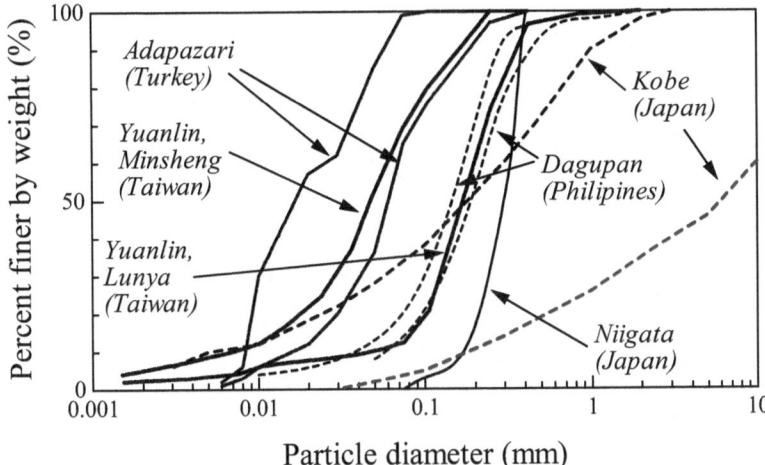

Fig. 6.24. Grain size distribution curves of liquefied sand in five cities (Yasuda el at., 2001a)

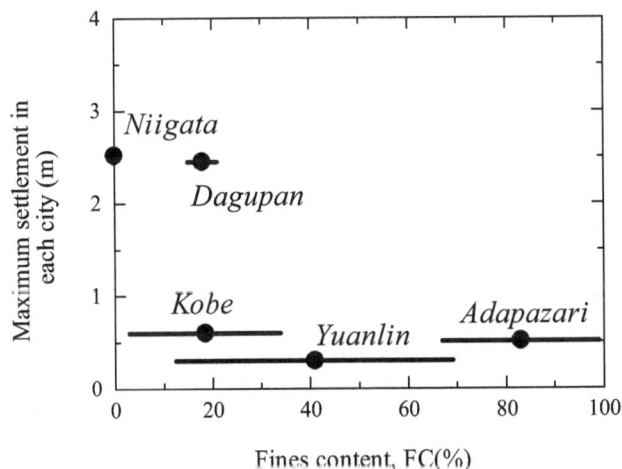

Fig. 6.25. Relationship between fines content and actual maximum settlement (Yasuda el at., 2001a)

Stress-Strain Curves of Silty Sand obtained by Laboratory Tests

Yasuda et al. (1998, 1999a) conducted many cyclic torsional shear tests to study the stress-stain relationship of liquefied sands as mentioned in 6.3.2. Figure 6.26 shows a summary of these tests with the relationships among the reduction rate of shear modulus, fines content and F_L. As shown in this figure, the reduction rate of shear modulus ratio is strongly influenced by the fines content. This implies that settlement or floatation of structures in silty ground is smaller than that in clean sandy ground.

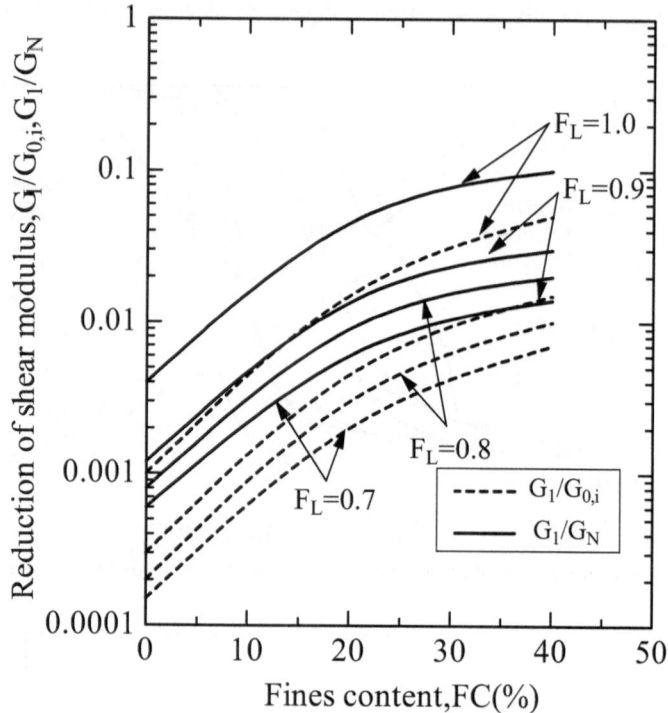

Fig. 6.26. Relationship between shear modulus ratio and fines content (Yasuda el at., 1999b; Courtesy of the JSCE)

6.4. Evaluation Methods for Liquefaction-induced Deformation of Structures

6.4.1. RAFT FOUNDATIONS

Empirical Method

Yoshimi and Tokimatsu (1977) collected data on the settlement of buildings during the 1964 Niigata earthquake, and found the relationship shown in Figure 6.27 between the width ratio and settlement ratio. Settlements of oil tanks during the 1983 Nihonkai-chubu earthquake in Japan were compared with the depth of liquefaction and diameter of the tanks by Yasuda and plotted in Figure 6.27. Settlement of buildings and tanks can be estimated roughly by these relationships.

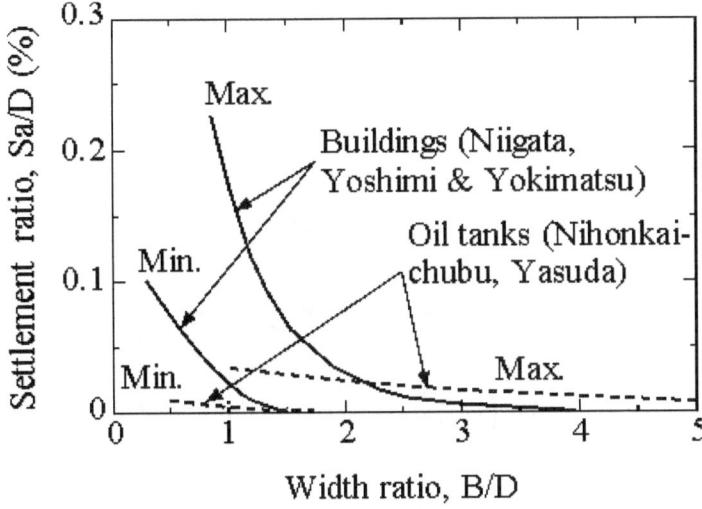

B : Width of building (m)
D : Depth of liquefied sand (m)
Sa : Average settlement (m)

Fig. 6.27. Relationship between width ratio and settlement ratio (Yasuda and Berrill, 2000; Reproduced with permission from Technomic Publishing Co.)

Analytical Methods

The settlement of raft foundations can be evaluated by a seismic response analysis in which liquefaction is considered. In 1992, joint analyses of a building that settled during the Niigata earthquake were carried out by eight program codes (JGS, 1991). Estimated settlements were smaller than the actual settlement at that time. Subsequently, those computer codes have been modified to include large strains. However, further joint analyses have not been conducted, and it is not clear yet whether these codes can evaluate large settlements of the order of 2 to 3 meters accurately.

Alternatively, the residual deformation method can be applied to the settlement of raft foundations. For example, the residual deformation method "ALID" was used to estimate the settlement of a footing of a power transmission tower (Yasuda et al., 2001c). Figure 6.28 shows the results of the analysis. In this method, it is assumed that residual deformation occurs in liquefied ground due to a reduction of shear modulus. Relationships among the shear modulus ratio, $G_1/G_{0,i}$ the factor of safety against liquefaction, F_L, and fines content F_C (percentage of particles smaller than 75μm) shown in Figure 6.26 were used. In the analysis, the finite element method is used twice as follows: (1) In the first step, the deformation of the ground is calculated by the finite element method using the shear modulus before the earthquake. (2) The finite element method is applied again using a reduced shear modulus, due to liquefaction., (3) The difference in the deformation measured by the two analyses is assumed to equal the residual ground deformation.

Settlement: 1.46m

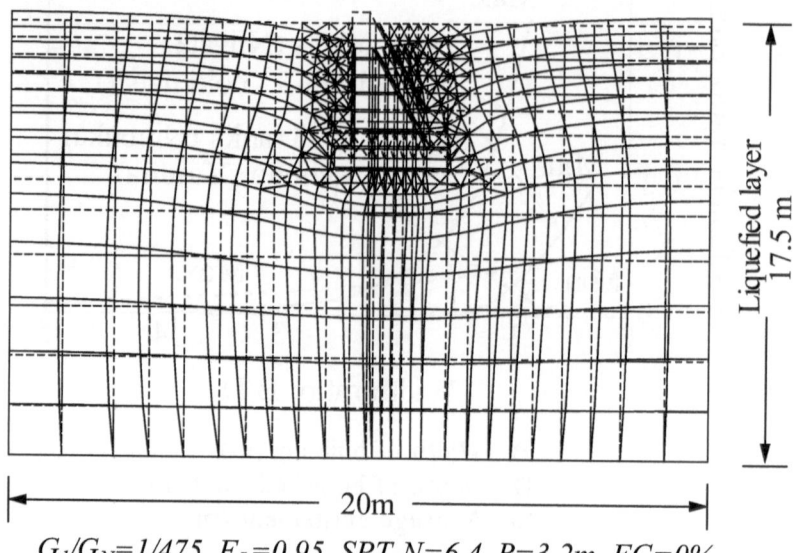

$G_I/G_N=1/475,\ F_L=0.95,\ SPT\text{-}N=6.4,\ B=3.2m,\ FC=0\%$

Fig. 6.28. Settlement of a footing of a transmission tower, analysed by residual deformation method "ALID" (Yasuda el at., 2001b)

6.4.2. PILE FOUNDATIONS

Concept of Design

Figure 6.29 demonstrates the concepts of pile behaviour, with time histories of ground displacement, pile head displacement, pile bending moment, pore pressure and soil spring stiffness during the process of liquefaction and associated ground flow (Yasuda and Berrill, 2000). Amplitude of the displacement and the bending moment in the pile increase rapidly just or slightly before the occurrence of liquefaction. Then these amplitudes decrease after the initiation of liquefaction. The soil spring stiffness decreases with liquefaction. Liquefaction-induced ground flow or settlement of structures with raft foundations begins just after the occurrence of liquefaction. Ground displacement due to the flow increases gradually. Then the flow stops after a short time due to dissipation of the excess pore water pressure and/or topographical balance of the soil mass. The displacement and bending moment of the pile also increase gradually and reach maximum values.

As shown in this conceptual diagram, the bending moment of the pile suffers large values at two distinct times: during the initial occurrence of liquefaction (Time "A") and during the subsequent flow (Time "B"). Detailed dynamic response analyses may be used to estimate these values. However, the displacement and bending moment at these stages are evaluated independently in recent design methods, because the behaviour of the pile and the ground are grossly different at the two points in time.

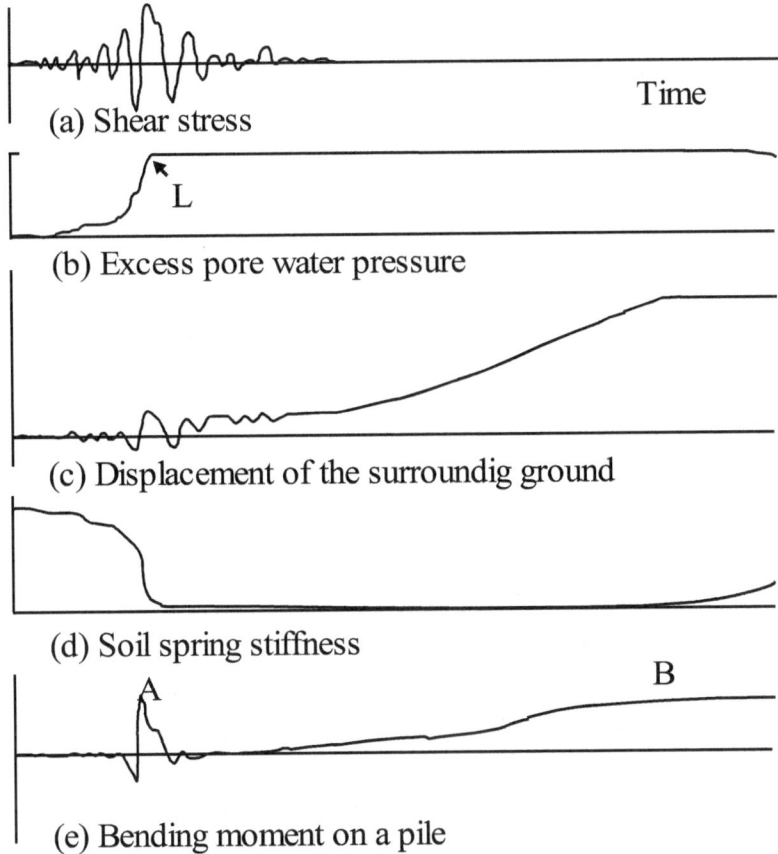

(a) Shear stress   Time

(b) Excess pore water pressure

(c) Displacement of the surroundig ground

(d) Soil spring stiffness

(e) Bending moment on a pile

Fig. 6.29. Concept of time histories of shear stress, bending moment and other items (Yasuda and Berrill, 2000; Reproduced with permission from Technomic Publ. Co.)

Recent Design Methods

In recent practice for the design of pile foundation in liquefied ground, two grades of method are used: i) Dynamic response analysis, and ii) The static evaluation method. Several methods of dynamic response analysis considering liquefaction have been developed in recent years. However, the dynamic response analysis of pile foundations considering liquefaction is still complex and not easy to use for design. Therefore, static evaluation methods are used in recent design.

As shown in Figure 6.30, there are two approaches in the static method: the seismic coefficient method and the seismic deformation method. In the seismic coefficient method, reduction factors for bearing capacity and for soil spring stiffness coefficients are different among current design codes in Japan, as shown in Figure 6.31. The reduction rate is affected by several factors, such as density and grain size of the soil, severity of liquefaction, differential displacement between the pile and surrounding soil etc. Further studies of the effect of liquefaction on soil spring stiffness are necessary.

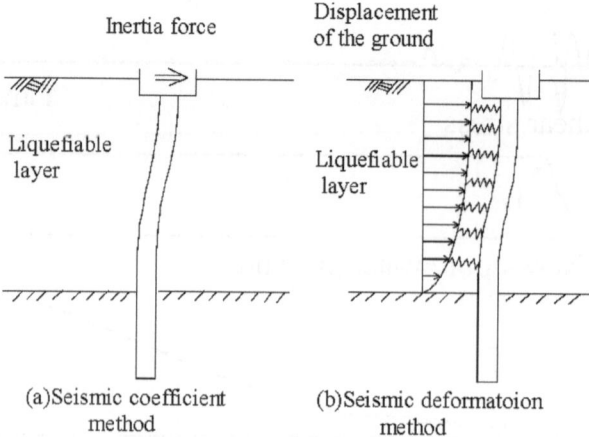

Fig. 6.30. Two types of design method for static analyses

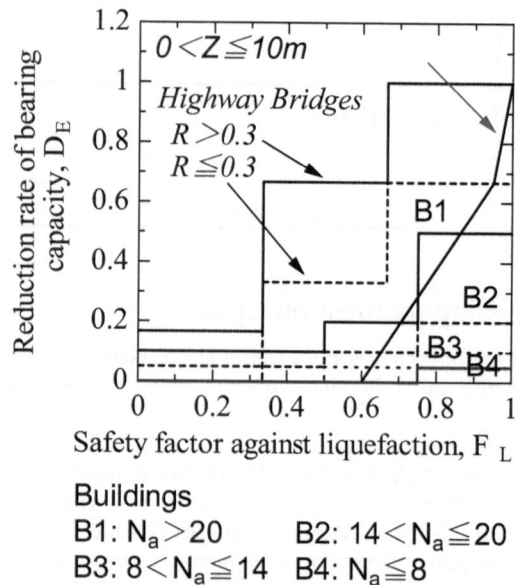

Fig. 6.31. Comparison of reduction rate for bearing capacity among three design codes
in Japan

6.4.3. EMBANKMENTS

Design Methods

Two grades of design methods have been used for embankments: i) detailed seismic
response analyses considering the effect of liquefaction, and ii) simple analyses. The
detailed seismic response analyses have been carried out for special embankments, such

as high filled dams. However, like raft and pile foundation analyses, seismic response analyses are still complex. On the contrary, some simple analytical methods have been proposed recently and applied to normal embankments, such as river, road and railway embankments. One simple method proposed by the author and his colleague is introduced below.

An Example analyzed by a Simplified Method

The simplified method "ALID" mentioned before was applied to estimate the settlement of river levees damaged or non-damaged during the 1993 Hokkaido-nansei-oki and the 1995 Hyogoken-Nambu (Kobe) earthquakes, and tested by centrifuge equipment. Figure 6.32 shows the analyzed results for the levee at Site No.9 of Shiribeshi-toshibetsu River. The levee body settled and the liquefied soil under the levee body was pushed out, as shown by the heave of the ground at the toe of the dike in Figure 6.32. Figure 6.33 compares the analyzed settlements and actual settlements. Though the data are scattered, the analyzed settlements agreed fairly well with the actual settlements during the past two earthquakes.

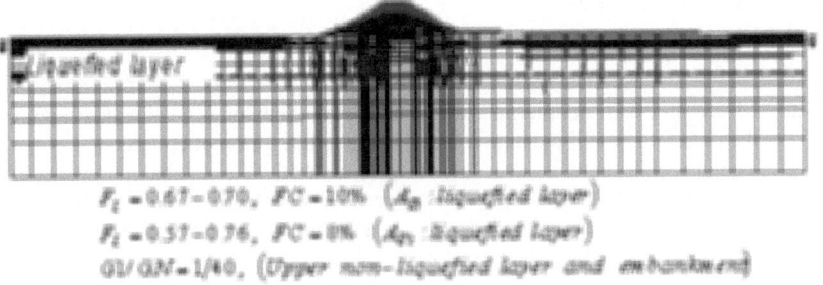

Fig. 6.32. Deformation of the river levee at Site No.9 analysed by "ALID" (Yasuda el at., 2001c)

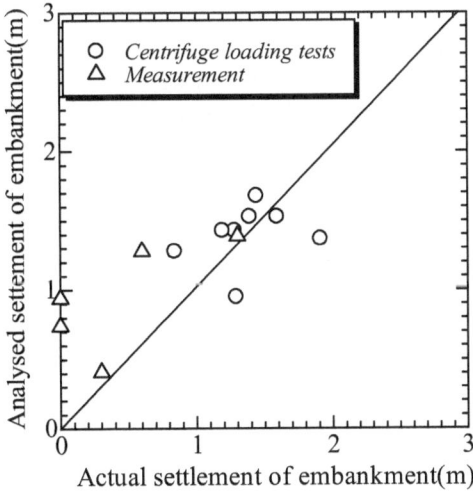

Fig. 6.33. Relationship between actual and analysed settlements

6.5. Countermeasures against Liquefaction-induced Damage of Structures

6.5.1. CURRENT COUNTERMEASURES

Current countermeasures against liquefaction are shown in Table 6.2. These methods are classified into two categories (JGS, 1998): i) improve the liquefiable soil to prevent liquefaction, ii) strengthen structures to prevent their collapse if the ground should be liquefied. In the first category, ground is improved to increase liquefaction strength by the following factors: ℵ high density, ℑ not-liquefiable grain size, ℜ stable skeleton or ℘ low saturation. Other methods to prevent liquefaction are: ⊗ immediate dissipation of increased excess pore pressure, ⊕ reduction of shear stress by increasing confining pressure, ∅ reduction of shear stress by building an underground wall. Appropriate countermeasures in the second category differ by the type of structure. In the countermeasures shown in Tables 6.2 to 6.4, the additional pile method has been applied for bridge foundations, but other methods have been applied for few structures only.

6.5.2. RECENT PROBLEMS

After the Kobe earthquake two problems in the current countermeasures were pointed out:

1. Effectiveness of countermeasures under strong shaking must evaluate not only the occurrence of liquefaction but also the deformation of structures.
2. The gravel drain method is not effective under the strong shaking evaluated by the current design method. However, it is necessary to discuss whether the gravel drain method is in reality useless under strong shaking, and if not, appropriate design method must be developed.

6.6. Liquefaction-induced Flow of the Ground

6.6.1. CONCEPT OF DESIGN METHOD

Concept of Design Methods

Liquefaction-induced ground flow can be divided into two classes, as illustrated in Figure 6.34: ℵ ground flow on gentle slopes, and ℑ ground flow behind quay walls. There are four approaches to allowing for the effect of ground flow in the design of piled foundations:

a) Evaluating the deformation of both piles and ground simultaneously,
b) Evaluating the pressure acting on piles due to the ground flow first, then evaluating the resulting deformation of the piles,
c) Estimating ground displacement first, and then evaluating the deformation of piles (seismic deformation method)
d) Estimating deformation, assuming that the liquefied ground behaves as a viscous fluid.

Table 6.2. Current countermeasures against liquefaction (modified from JGS, 1998)

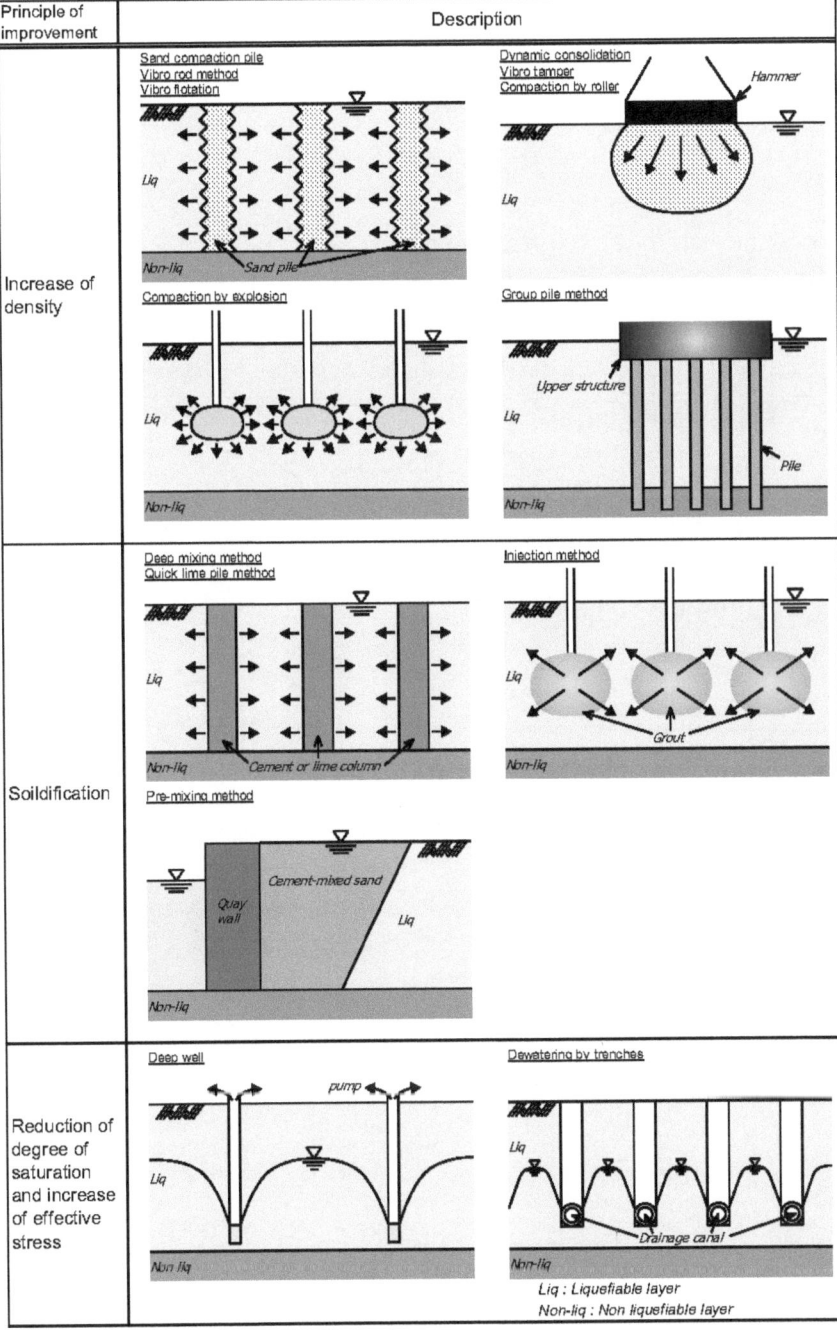

Principle of improvement	Description
Increase of density	
Soildification	
Reduction of degree of saturation and increase of effective stress	

Liq : Liquefiable layer
Non-liq : Non liquefiable layer

Table 6.3. Current countermeasures against liquefaction (modified from JGS, 1998)

Table 6.4. Current countermeasures against liquefaction (modified from JGS)

Principle of improvement	Description
Structural counter-measure	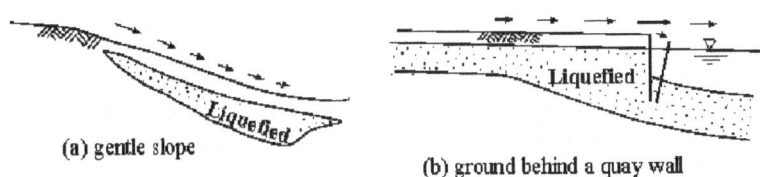 Lift prevention pile or sheet pile / Constraint of surroundings / Absorption of ground deformation by flexible joint / Provision of supplemental foundation for mat foundation / Reinforcement of mat foundation by geogrid / Sheet piling for embankment

Liq : Liquefiable layer
Non-liq : Non liquefiable layer

(a) gentle slope

(b) ground behind a quay wall

Fig. 6.34. Patterns of liquefaction-induced flow

The first approach is logically correct, but it is not an easy one, even by the latest effective-stress response methods of analysis, because of the difficulty of incorporating large ground displacements and interaction between the ground and structure. The second approach is simple and was introduced immediately after the Hyogoken-Nambu earthquake in the new specification for highway bridges in Japan (The Japanese Road Association, 1996). The third approach is also simple and can be applied to complex topographical conditions of the ground and quay walls. This approach was also developed after the Hyogoken-Nambu earthquake, and has been adopted recently in the design codes for Railways and High Pressure Tanks in Japan. Application of the fourth approach is still experimental, and it has not yet been introduced into design codes.

Earth Pressure Method

In the new design code for highway bridges in Japan (The Japanese Road Association, 1996), the forces, shown in Figure 6.35, are applied to the foundation:

$$q_{NL}=C_sC_{NL}K_p\gamma_{NL}X \ (0 \le X \le H_{NL}) \tag{6.6}$$

$$q_L=C_sC_L\{\gamma_{NL}H_{NL}+\gamma_L(X-H_{NL})\} \ (H_{NL}< X \le H_{NL}+H_L) \tag{6.7}$$

where, C_s: correction factor for the distance, S, from quay wall, 1.0 for $S \le 50$ m, 0.5 for $50 < S \le 100$ m, C_{NL}: correction factor in non-liquefiable layer, 0 for $P_L \le 5$, $(0.2P_L-1)/3$ for $5 < P_L \le 20$, 1 for $P_L > 20$, C_L: correction factor in liquefiable layer, $=0.3$, K_p: coefficient for passive pressure, X: depth (m)

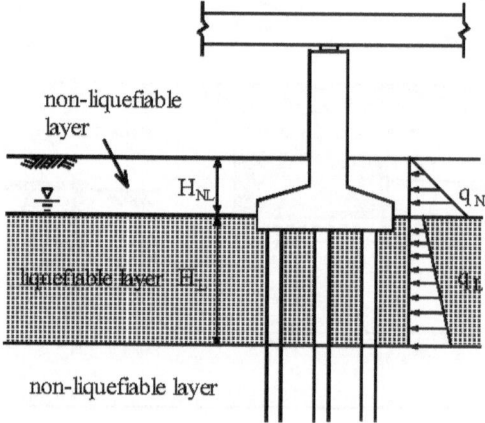

Fig. 6.35. Earth pressure considered in the specification for highway bridges (JRA, 1997)

It must be noted that these methods were derived from inverse analyses of the piles damaged during the Hyogoken-Nambu earthquake only, and the coefficients C_S and C_N have not been derived from analyses or tests. Therefore further study is necessary to fully establish and validate this method.

Seismic Deformation Method

In the seismic design method, the displacement of the ground during flow must be estimated first. Then, horizontal forces are applied to the pile through soil springs. Therefore, both ground displacement due to flow and the value of soil-spring stiffness for the liquefied ground must be evaluated in some way. There are three classes of method to estimate the flow-associated displacement: i) by use of an empirical formula,

ii) by a residual deformation analysis, and iii) by an effective stress response analysis. Iai et al. (1998) succeeded in analyzing the deformation of quay walls and the ground by use of the detailed seismic response analysis code "FLIP." Yasuda et al. (1999) tried to analyze the ground deformation by a simple residual deformation method, "ALID" same procedure as 6.4.1.

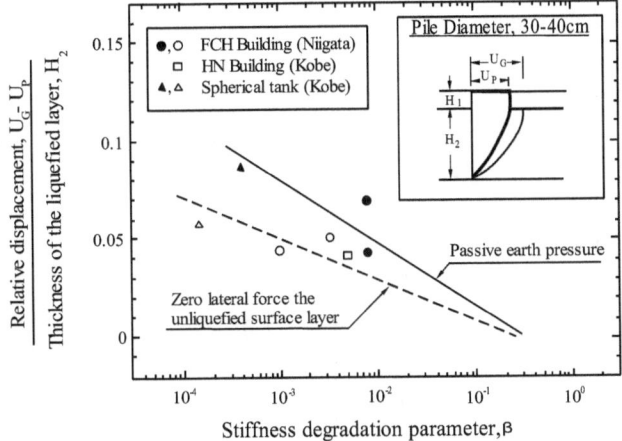

Fig. 6.36. Relationship between stiffness degradation parameter and the ratio of relative displacement to thickness of liquefied layer (Ishihara and Cubrinovski, 1998; Reproduced with permission from Balkema)

The properties of equivalent soil springs to represent liquefied ground during flow have been studied by conducting inverse analyses on damaged piles and by shaking table tests. Ishihara et al. (1998) made inverse analyses of damaged piles during the 1964 Niigata and 1995 Hyogoken-Nambu earthquakes. Figure 6.36 summarizes the relationship between the stiffness degradation parameter, β and the ratio of relative displacement and thickness of the liquefied layer. Estimated β, which is same as the reduction rate of soil-spring stiffness shown in Figure 6.26, was found to be about 1/100 to 1/10000, and decreased with the ratio of relative displacement and thickness of the liquefied layer.

6.6.2. COUNTERMEASURES AGAINST THE FLOW

Countermeasures against liquefaction-induced flow have been studied since the 1995 Kobe earthquake. Some methods have been applied to quay walls and express highways in Tokyo, Kobe and Osaka. Table 6.5 summarise the applied countermeasures together with some ideas which have not been applied yet (Kanatani et al., 2000).

6.7. Concluding Remarks

New design methods for liquefaction which have been developed recently were introduced in this chapter. The evaluation of liquefaction-induced deformation of structures seems very important for rational design. So, studies on the methods to evaluate deformation must be studied more, together with discussions on the allowable value of deformation of each structure.

Table 6.5. Countermeasures against liquefaction-induced flow behind the quay wall
(Kanatani el at., 2000)

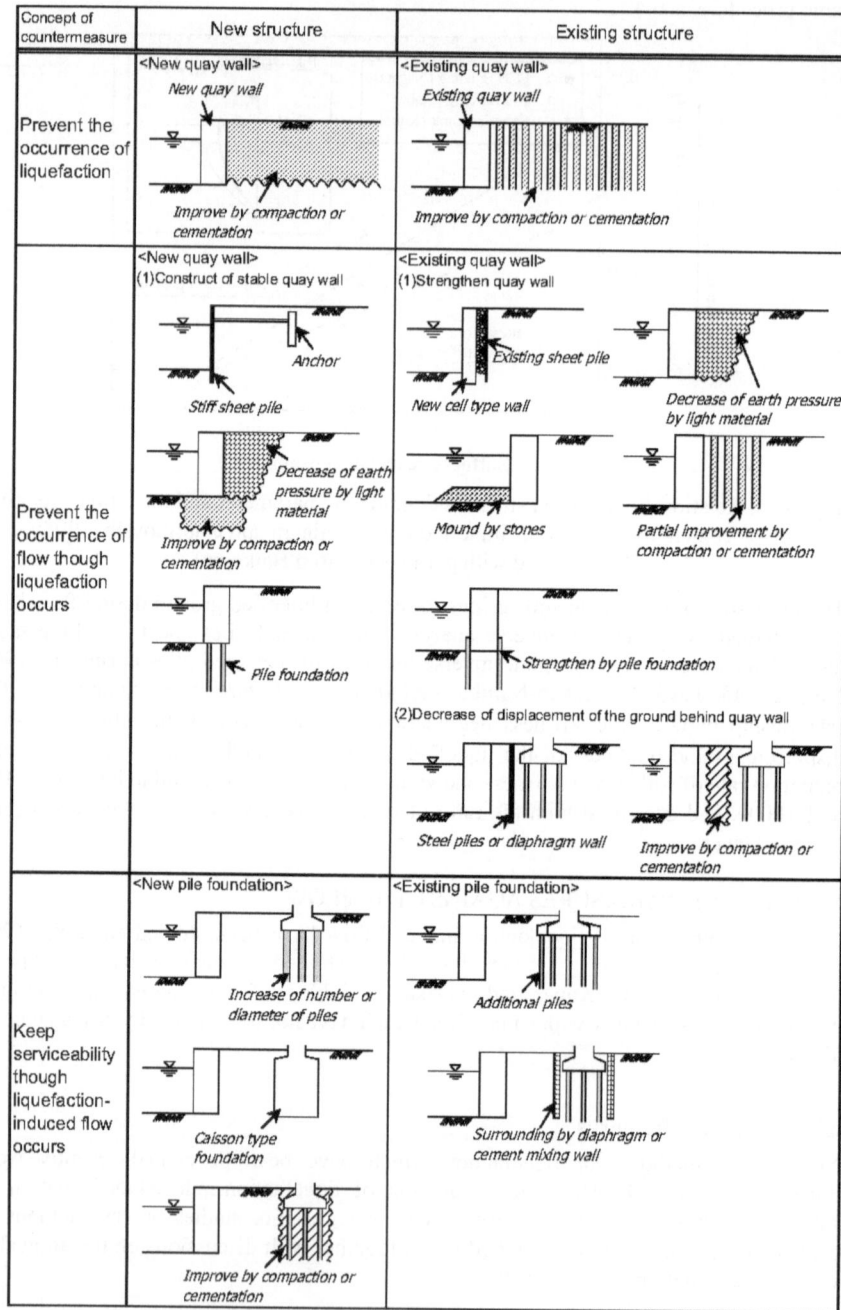

CHAPTER 7
SEISMIC ZONATION METHODOLOGIES WITH PARTICULAR REFERENCE TO THE ITALIAN SITUATION

Albero Marcellini and Marco Pagani
Istituto per la Dinamica dei Processi Ambientali, CNR, Milano and Dipartimento di Scienze della Terra, Università degli Studi di Milano

7.1. Introduction

Experimental evidences of both large and moderate earthquakes show a high degree of variability of damage distribution. Since the mid of the fifties of past century, scientists realised the existence of a strong influence of soil characteristics and of near field source effects on the recorded values of strong motion. These observations constitute the premises of microzonation studies; the intention is to provide a tool to prevent damages by means of detailed assessing of design forces and land use planning in seismic areas.

After the adoption of some over-simplified approaches in the fifties using microtremor analysis (Kanai, 1957), in the sixties Medvedev (1977), proposed a zonation method based on an empirical correlation between the seismic impedance ratio and the variation of macro seismic intensity. In the early eighties, the microzonation approaches became more sophisticated partly owing to the boost of nuclear power plants installations and the deployment of dense accelerometric arrays. Since then, the methods were implemented and nowadays Figure 7.1 well represents a scheme of microzonation methodology commonly adopted. As it is possible to see it encompasses the whole topics necessary to perform a microzonation that basically can be subdivided into three major items:

(1) Evaluation of the expected input motion
(2) Site effects analysis
(3) Preparation of microzonation maps and recommendations for practical application

Now, whenever the scheme of Figure 7.1 is worldwide applied, the relative importance one component has with respect to the others is subjected to large variability (Marcellini et al., 2001a) due both to the physics of the phenomenon and to goal of the study. Actually the following factors play the key role: a) the administrative situation of the zone (regulations and seismic codes are significantly different country to country and in some cases even region by region) b) the size of the expected earthquake (generally the stronger the earthquake the larger becomes the importance of the source with respect to the amplification factors), c) the nature of soils (for instance, the presence of potentially liquefying soils urges detailed and expensive geotechnical investigations).

This article focus on the experiences gained in Italy during the last ten years. In other words, the microzonation suggestions of this paper should be considered particularly relevant for seismic areas subjected to moderate earthquakes, that is, earthquakes with magnitude lower than 7.

A. Ansal (ed.), *Recent Advances in Earthquake Geotechnical Engineering and Microzonation*, 231–252.
© 2004 *Kluwer Academic Publishers.*

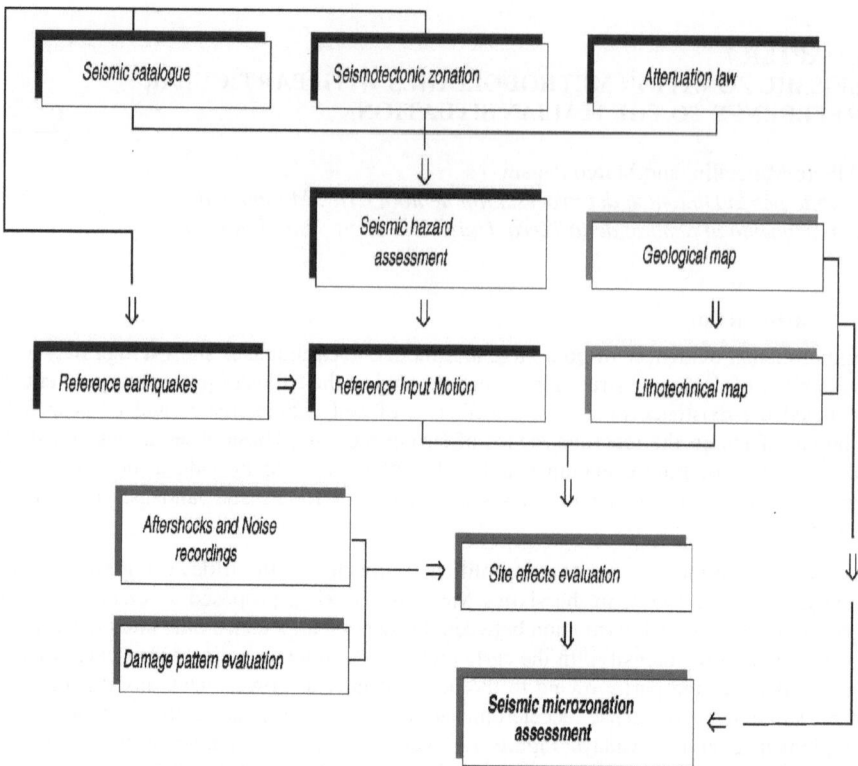

Fig. 7.1. Microzonation scheme (after Marcellini et al., 2001a; Reproduced by
permission of Italian Geotechnical Journal - Patrone Editore)

Before tackling the strictly scientific aspects we want stress the fact that microzonation
falls into the category of "applied research", that is, to perform a microzonation that will
not (or cannot) applied is waste of time. It means that the typologies of expected results
must be defined with local administrators (or in some cases central government). As an
example we report three typologies of results.

Figure 7.2 shows the microzonation outcomes of three municipalities situated along the
Adriatic Sea coast in the Emilia-Romagna region (Daminelli et al., 2000). These
municipalities are located within a zone subjected to moderate earthquakes (maximum
expected magnitude around 6.5). Soils data available for microzoning consisted on a
detailed geological survey and a number of CPT tests (around 100). Given the lack of
geophysical investigations, shear wave velocities have been estimated using statistical
correlations with lithology and CPT data: as a result the obtained amplification values
are affected by a consistent degree of uncertainty. The microzonation representative
parameter was the ratio of Housner Intensity (with integration limits between 0.1 and
0.5 s) between the site and the outcropping bedrock. Housner intensity has been
selected to address ground shaking because of its greater stability with respect to other
parameters like PGA, strongly unstable in particular in the high frequency range.
Figure 7.2 shows an example of the computed maps.

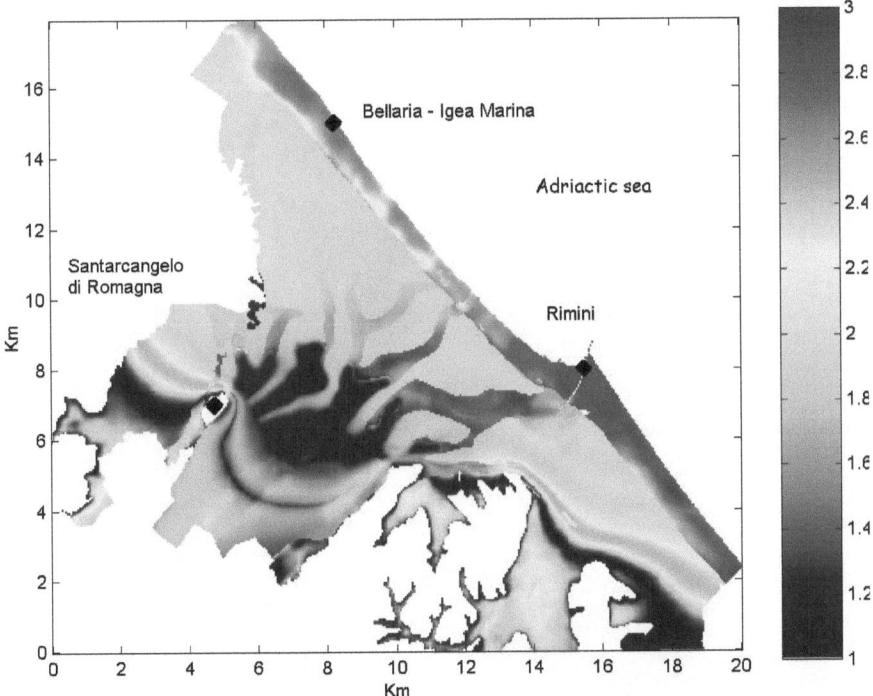

Fig. 7.2. Microzonation of Rimini, Santarcangelo di Romagna and Bellaria: amplification map in terms of Housner intensity ratio (0.1-0.5s). Higher seismic amplifications decrease moving toward the mountainous area (in the figure from NE to SW)

In other cases microzonation maps in terms of response spectra, as we did for the municipalities of Gatteo, Savignano and S. Mauro Pascoli (see Figure 7.3) are more suitable for local administrators (Marcellini et al., 1999). To note that these cited microzonations were performed not after an earthquake.

The microzonation of Fabriano, situated in the Marche Region at the foothills of the Apennine chain, has been performed just after the 1997 Umbria-Marche earthquake. Boosted by the necessity to start the reconstruction as soon as possible, the request was a clear and compulsory zonation of the area of the town. Fortunately supported by a lot of data and experimental evidences (pattern of damage), it was possible to provide the planners the map reported in Figure 7.4. The ground motion parameter (F.A.) was not selected by experts and scientists involved in the project, but were obliged by the administrators in order to translate it immediately into a seismic coefficient multiplicative of the standard design forces provided by the national seismic code.

Henceforth through the Fabriano case history, we will deal with the different aspects of microzonation.

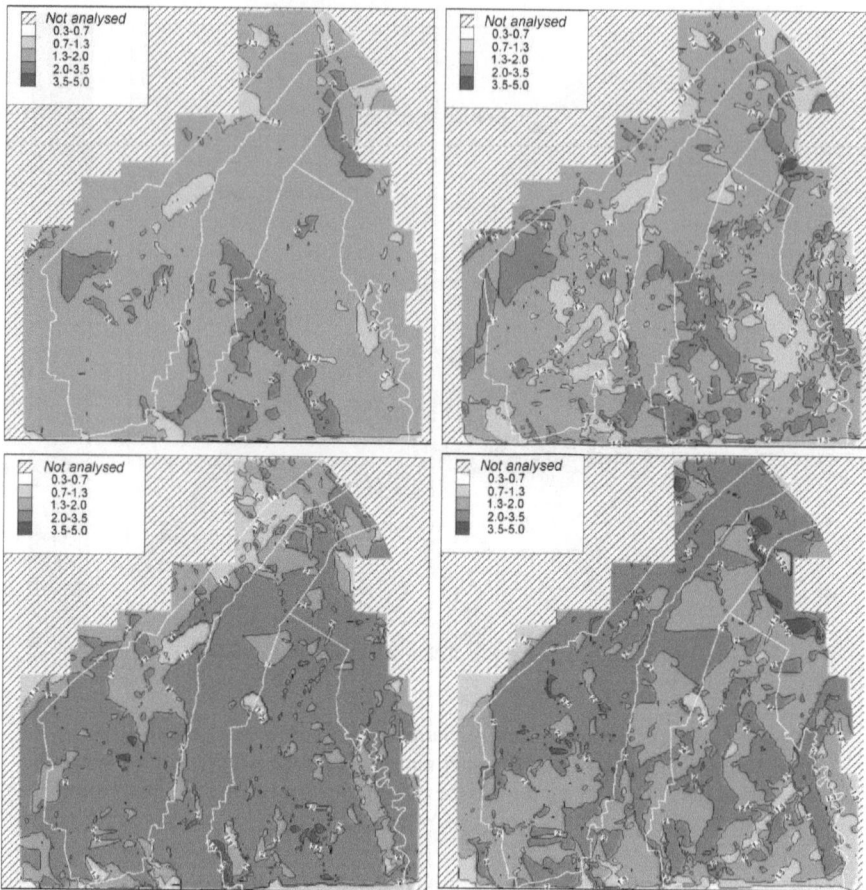

Fig. 7.3. Microzonation of Gatteo, Savignano and S. Mauro Pascoli: Amplification maps. Top left PGA/PGA_H, top right S_a/S_{aH} (T=0.2s), bottom left S_a/S_{aH} (T=0.5s) and bottom right S_a/S_{aH} (T=1.0s). PGA_H and S_{aH} are Peak ground acceleration and spectral acceleration (ξ=5%) computed by hazard analysis (RP=474 yr) whereas PGA and S_a were obtained after the evaluation of site effects. It is possible to observe that the amplification pattern changes dramatically as a function of T.

7.2. Evaluation of the Expected Input Motion

We define the reference motion at a site as the ground motion evaluated without taking into account site effects (as an example reference motion could be represented by accelerograms derived by the response spectrum computed via standard probabilistic hazard assessment, where soil conditions are only roughly accounted for by specific attenuation relationships). The reference motion (usually in terms of accelerograms) to use in microzonation studies for site effects evaluation is one of the most crucial problems and constitutes the link between regional and local hazard.

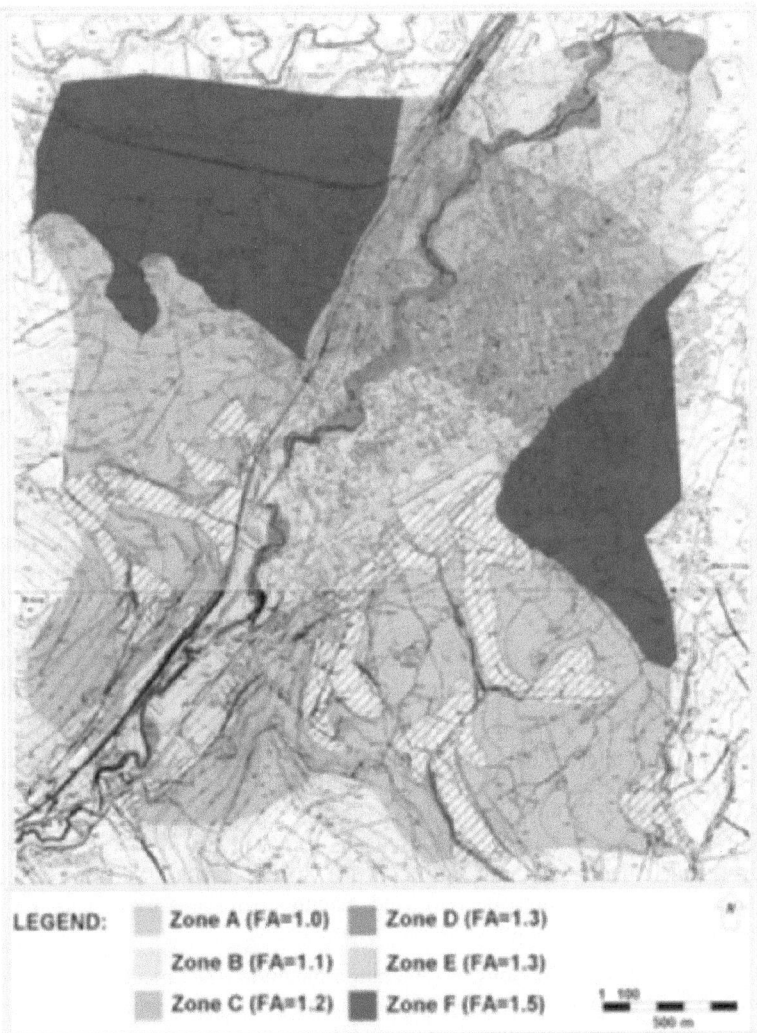

Fig. 7.4. Fabriano zonation map; FA represent an amplification coefficient, eventually used by the designer as multiplicative factor of design forces (after Marcellini et al., 2001a, Reproduced by permission of Italian Geotechnical Journal - Patrone Editore).

There are at least three major approaches that can be adopted for the evaluation of reference motion: the probabilistic approach, the stochastic and the deterministic one.

We usually rely on a two-step procedure to evaluate the reference motion. The first step consists on the evaluation of expected acceleration response spectrum (5% damping) using standard probabilistic seismic hazard assessment with response spectra attenuation laws, horizontal component. The second step lies on the determination of accelerograms from the above obtained response spectrum. There are different procedures to generate artificially accelerograms whose response spectra match a target response spectrum. As a personal opinion, we prefer to use accelerograms actually recorded, selected by means of a simple procedure here synthesised: (1) to browse into the accelerograms database (2) to choose the accelerograms recorded at or nearby the site (if available; in case not, to choose the accelerograms produced by earthquakes originated in a similar seismogenic situation), (3) to compute the response spectra of the sorted accelerograms, (4) to adopt as reference motion the accelerograms whose spectra best fit the target spectrum (hazard spectrum).

The deterministic approach computes the expected motion at the given site by means of a kinematic or sometimes dynamic description of the earthquake source whereas "intermediate" methods generally require stochastic approaches to compute reference motion.

Through the case history of Fabriano microzonation, we will present three different procedures to compute reference motion: a) deterministic, b) stochastic, c) probabilistic and we will discuss advantages and disadvantages of each method.

7.2.1. DETERMINISTIC APPROACH

It is the well-known approach adopted to compute synthetic seismograms. Basically it requires: 1) the definition of seismic source parameters (in general kinematics models are adopted); 2) the application of Green's function of elastodynamic; 3) the use of the wave equation in elastic media, or pure viscoelastic media (attenuation taken into account using the quality factor Q); 4) the application of the representation theorem. Different authors applied deterministic approaches in seismic microzonation (Panza et al., 1996).

The deterministic approach has been applied to Fabriano microzonation by Priolo (2001). The first step consisted on the determination of the "scenario" earthquake; in the present case it has been decided to consider the strongest earthquake that hit Fabriano in the past. Castelli and Monachesi (2001) on the basis of historical investigations defined the event of April 24, 1741 (I_{MCS}=9) as the "scenario" earthquake; through empirical correlation the estimated magnitude for this event was 6.2. The epicentre was placed at some 8km North -East of the town.

Priolo (2001) used the wave number integration method, a 3-D method that solves the full equation of seismic wave propagation through a layered medium, to compute seismic signal in Fabriano under the assumption of the repetition of the 1741 event. Source parameters adopted were 13 MPa stress-drop and a seismic moment (M_o) of $3.1 \cdot 10^{18}$ Nm together with the assumption of a normal fault mechanism (ϕ=150°, δ=40°, λ=-90°) such as the earthquake recently occurred in 1997 in Central Italy. The crustal model consisted of 14 layers (in the uppermost 300m the average values of V_P and V_S are 2500 and 1400 m/s, respectively).

The signals at Fabriano have been computed using both point source and extended source. Figure 7.5 shows some of the results obtained by Priolo considering a point source: to note the significant difference between horizontal and vertical components and the relevant influence of the depth. Considering a realistic hypocentral depth (6 km) the computed horizontal PGA in Fabriano (8.24km from the source) has been estimated to be 0.21g.

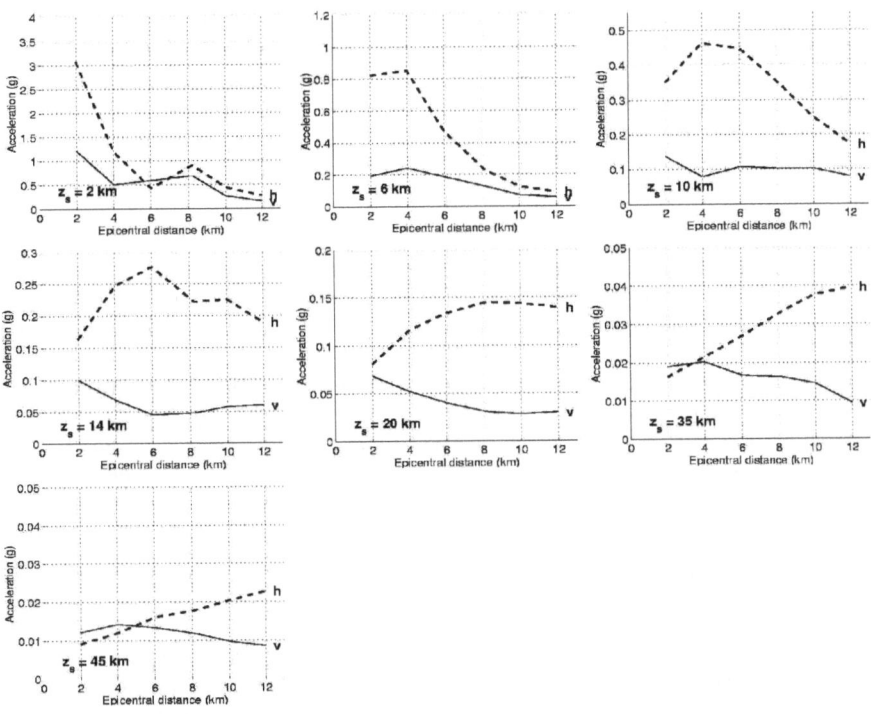

Fig. 7.5. Deterministic simulation using point source model: computed PGA values against epicentral distances and source depths (z_S): h - horizontal component; v - vertical component (after Marcellini et al., 2001b; Reproduced with permission from Elsevier).

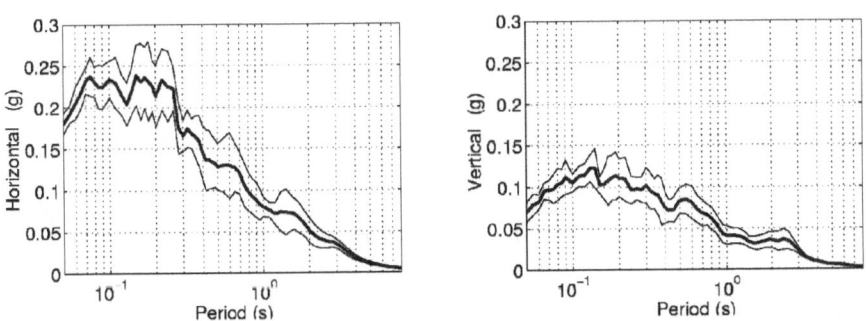

Fig. 7.6. Acceleration response spectra (5% damping) computed at Fabriano, using an extended source model (after Marcellini et al., 2001b; Reproduced with permission from Elsevier).

Using the same normal fault mechanism Priolo tackled the extended source simulation by decomposing the entire fault area (20 x 12 km^2) into 15 elementary sub sources of 4 x 4 km^2 and M_o=2.1x10^{17} Nm, regularly distributed on the fault. The simulation has been performed considering several hypothesis of rupture nucleation point, with a velocity rupture of 2.8 km/s. Results obtained using an extended source, as it is expected, show lower acceleration values, but the differences decrease in terms of velocity and displacement (Priolo, 2001). Figure 7.6 shows the computed response spectra, 5% damping, considering a bilateral source that nucleates at the epicentre of the 1741 earthquake.

7.2.2. STOCHASTIC APPROACH

The stochastic nature of high-frequency ground motion is clearly visible on the acceleration time-histories commonly recorded during an earthquake (Hanks and McGuire, 1981), and different empirical approaches can be used to tackle this aspect. They basically consist in simulating time sequences that match existing data with respect to amplitude, frequency content and signal duration. However, in many cases simple propagation and source mechanism deterministic models, can successfully predict the acceleration amplitude Fourier spectrum. Therefore, in order to simulate the acceleration time histories to use as "reference motion" in seismic microzonation analysis, it seems reasonable to take into account of:

1. stochastic properties of the recorded ground motion;
2. simplified description of the physical properties of source and path to evaluate a reliable far field acceleration spectrum.

An approach of this kind, has been introduced by Boore (1983), where point 1) is achieved by using a time sequence of band-limited random white Gaussian noise and point 2) by considering the far field Brune spectrum (1970, 1971) multiplied by three attenuation terms representing the geometrical spreading, the whole attenuation path and the high frequency decay of the acceleration spectrum (due to same kind of source effect or site effect according to Papageorgiou and Aki (1983) and Hanks (1982) interpretations, respectively). A modified version of this procedure has been used in this work and it is here briefly summarised (see Tento (1999) for further details).

The stochastic properties of the acceleration time history are simulated by means of a time sequence of band-limited random white Gaussian noise windowed by (Saragoni and Hart, 1974):

$$w(t) = at^b e^{-ct} H(t) \tag{7.1}$$

where $H(t)$ is the Heaviside function and coefficients a, b, and c are chosen to normalise the resulting time sequence and to match the Saragoni and Hart window. The above coefficients are related to strong-motion duration T_d that is established by $T_d = 1/f_c$ (f_c is the earthquake corner frequency).

The Fourier spectrum of the resulting signal is then multiplied by the model spectrum:

$$A(f,R) = S(f) \frac{e^{-k\pi f} e^{-\frac{\pi R}{\beta Q_\beta}}}{R} \tag{7.2}$$

where $S(f)$ is the acceleration far field source spectrum, k and Q_β represent the anelastic attenuation and the scattering mechanism respectively (see Rovelli et al. (1988) among the others), R is the hypocentre distance and the $1/R$ factor models the geometrical spreading, β represents the S wave velocity. The source spectrum predicted by the Brune model (1970, 1971) is taken as average far field spectrum, thus:

$$S(f) = \frac{F_s \langle R_{\theta,\phi} \rangle P}{4\pi\rho\beta^3} M_0 \frac{(2\pi f)^2}{1 + \left(\dfrac{f}{f_c}\right)^2} \tag{7.3}$$

where F_S, $\langle R_{\theta,\phi} \rangle$, P, M_0 and ρ represent the free surface amplification factor, the average radiation pattern, the partition energy factor into two horizontal components, the seismic moment and the medium density. The Brune stress drop is related to M_0 and f_c by $\Delta\sigma_B \propto M_0 f_c^3$.

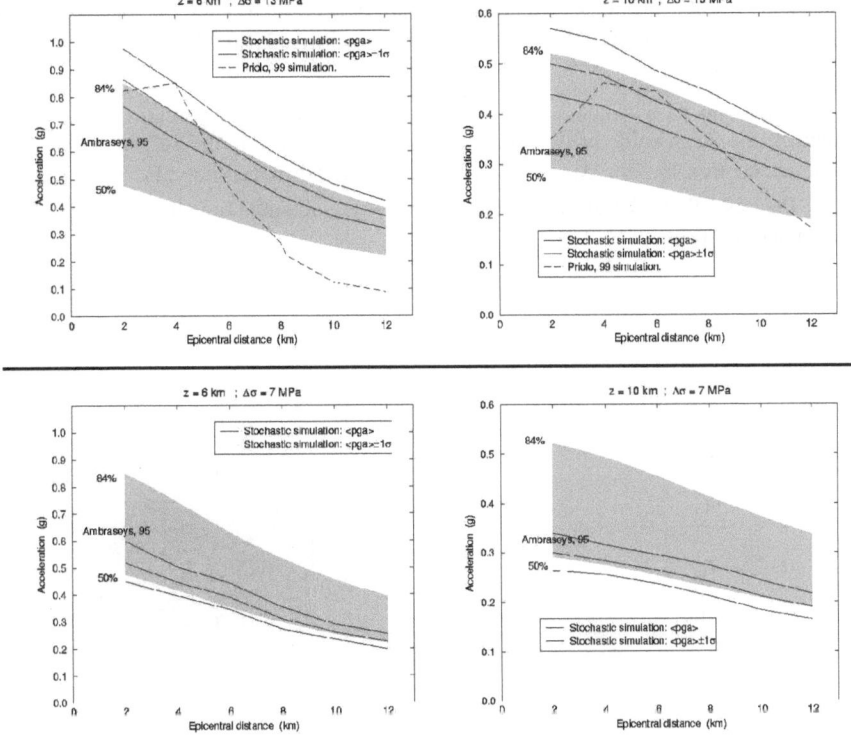

Fig. 7.7. Predicted PGA by means of stochastic simulation: mean value and mean plus and minus one standard deviation are compared with Priolo point source simulation (Priolo, 2001). Ambraseys 50 and 84 percentile attenuation laws (shaded area) are also shown for comparison (Ambraseys, 1995) (after Marcellini et al., 2001b; Reproduced with permission from Elsevier).

The above procedure has been applied to Fabriano microzonation by using a point source model of the April 24, 1741 M=6.2 earthquake, in order to compare the results obtained by this method with Priolo point source simulation (Priolo, 2001) assuming, the same value of the seismic moment and two stress-drop values (13 and 7 MPa). The attenuation terms Q_β and k were fixed, on the basis of the simplified crustal model of Priolo, to 100 and 0.01s whereas β and ρ values adopted were 3.5 km/s and 2.8 gr/cm^3. The simulation was performed for two values of hypocentre depth (6 and 10 km) and epicentral distances ranging between 2 and 12km. The results are shown on Figure 7.7, where the Priolo (2001) simulation and the 50 and 84 percentile values of the Ambraseys (1995) attenuation law are also plotted for comparison.

Contrary to the deterministic point-source simulation, the stochastic procedure shows a monotonic decay of the PGA with distance. For 6km hypocentre depth, in the examined range of distances, the stochastic simulation shows a less rapid decay with respect to the deterministic one. For 10km hypocentre depth, this discrepancy becomes noticeable only for distances greater than about 10km. In Fabriano, by considering the hypocentre depth of 6 km, the stochastic model predicts a 2.4 times greater PGA value than Priolo result.

The comparison with Ambraseys (1995) attenuation law is really impressive. At all distances, the mean value of the predicted PGA is very close to the 84 percentile value of the attenuation law for $\Delta\sigma_B = 13$ MPa and to the 50 percentile value for $\Delta\sigma_B = 7$ MPa. Figures 7.9 and 7.8 show some examples of the obtained time histories and of the corresponding spectra.

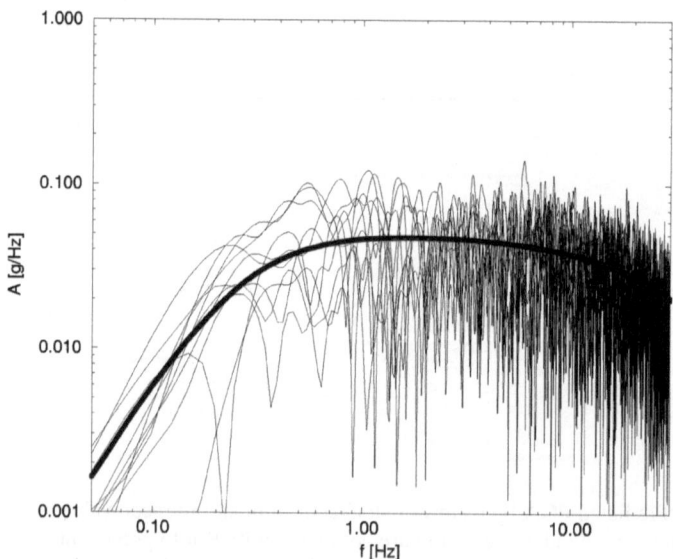

Fig. 7.8. Acceleration Fourier spectra (thin lines) of the time histories showed in Figure 7.9 and Fourier spectrum of the adopted model (thick line) (after Marcellini et al., 2001b; Reproduced with permission from Elsevier).

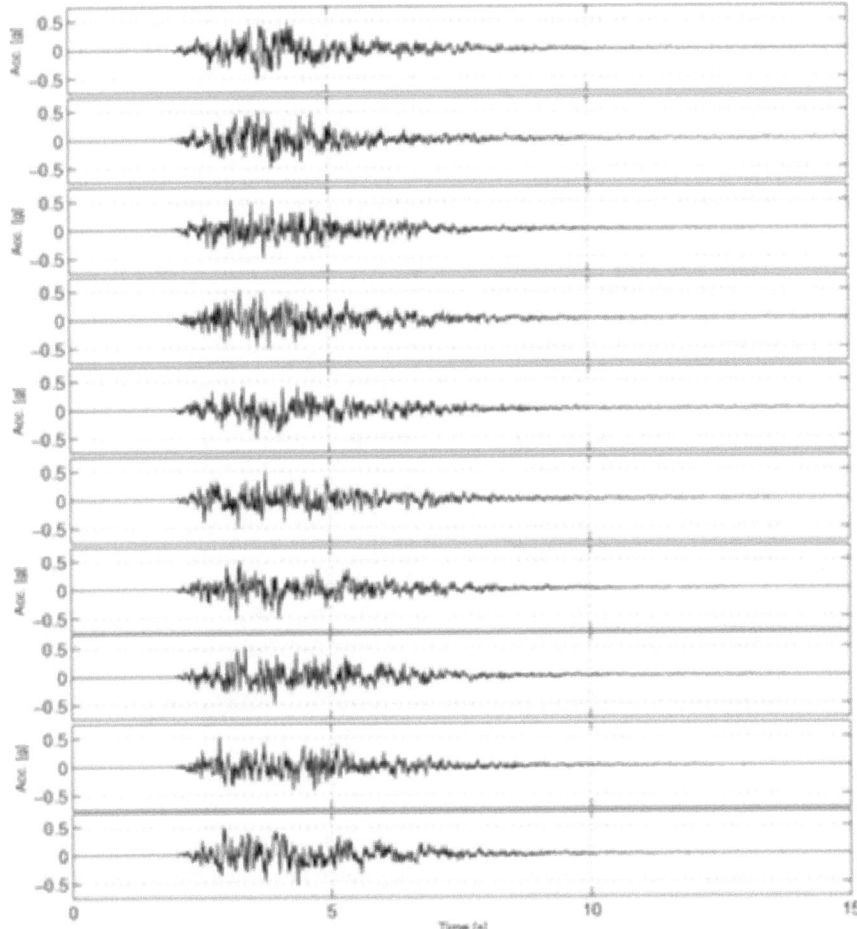

Fig. 7.9. Acceleration time histories obtained by stochastic simulation with $3.1 \cdot 10^{18}$ Nm seismic moment, 13 MPa stress drop and 6 km source depth; Q_β and k have been fixed to 100 and 0.01s, respectively (after Franceschina et al., 2001; Reproduced by permission of Italian Geotechnical Journal - Patrone Editore).

7.2.3. PROBABILISTIC APPROACH

One of the main advantages of this approach is to assess a reference motion compatible with the hazard at regional level; basically the procedure is a two-step approach: 1) evaluation of seismic hazard at the reference site; 2) assessment of the reference motion.

Seismic hazard evaluation

The approach here adopted is the classical one (the so called "Cornell approach"; Cornell, 1968) that consists on seismic source identification, temporal behaviour described by a Poisson model and a negative exponential distribution for the magnitude. Seismic hazard is computed at the site in terms of probability of non exceedance of the strong motion parameter used in the adopted attenuation law.

A 474 years return period is considered as a standard for seismic classification and building code purposes, generally PGA is assumed as the strong motion parameter. As far as Fabriano is concerned seismic hazard has been computed by Peruzza (2000) using SEISRISK III computer code (Bender and Perkins, 1987). Seismic source zones have been drawn following Scandone et al. (1990) and historical seismic catalogue

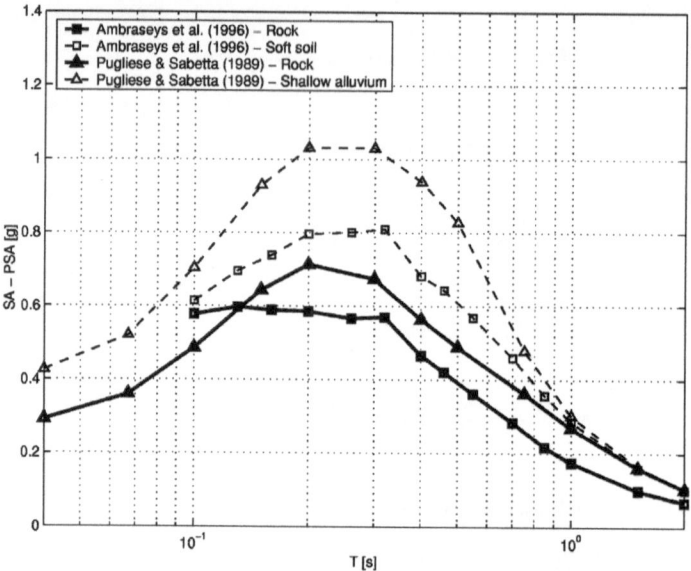

Fig. 7.10. Hazard response spectra computed via probabilistic approach (RP=474 yr, 5% damping) using different attenuation laws (after Marcellini et al., 2001b; Reproduced with permission from Elsevier)

Reference motion

As noted previously, different approaches can be used to obtain accelerograms from uniform hazard response spectra but because there is not a biunique correspondence between response spectrum and accelerogram, some statistical procedure must be applied.

An appealing procedure is the generation of artificial accelerograms that match the target response spectrum. One of these methods is the Auto Regressive Moving Average (ARMA) technique, which is illustrated for example in Ciampoli (1997).

As discussed above we prefer to adopt real accelerograms; the procedure followed to match the accelerograms with response spectra is very simple (a description can be found in Tento, 1999). It is important to note that this type of procedure suggests the adoption of several accelerograms as a reference motion. In the case of Fabriano ten accelerograms have been considered (Figure 7.11); they were taken from the Italian accelerograms database.

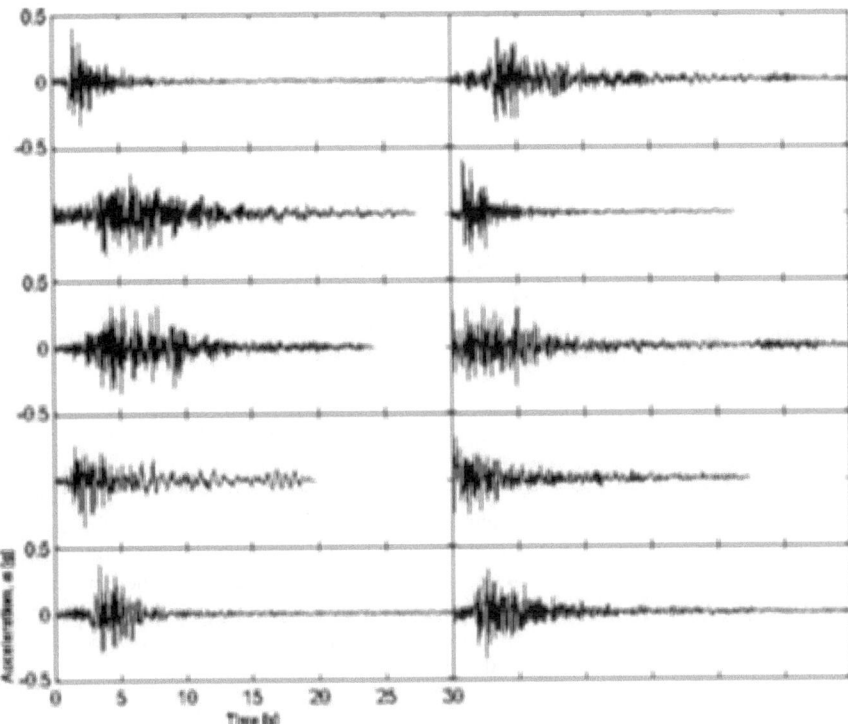

Fig. 7.11. Selected accelerograms obtained from the uniform probability response spectrum.

7.2.4. DISCUSSION

We presented three different approaches to evaluate the reference motion: the choice of one method instead of another is not a simple matter. On the other hand what seems a good solution, that is, the use of all the three approaches brings to a non-uniqueness of results. The three types of reference motion, when used to estimate site effects can produce contradictory results. As shown in Figure 7.12, the results obtained using a simple 1-D model in a site at Fabriano (where the reference motion has been computed with the three mentioned approaches) with lithology and V_S profile shown in Figure 7.13, are strongly dependent on the type of reference motion. Generally seismologists do like the deterministic approach because the physical background sounds better; the objection raised by risk experts is that the motion obtained is subjected to strong variations depending on source or path parameters.

Therefore the deterministic approach is not recommended, unless the source and path parameters can be adequately modelled by suitable probabilistic distributions, and it is well known that right now probability functions of stress drop, seismic moment or Q value in a given area are not available: it is enough difficult to estimate with sufficient accuracy average values.

A. Marcellini and M. Pagani

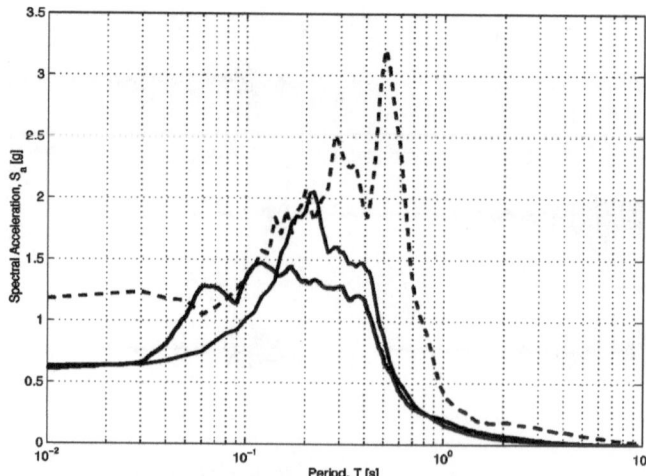

Fig. 7.12. Computed response spectra for the site shown in Figure 7.13 under different input motions (continuous line: input motion obtained following a probabilistic approach; dashed line: input motion obtained by deterministic computation; dark continuous line: input motion computed using a stochastic approach). The input motion is the ground shaking at bedrock as obtained after a deconvolution of reference motion (after Franceschina et al., 2001; Reproduced by permission of Italian Geotechnical Journal - Patrone Editore).

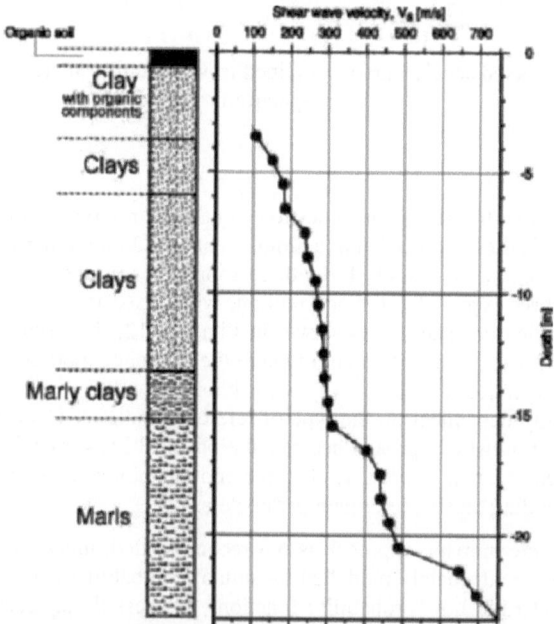

Fig. 7.13. Site MS2 in Fabriano: stratigraphic profile and measured shear waves velocity (m/s) obtained by down-hole measurements.

Probabilistic approach fulfils the requirements of compatibility with regional hazard (that means compatibility with seismic codes). The computed response spectrum is uniform probability ensuring equal risk to building of different characteristic (mass, stiffness and damping). However there are two weak points in this approach, the first is the arbitrariness when choosing the procedure to derive the signal from the spectrum (real or artificially generated accelerograms?) and about the number of accelerograms to use (1, 10 or more?). The second is due to the probabilistic approach itself. The uniform hazard response spectrum obtained for a given return period overestimates short periods and underestimates long periods (of the spectrum) compared to the spectrum expected from an earthquake with the same return period (Marcellini et al., 1994). This fact is unavoidable and is dependent on the β of Gutenberg-Richter relation, actually only if β is equal to 0 this mismatch will disappear, but in practice it never happens.

The stochastic approach here presented could be thought of as a compromise: it is more "physical" than the probabilistic approach, it does not require input parameters highly unpredictable as the deterministic approach, but, once again it does not fully satisfy requirements of risk experts, that is to say what is the probability of the evaluated ground motion for a given return period.

How to deal with this puzzle? There is not a full agreement among the experts. The choice depends on the goal of the study, the maximum expected magnitude and the soil type. As an example, for microzonation of a municipality in an area subjected to moderate seismicity, we found the probabilistic approach more suitable, but using the deterministic one for specific conditions such as in the presence of loose soils susceptible to liquefaction or ground failures. Deterministic approach can also be used to estimate the "upper boundary" of the expected motion. In case of strong earthquake deterministic approach becomes more important, due also to the importance of the directivity effects. To design a special industrial plant, the deterministic approach assumes also a crucial role; in particular it helps to estimate the motion in the long period range, where the probabilistic approach is not too much reliable. As a matter of fact the experience plays a significant role in the choice of the approach to adopt.

7.3. Site Effects Evaluation

The importance of site effects is very well known. Due to the fact that this argument is tackled in a number of papers of the present book we limit this paragraph to the presentation of some results of site effects analysis obtained in the framework of Fabriano microzonation (Tento et al., 2001)

In particular we focus on the results obtained with a dense seismic temporary network installed just after the quake (Figure 7.14). Both aftershocks and noise were recorded in order to assess the different soil amplification factors. Figure 7.15 shows that receiver function and Nakamura methods exhibit a good agreement and they agree also with spectral ratios as far as the fundamental period of soils is concerned (to note that, in the light of preparation of the microzonation map, we attribute greater importance to the spectral ratios).

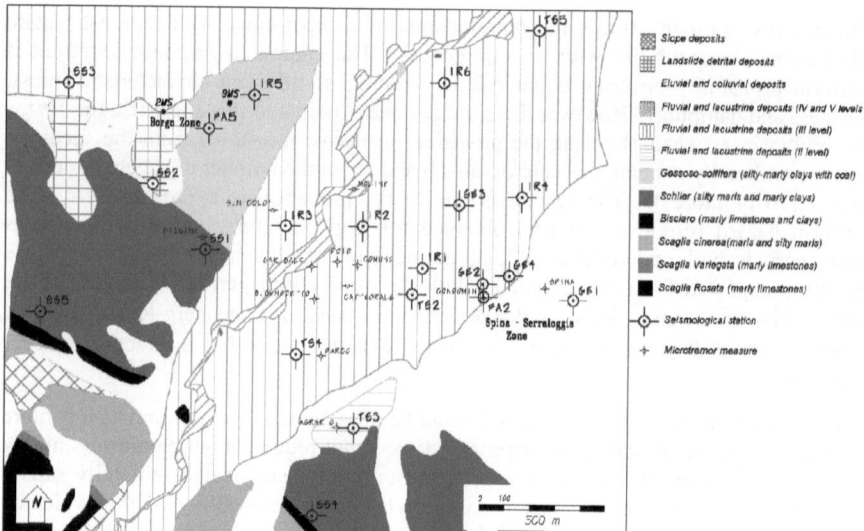

Fig. 7.14. Fabriano: geological map, locations of seismological stations and of microtremor measurements

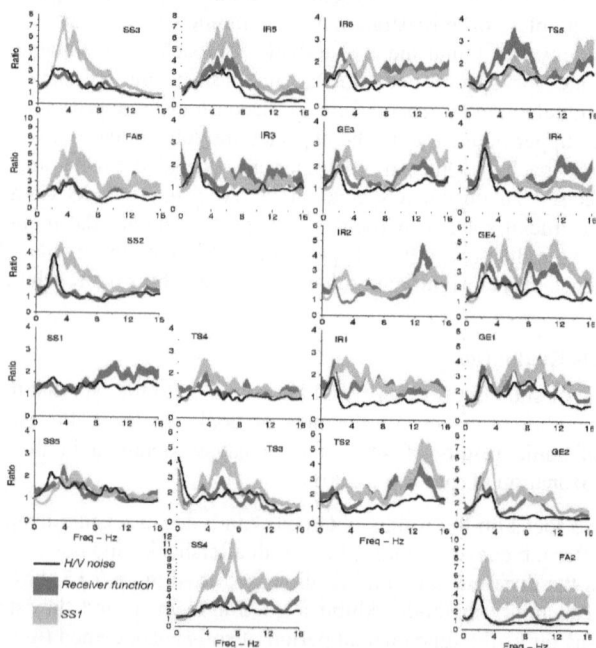

Fig. 7.15. Fabriano: Receiver Function (dark grey) and spectral ratios (light grey) evaluated using weak motion recordings and H/V ratios as obtained by noise (black line). The line thickness (dark and light grey lines) represents the 95% confidence interval.

Concerning aftershocks we remind that Fabriano is situated some 30km North with respect to epicentre area. A great concern was (before analysing the data) about the influence of the back-azimuth. Figure 7.16 presents spectral ratios computed by rotating the horizontal components in the longitudinal and transverse directions of the valley (longitudinal means approx at N45°E as represented in Figure 7.14).

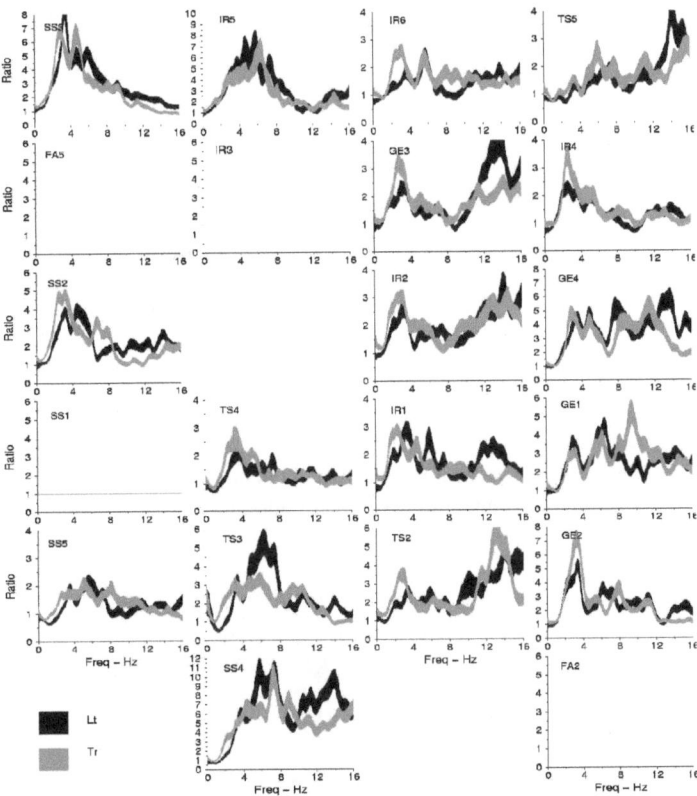

Fig. 7.16. Fabriano: Spectral ratios. Horizontal components of the recorded signals are rotated parallel (dark grey) and perpendicularly to the valley (light grey). The line thickness (dark and light grey lines) represents the 95% confidence interval.

Sometimes the crude observation of data bring to interesting results: it is the case shown in Figure 7.18 where it is represented the incidence angle at the station simply computed using the horizontal-vertical ratio of the first pulse: the comparison with figures 7.14 and 7.16 evidences a significant correlation between low incidence angle and softer materials, also characterized by higher spectral ratios.

A very crucial aspect of site effects in microzonation is the density of points to consider in the analysed area because to translate point information into areal information it is necessary to adopt some interpolation procedure. Figures 7.17 represent a map of predominant periods obtained by integrating weak motion data and microtremors measurements.

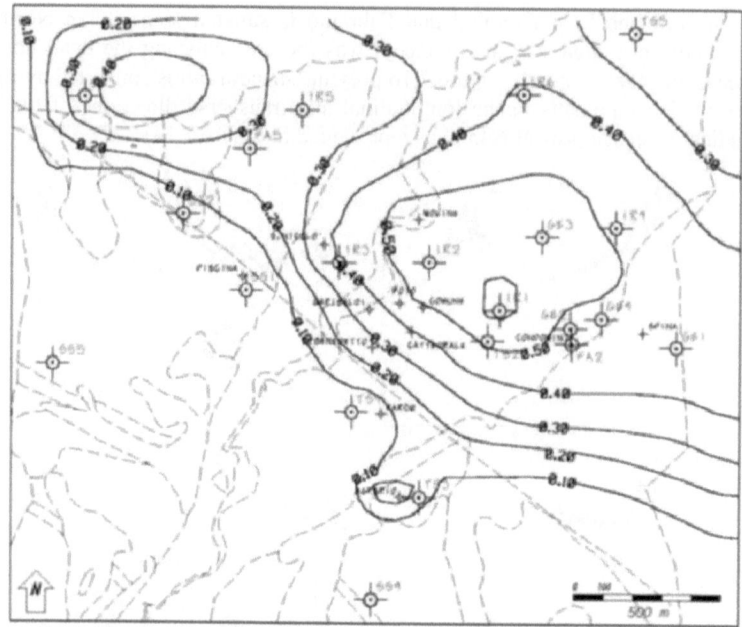

Fig. 7.17. Fabriano: soils natural period map obtained by microtremors (Nakamura method) and weak motion data

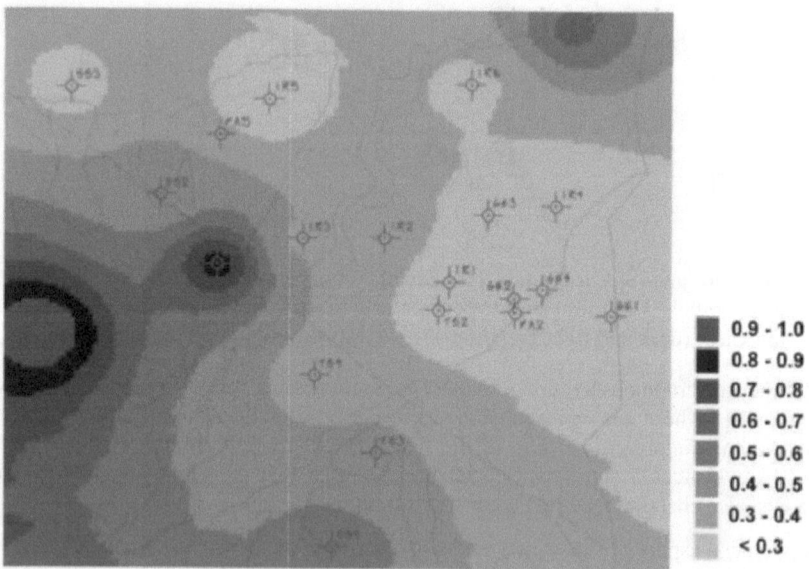

Fig. 7.18. Fabriano incidence angle map as evaluated by the analysis of the P-wave first arrival of the recorded earthquakes. Symbols represent recording stations sites as in Figures 7.14 and 7.17

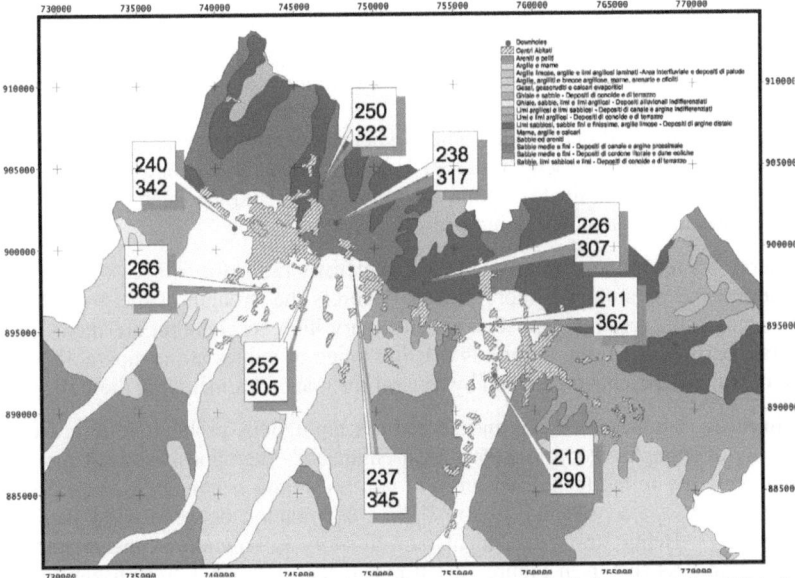

Fig. 7.19. Forlì-Cesena area: seismic hazard assessment. Each box contains PGA (gal) values computed considering and without considering local amplifications (top and bottom values in each box, respectively). The geology of the area is shown in the background.

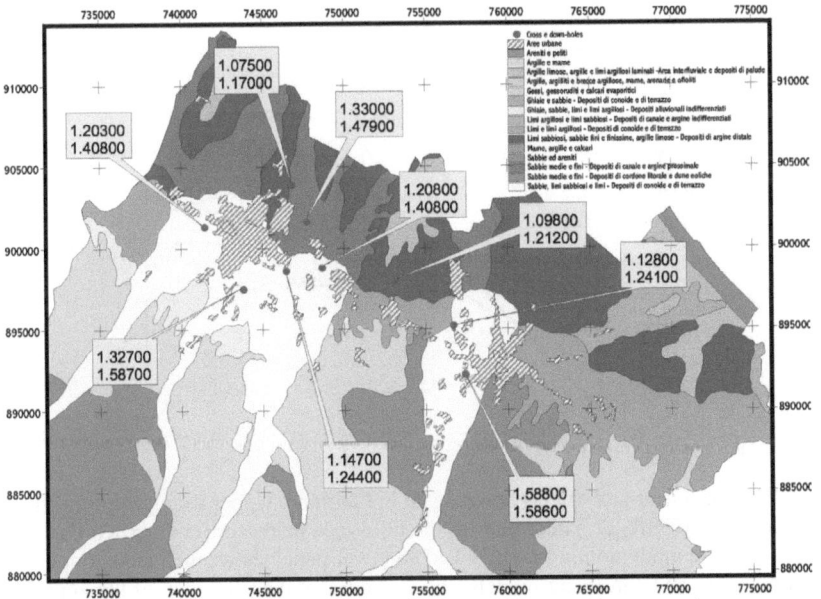

Fig. 7.20. Forlì-Cesena area: ratios of Housner intensity (SI) between values computed at each down-hole site and SI at the outcropping bedrock (top values in each box), bottom values in each box represent ratios of average spectral acceleration

7.4. Final Remarks

We want to stress again about the choice of the parameter to represent the expected ground motion. Figures 7.19 and 7.20 refer to the microzonation of the area of Forlì-Cesena (Emilia-Romagna Region). Figure 7.19 shows computed PGA values at the surface and at the bedrock for few sites (where down-hole measures were available) whereas Figure 7.20 represents ratios of Housner Intensity (SI) and ratios of average spectral acceleration (evaluated within the interval 0.1-2.5s), between values computed at the surface and at the outcropping bedrock.

In general, we recommend to adopt Housner Intensity ratios for two reasons: the first is that, an integral parameter is more likely to express the strength of the ground motion (is a common observation that several times very high acceleration are recorded after moderate earthquakes), the second is that the range of values are comparable with the range of soil coefficients prescribed by several seismic codes including the Italian one.

Non-linearity is another crucial and sometimes ambiguous problem in microzonation (generally overemphasized by geotechnical engineers - the opposite by seismologists). Without entering in the technical nature of non-linear soil behaviour already described in other papers, we would like to stress the close relation between input motion and non-linearity. Being strain dependent, non-linearity is function of the magnitude of input motion. As shown in Figure 7.21 it means that it depends from hazard analysis. In turn, hazard values depend on the choice of return periods, or in other words from the probability of non-exceedance of a given ground motion value. In conclusion the choice of a level of risk could determine if the behaviour of a given soil must be considered linear or non- linear.

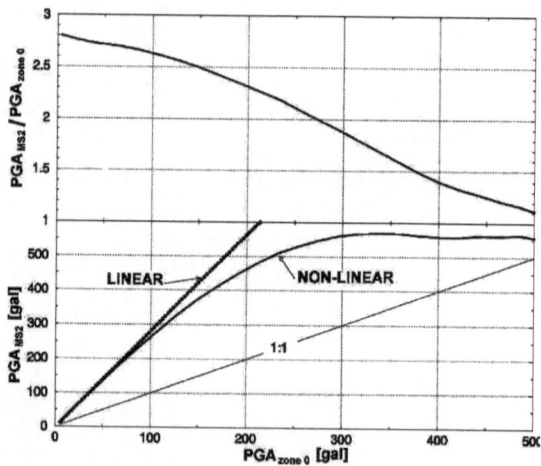

Fig. 7.21. Non-linear soil effects at MS2 site in Fabriano. Upper panel: the amplification ratio ($PGA_{MS2}/PGA_{Zone\ 0}$) decreases as the $PGA_{Zone\ 0}$ increases (in other words, as the PGA of the input motion increases). The lower panel shows the PGA computed adopting linear and non-linear models: it is evident that the behaviour of the upper panel curve depends on non-linearity. This graph well explains that the decreasing of amplification factor in terms of PGA does not imply a diminution of risk (after Tento et al., 2001; Reproduced by permission of Italian Geotechnical Journal - Patrone Editore).

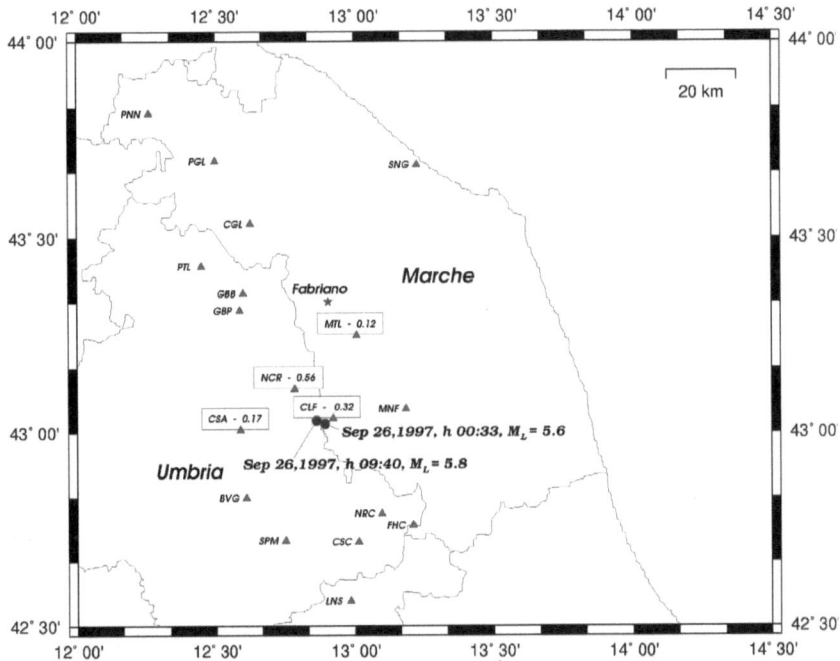

Fig. 7.22. Epicentre locations of the September 26th 1997 00.33 GMT and 09:40 GMT shocks (Amato et al., 1998). Small boxes report values of recorded PGA (g) (after Franceschina et al, 2001; Reproduced by permission of Italian Geotechnical Journal - Patrone Editore).

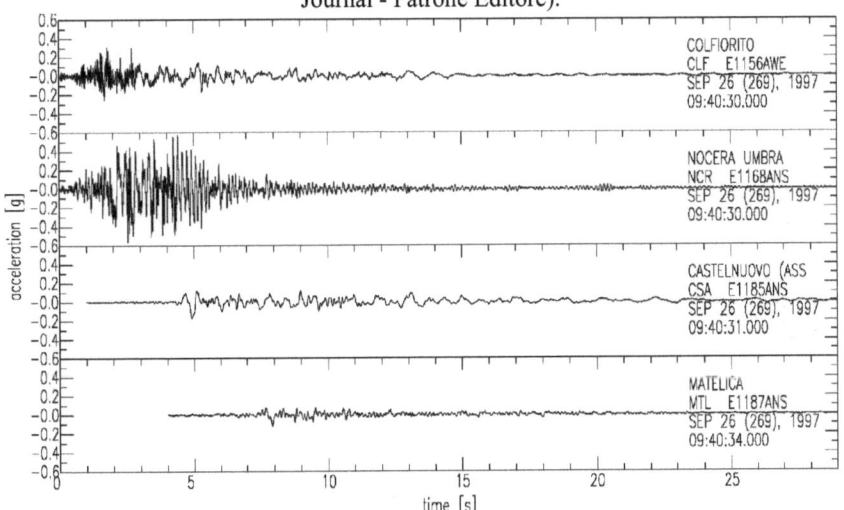

Fig. 7.23. Accelerometric recordings of the September 26th 1997, M_L=5.8 earthquake, 09:40 GMT (after Franceschina et al, 2001; Reproduced by permission of Italian Geotechnical Journal - Patrone Editore).

A final remark is devoted to the nature of ground motion. Figure 7.22 shows the accelerometric network that recorded the M=5.8 earthquake of Central Italy. It is possible to note the high spatial variability of recorded ground motion (Figure 7.23): compare Nocera and Matelica records, situated at some 20km apart. This fact underlines once again that source and directivity factors require more attention by investigators (at least the same devoted to soils).

Acknowledgements

The authors are indebted with R. Daminelli for preparation and correction of this manuscript. We would like to thank all the participants of the GNDT-Umbria-Marche microzonation project and in particular: G.L. Franceschina, E. Priolo, A. Tento

CHAPTER 8
SEISMIC MICROZONATION: A CASE STUDY

Atilla Ansal, Yeşim Biro
Boğaziçi University, Kandilli Observatory and Earthquake Research Institute
Ayfer Erken, Ümit Gülerce
Istanbul Technical University, Civil Engineering Faculty

8.1. Introduction

The earthquake damages are controlled basically by three interacting factor groups; earthquake source and path characteristics, local geological and geotechnical site conditions, structural design and construction features. Seismic microzonation can be considered as the assessment of the first two groups of factors. In general terms, it is the process for estimating the response of soil layers under earthquake excitations and thus the variation of earthquake characteristics on the ground surface. Seismic microzonation is the initial phase of earthquake risk mitigation and requires multidisciplinary approach with major contributions from geology, seismology, geotechnical and structural engineering. The final output should contain recommendations suitable for application by local administrators, urban planners and engineers.

The key issue affecting the applicability and the feasibility of any microzonation study is the usability and reliability of the parameters selected for microzonation. These parameters need to be meaningful for city planners as well as for public officials and should not lead to controversial arguments among the property owners and city administrators.

It was shown over and over again (Gazetas et al., 1990; Faccioli, 1991; Ansal, 1994; Bard, 1994; Chavez-Garcia et al., 1996; Chin-Hsiung et al., 1998; Gueguen et al., 1998; Kawase, 1998; Ansal, 1999; Athanasopoulus et al., 1999; Hartzell et al., 2001) based on the encountered earthquake damage and strong ground motion records that there are numerous source and site factors (i.e. near field effects, directivity, duration, focusing, topographical and basin effects, soil nonlinearity, etc.) that are important in assessing ground motion characteristics. The national seismic zoning maps are generally prepared in small scales such as 1:1,000,000 or less neglecting all these factors and independent of geological and geotechnical site conditions. However, seismic microzonation requires 1:5,000 or even 1:1,000 scale studies taking into consideration both earthquake source and regional geological and geotechnical site conditions in order to be used for urban and landuse planning. Thus, detailed seismological, geological and geotechnical studies are necessary to establish seismic microzonation maps.

A Seismic microzonation study consists of three stages: (1) estimation of the regional seismic hazard, (2) determination of the local geological and local geotechnical site conditions (3) assessment of the probable ground response and ground motion parameters on the ground surface.

There may be differences among the adopted procedures with respect to these three stages (Marcellini et al., 1995a, 1995b; Lachet et al., 1996; Fäh et al., 1997; Ansal, et al.,

A. Ansal (ed.), Recent Advances in Earthquake Geotechnical Engineering and Microzonation, 253–266.

2003; Ansal, 2002). These differences mostly arise from different intentions that produced microzonation maps and different levels of accuracy achieved based on the available input data in terms of local geological and geotechnical site conditions. One preference may be to produce microzonation maps to be used mainly for city and landuse planning. A second preference is to use the microzonation maps to estimate the possible earthquake characteristics for the assessment of structural vulnerability in an earthquake scenario study. A third preference may be to provide input for the earthquake design codes. The methodology presented in this chapter is more directed towards the first preference considering a microzonation case study performed in Silivri Municipality in Istanbul, Turkey, particularly relevant for regions characterized by high level of seismicity.

8.2. Regional Seismicity

The first phase of any seismic zonation study is the estimation of the regional seismic hazard based on probabilistic analyses using the available seismic and geological database (Mcguire, 1995; Frankel et al., 2000). Ground motion characteristics can be determined for a specific return period or exceedance probability depending on the purpose of the study. The sophistication of the adopted approach can vary between a single areal source models to very detailed multi-source models (e.g. Erdik, et al., 2003).

In areas with active seismicity and complex tectonic formations, it may be realistic to assume a single tectonic areal source with a fixed radius around the investigated area to determine the earthquake recurrence relationship for calculating exceedance probabilities with respect to earthquake magnitudes for the purpose of defining a probable earthquake magnitude. A simple single areal source model was adopted to estimate the regional earthquake hazard for Silivri.

The proposed seismic hazard analysis may be considered as composed of four consecutive stages that can be assumed independent and thus can be evaluated separately (Ansal and Iyisan, 1998). The first stage is the estimation of the probable earthquake magnitude based on seismological and geological data for the region. The second stage is the estimation of the source distance of the probable earthquake. The third stage is the estimation of the earthquake characteristics at the competent soil or rock outcrop based on appropriate attenuation relationships. The fourth stage is the estimation of earthquake characteristics on the ground surface based on the local geotechnical site conditions. Each of these stages involves various degrees of uncertainties; therefore probabilistic approaches were adopted at each stage to account for the variability by selecting identical exceedance probabilities at all stages to keep the overall risk level constant.

In the first stage of seismic microzonation studies for Silivri, Gutenberg-Richter recurrence relationship with a Poisson model was adopted using the available instrumental and historical seismic databases for evaluating the seismic hazard as shown in Figure 8.1(a) for a source radius of 100 km.

Earthquake records for the historical era (approximately between 496 B.C. and 1916) with intensity $I_o > V$, for Silivri region were compiled based on available earthquake catalogues (Ergin et al., 1967; Sipahioğlu, 1984; Ambraseys and Finkel, 1995; Papazachos et al., 1997). Since the records for this period are in terms of intensities, the relation developed for Turkey by Ansal (1997);

$$M = 0.594\,I_o + 1.36 \tag{8.1}$$

was used to convert the intensities (I_o) to magnitudes (M_s). The seismic hazard analyses were performed in terms of these calculated magnitudes. Due to the nature of the historical earthquake records, it would be more reliable to base the seismic hazard analyses on medium strong and strong earthquakes. Thus, only earthquakes with intensity $I_o \geq VI$ have been used in the analyses.

The earthquake records for the instrumental era (approximately between 1904 and 2002) were compiled based on available earthquake catalogues (BUKOERI, 2002; Ergin et al., 1967, 1971; Gençoğlu et al., 1990; Güçlü et al., 1986).

One shortcoming with the use of instrumental records is the limited time interval for the compiled data that would not represent the tectonic regime going on for millions of years. The other weakness is the non-uniform statistical distribution of earthquakes with respect to their magnitudes and the presence of relatively large number of small events that affects the selected probability distribution function. On the other hand, historical data compiled for a longer time interval may not be very accurate with respect to epicentre locations, dates, and intensities.

It would be more reliable to utilise the earthquake records from historical and instrumental era together to perform seismic hazard analyses. A weighted averaging procedure was adopted by assigning weights of 40% and 60% for historical and instrumental records, respectively. Thus, the magnitude of a probable earthquake corresponding to the average exceedance probability of 10% for a period of 50 years was estimated as M=6.7, as shown in Figure 8.1b.

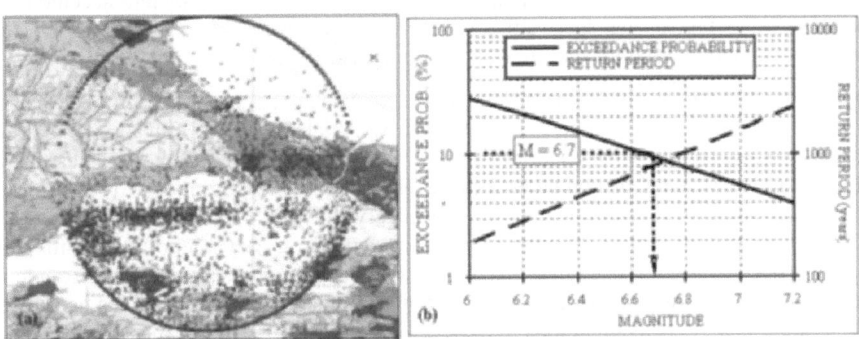

Fig. 8.1. Seismic activity within 100 km of Silivri and evaluation of seismic hazard for based on single areal source model

The next stage is the estimation of the possible epicentre locations for this probable earthquake. A statistical analysis was conducted for this purpose by assuming that all epicentres of past earthquakes are possible epicentres for future earthquakes (Ambraseys et al., 1996; Kafka and Walcott, 1998). The statistical distribution of epicentre distances of instrumentally recorded earthquakes with magnitude M>2 around Silivri was determined as shown in Figure 8.2(a).

One way to obtain the statistical distributions for probability analysis is to use the probability density functions. The Beta probability density function was observed to give the best fit for the distribution of epicentre distances. Another way is to define the observed distribution of data as a discrete statistical density distribution for evaluating the probabilities. Figure 8.2(b) shows both methods and demonstrates the similarity in both procedures. By this approach, the epicentre distance for M=6.7 earthquake corresponding to 10% exceedance probability around Silivri is determined as 24 km, with a 10% probability of being less.

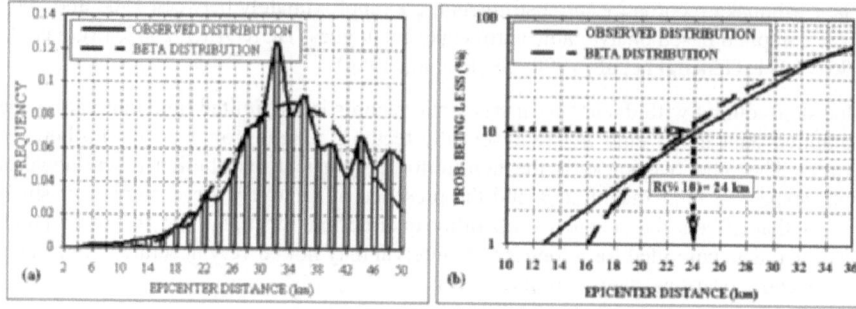

Fig. 8.2. Estimation of the probable epicentre distance

The third step is the estimation of the earthquake characteristics at the bedrock or competent soil outcrop using a suitable attenuation relationship in accordance with the geological and tectonic features of the region. One option that may be suitable for microzonation studies is to use a contemporary attenuation relationship to estimate the spectral accelerations at the competent soil outcrop. Once the probable acceleration spectrum is obtained then spectrum compatible outcrop acceleration time histories can be calculated to be used for the site response analyses.

In the case of Silivri study, the acceleration response spectrum was determined using the spectral attenuation relationship proposed by Abrahamson and Silva (1997). The simulated acceleration time histories were calculated based on the method suggested by Deodatis (1996) using the computer code developed by Papageorgiou et al. (2000). The target acceleration spectra compatible time history (TARSCTHS -Papageorgiou et. al, 2000) is a ground motion simulation program generating a synthetic time history of ground accelerations. The time domain simulations are non-stationary with random phases. Six spectrum compatible acceleration time histories were calculated as shown in Figure 8.3. The simulations were performed taking into account the possible earthquake magnitude and (a) using the estimated epicentre distance as a parameter for duration control for the first three and (b) by specifying the duration as 30 seconds for the remaining three records.

The acceleration response spectra of the simulated acceleration time histories are in very good agreement with the target spectra for the high period range but with some discrepancies in the low period range as shown in Figure 8.4. However, since this low period or high frequency range of the acceleration time histories would not produce significant effects on the site response analyses, the calculated acceleration time histories were considered suitable.

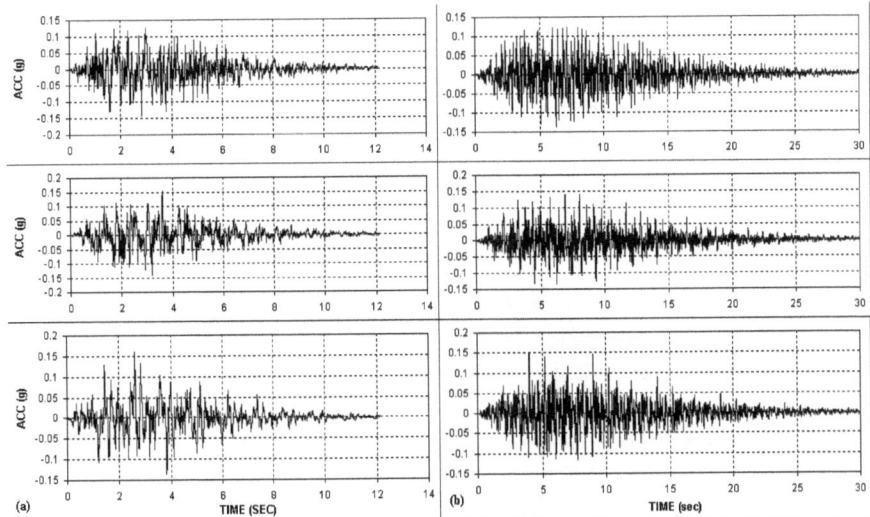

Fig. 8.3. Six simulated acceleration records calculated for site response analyses

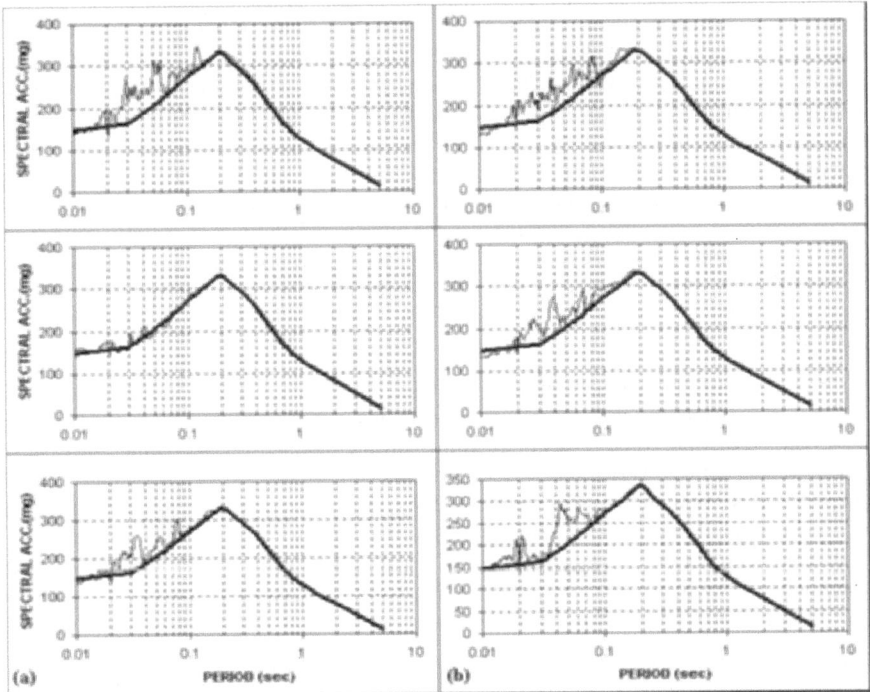

Fig. 8.4. Comparisons of the acceleration response spectra of the six simulated spectrum compatible acceleration time histories with the acceleration spectra calculated by Abrahamson and Silva (1997) spectral attenuation relationship

8.3. Geological and Geotechnical Site Conditions

The second stage in seismic microzonation would start with the assessment of the local geological formations and with the mapping of the surface geology based on available information, site surveys, site investigations, and soil explorations. The purpose is to determine the boundaries and the characteristics of the geological formation and to prepare a geology map at a scale of 1:5000 or larger as shown for Silivri in Figure 8.5.

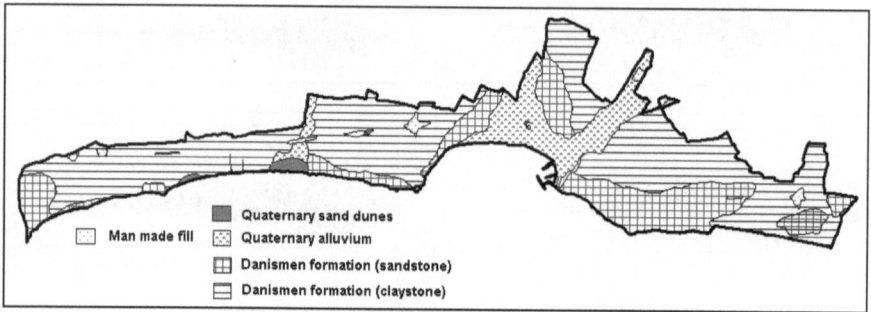

Fig. 8.5. Geological formations in Silivri

This map clearly indicates the geological formations and their variation in Silivri. However, it is important, as pointed out by Willis et al., (2000), to base the site classification on measured characteristics of geologic units taking into consideration the possible variations in each unit. The deviations from the mean values obtained for each geological unit may exceed the permissible limits to justify its use for assessing the effects of local soil conditions. It should be noted that the studies in Silivri also demonstrated that the existing geological units are not homogenous and significant changes in their properties could be observed from one point to another, even in the same formation. Therefore, considering the geological units as the only criteria in seismic microzonation is not appropriate. The major purpose of the geology map may be regarded as the basic information to plan the detailed site investigations and to control the reliability of the results obtained by site characterisations and site response analyses.

Wills and Silva (1998) suggested using average shear wave velocity in the upper 30 m as one parameter to characterise the geological units while also admitting the importance of other factors such as impedance contrast, 3-dimensional basin and topographical effects, and source effects such as rupture directivity on ground motion characteristics. In their compiled database, they have encountered significant variations in the equivalent shear wave velocities especially in the case of alluvium deposits. Wills and Silva (1998) suggested using shear wave velocity for classifying site conditions rather than geological units, even though the determination of shear wave velocities requires extensive field investigations.

During site characterisation it is necessary to determine the variations in soil stratification and engineering properties of soil and rock layers encountered at the site preferably based on in-situ tests and laboratory tests conducted on the samples obtained during the soil exploration. The purpose is to determine the site conditions as accurate as possible in order to realistically assess the site response characteristics.

The site characterisation in Silivri was conducted by adopting a grid system. The investigated area was divided into cells of 500×500 meters. A representative hypotetical soil profile was determined for each cell based on all available borehole data, in-situ and laboratory test results. The purpose is to estimate the effects of site conditions at a scale of 1:5000 by assigning these representative boreholes at the centre of each cell, as shown for Silivri in Figure 8.6.

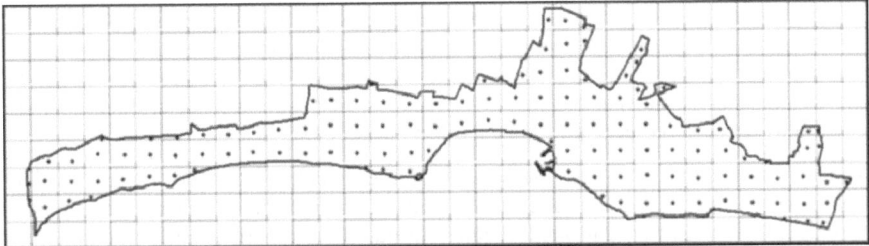

Fig. 8.6. Site characterisation by a grid system for Silivri

There are basically two reasons behind this approach (1) to utilise all the available data in each cell in order to have more comprehensive and reliable information about the soil profile; (2) to eliminate the effects of different distances among boreholes or site investigation points during GIS mapping.

The results obtained were mapped using GIS techniques by applying linear interpolation among the grid points, thus enabling a smooth transition of the selected parameters. Soft transition boundaries were preferred to show the variation of the mapped parameters. Better defined clear boundaries were not used and are not recommended due to the accuracy of the study and in addition to allow some flexibility to the urban planners and to avoid misinterpretation by the end users that may consider the clear boundaries as accurate estimations of the different zones.

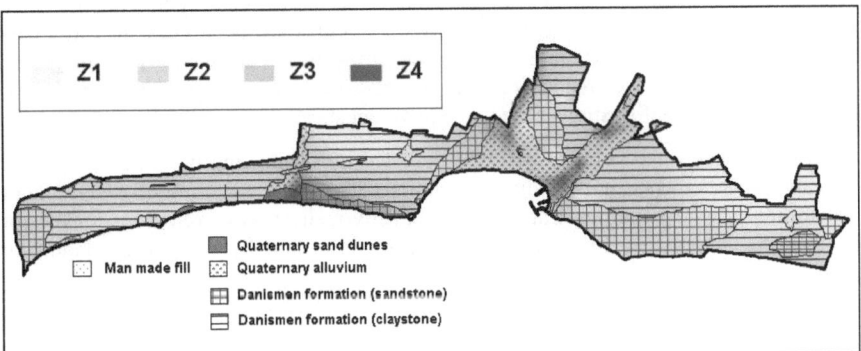

Fig. 8.7. Zonation with respect to site classification defined in Turkish Earthquake Code (1997) in comparison to the surface geology for Silivri

The objective of site characterisation is the determination of each representative soil profile estimated for each cell with respect to a site classification criterion. Borehole samples and geological studies were used to classify the site conditions in Silivri according to the Turkish Earthquake Code (1997). The zonation map with respect to the site classes specified in the Turkish Earthquake Code (1997) is shown in Figure 8.7. There appears to be a reasonable agreement with respect to surface geology. Softer sites Z4 and Z3 were mostly in the areas that were classified as Quaternary sand dunes and Quaternary alluvium deposits while Z1 and Z2 are in areas that were classified as claystone and sandstone formations.

It is important to note that these results are based on representative site profiles that were estimated based on the available data and the accuracy of this site classification map may not be suitable for assessing the local site conditions to be used for structural design purposes. Site specific soil investigations are required to determine local site conditions for each building area to be used in the structural design.

The approach adopted in the assessment of the calculated zonation maps involves division of the area into three zones as (A, B, and C). Since the site characterisations, as well as all the analyses performed, require various approximations and some important assumptions, it was preferred not to present the numerical values for any parameter. In all cases, the variations of each calculated parameter are considered separately and their frequency distributions were determined for the whole microzonation study region. Thus the zone *A* shows the most favourable (in terms of earthquake hazard) 33% percentile (e.g. higher equivalent shear wave velocities, lower spectral amplifications and lower spectral accelerations), zone *B* the medium 34% percentile and zone *C* shows the most unsuitable 33% percentile (e.g. lower equivalent shear wave velocities, higher spectral amplifications and higher spectral accelerations). In this way it was possible to define three zones with different earthquake hazard levels in a relative way enabling the urban planners and municipal officials to have the flexibility in deciding the appropriate planning alternatives taking into consideration all other relevant factors besides earthquake hazard.

One popular criterion for site characterisation is to use equivalent (average) shear wave velocity defined as the weighted average of shear wave velocities of soil and rock layers in the top 30 meters. Equivalent shear wave velocities are being used in earthquake codes for the purpose of evaluating the design earthquake characteristics on the ground surface (Borchert, 1994). In addition, it is also possible to use empirical relationships to estimate spectral amplifications based on equivalent shear wave velocity.

Equivalent shear wave velocity can be calculated by conducting in situ seismic wave velocity measurements or by using correlations developed in terms of *SPT*-standard penetration or *CPT*-cone penetration tests. In Silivri, shear wave velocities for the soil layers were determined from seismic refraction tests and by using the relationship proposed by Iyisan (1996);

$$V_S = 51.5 \, N^{0.516} \tag{8.2}$$

where *N* is the SPT blow counts for 30 cm and V_S is the shear wave velocity in m/sec. This relationship is valid for all soil types for estimating shear wave velocities from SPT tests. There are large numbers of similar correlations in the literature. The Iyisan relationship was used because it was developed based on local data sets under similar site conditions.

The zonation with respect to the equivalent shear wave velocity in Silivri is shown in

Figure 8.8 in terms of three percentile groups in comparison with the surface geology. Zones determined as *A* are the areas with the higher 33% percentile of equivalent shear wave velocity and zones determined as *C* are the areas with the lower 33% percentile of equivalent shear wave velocity. The zonation with respect to equivalent shear wave velocity is in agreement with the geology where zone *C*, representing relatively higher earthquake hazard zones coincides mostly with the Quaternary alluvium deposits while zone *A* representing the upper 33% percentile of the equivalent shear wave velocity concides with the Danismen (sandstone & claystone) formations.

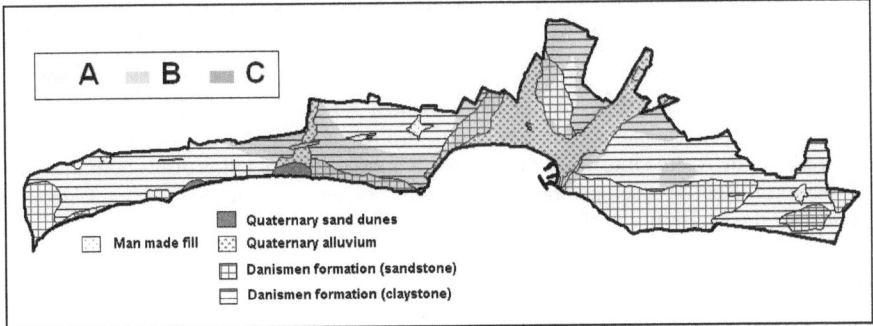

Fig. 8.8. Zonation with respect to equivalent shear wave velocity in comparison to surface geology for Silivri

8.4. Earthquake Characteristics on the Ground Surface

One option to estimate the earthquake characteristics on the ground surface is to use an empirical relationship in terms of equivalent shear wave velocities. In the case of Silivri microzonation study, the peak spectral amplifications based on equivalent shear wave velocity were calculated using the relationship given by Midorikawa (1987);

$$A_k = 68 \ V_S^{-0.6} \tag{8.3}$$

where A_k is the spectral amplification and V_S is the equivalent shear wave velocity in m/sec.

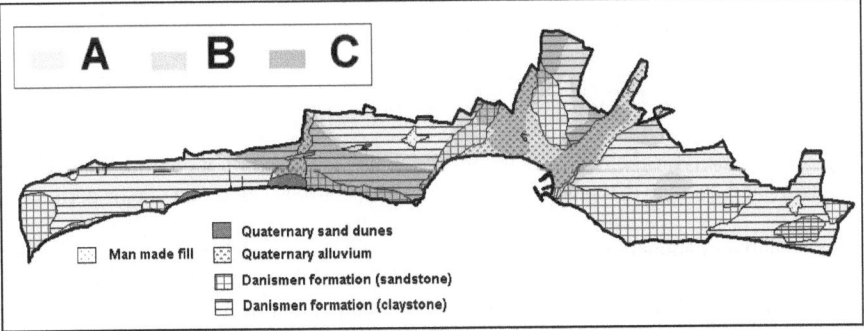

Fig. 8.9. Zonation with respect spectral amplifications calculated from equivalent shear wave velocities in comparison with the surface gology for Silivri

The peak spectral amplifications calculated based on Equation (8.3) were evaluated and the zonation with respect to peak spectral amplifications were also mapped based on three percentile groups, as shown in Figure 8.9 where C shows the areas with relatively higher spectral amplification and thus higher level of earthquake hazard.

The second and more comprehensive assessment of earthquake characteristics on the ground surface can be achieved by conducting site response analyses for all representative soil profiles. In the case of the Silivri study, site response analyses were conducted for the spectra compatible six simulated acceleration time histories determined based on the regional earthquake hazard study using the site response analysis code proposed by Schnabel et al. (1972) and later modified by Idriss and Sun (1992).

The variation of spectral accelerations calculated for six input motion as well as the average of all 6 response spectra are shown in Figure 8.10. As can be observed from Figure 8.10, there are significant differences among different soil profiles both in terms of the ordinates of the spectral accelerations as well as the predominant periods indicating the variability of the site conditions.

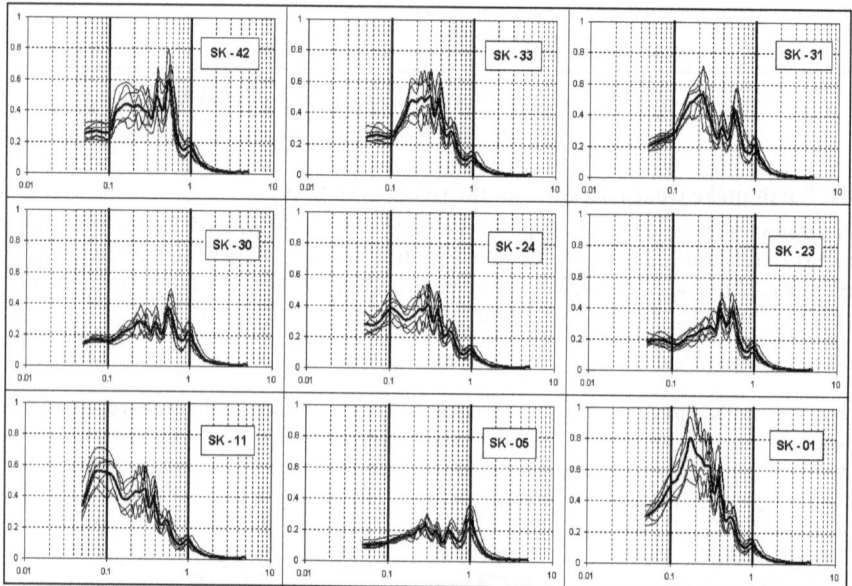

Fig. 8.10. Some typical elastic acceleration response spectra for 6 input motions and the calculated average that are plotted in terms of spectral acceleration versus period

Representative parameters reflecting the calculated site amplification characteristics could be peak spectral accelerations or average spectral accelerations calculated between the periods of 0.1-1.0 second. It was assumed that peak spectral accelerations could be used as a zonation criterion since a linear correlation was observed between these two parameters with relatively high regression coefficient as shown in Figure 8.11.

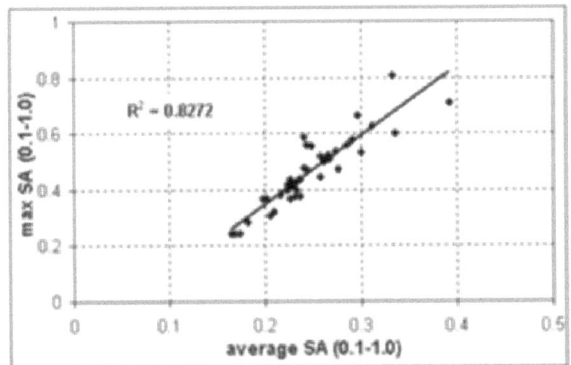

Fig. 8.11. Correlation between the calculated peak spectral accelerations and average spectral accelerations between 0.1-1.0 sec. periods

The average of the six peak spectral accelerations was used as the representative parameter for microzonation that was performed with respect to three percentile groups where *A* shows the lower 33% percentile and *C* shows the higher 33% percentile in terms of calculated spectral accelerations in comparison with the surface geology as shown in Figure 8.12. The zonation with respect to spectral accelerations is in general agreement with the surface geology where *C* shows the areas with relatively higher earthquake hazard. These areas mostly coincide with the Quaternary deposits however, as can be observed from Figure 8.12, there are also zones classified as *C* in regions that were defined as claystone formations in the surface geology map. These were the areas where the claystone formation shows different characteristic mostly due to weathering.

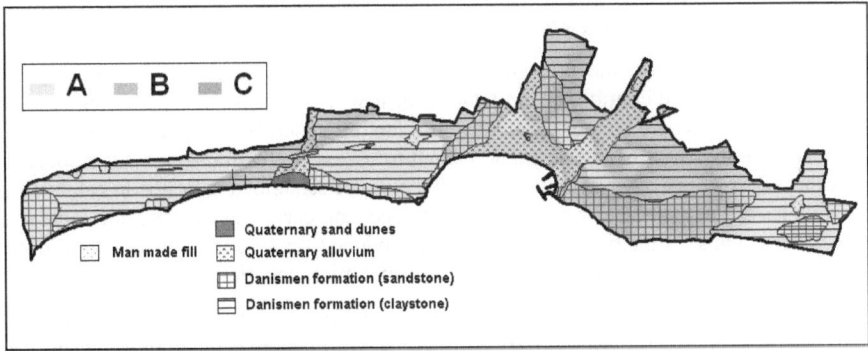

Fig. 8.12. Zonation with respect to peak spectral accelerations calculated by site response analyses in comparison with the geology in Silivri

Microtremor records are being widely used in the recent seismic microzonation studies. Microtremors are very low amplitude vibrations, which may only be measured by very sensitive seismometers. The methodology is relatively easy and economically feasible that enables quick assessment of predominant soil periods and approximate estimates of spectral amplifications (Nakamura, 1989).

Even though it is generally accepted that H/V ratios obtained from microtremor records would not lead to very reliable spectral amplification values (Bard, 1999, Lachet and Bard, 1994) they can still be taken into consideration when finalising the microzonation with respect to site amplification. Therefore, the results obtained from the microtremor study were utilised to map the variation of spectral amplifications for Silivri as shown in Figure 8.13.

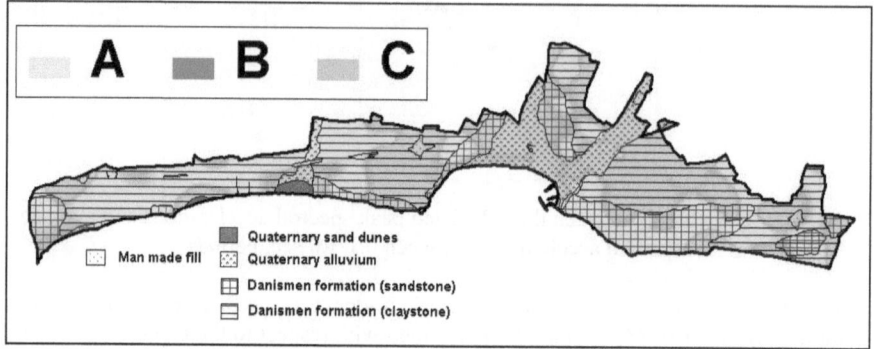

Fig. 8.13. Variation of spectral amplifications obtained from microtremor records in comparison with surface geology in Silivri

8.5. Seismic Microzonation with Respect to Ground Shaking

The basic intention of the site response analysis is to estimate the effect of local site conditions in assessing the site amplification with respect to ground shaking. It would be logical to base this decision on all the available results obtained from site characterisation based on equivalent shear wave velocity, site response analysis as well as from microtremor measurements conducted in the region.

As briefly described in the preceding sections, many parameters such as geological formations, local soil conditions, equivalent shear wave velocity, spectral amplification and their variations are among the controlling factors in seismic microzonation studies. A consistent approach has to be implemented to assess each parameter with respect to all other parameters. However, the main objective of seismic zonation is to establish a seismic hazard map in a large scale taking into account, earthquake source and local site conditions. Thus, estimation of the earthquake induced forces and their variation in the investigated area must be the main target in seismic microzonation.

Even though seismic microzonation contains important information for city and regional planning, considering different structures with different functions, site specific studies need to be performed at each site during the design stage to evaluate the local site conditions. A classification in terms of priority levels can be used to resolve this restriction and site characteristics determined from seismic microzonation studies may be used in the design process of the selected structures, e.g. one or two storey residential buildings. On the other hand, site specific studies, including in-situ and laboratory tests, must be obligatory in the assessment of the required parameters for the structures with higher importance levels.

The final seismic microzonation map with respect to ground shaking was obtained by evaluating all the available data based on the grid system given in Figure 8.6 where spectral amplification is estimated for each grid point based on the results obtained from equivalent shear wave velocity, site response analysis and microtremor H/V ratio taking into consideration the representative soil profiles for each cell. The final seismic microzonation map with respect to ground shaking is shown in Figure 8.13 in comparison with the surface geology. In general there is good correlation with respect to alluvial deposits but there are also significant differences for the sandstone and claystone formations as expected. The advantage of this assessment is the capability to consider the variations of the calculated spectral accelerations with respect to three different methodologies: equivalent shear wave velocity, site response analysis, and microtremor H/V ratio. Thus, it can be considered partly independent of the approach adopted to determine the site amplification characteristics.

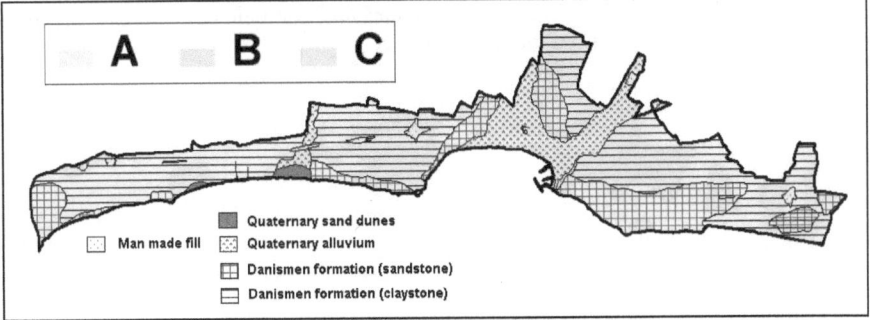

Fig. 8.14. Proposed seismic microzonation with respect to ground shaking in comparison to surface geology for Silivri

The interpretation of the seismic microzonation map with respect to ground shaking involves three zones that are classified relatively according to the expected ground motion characteristics taking into consideration source and site factors. For the purpose of urban planning areas classified as zone *A* are the areas that will be less affected with lower earthquake hazard. These are the areas that are most suitable for the critical structures and higher population densities. On the contrary areas classified as zone *C* are the areas that will be most adversely affected during possible earthquakes in the future. However, it is important to point out that this classification was conducted for the three zones that were defined relatively with respect to each other. Thus considering only the Silivri municipality, zones classified as *C* are the areas that would experience more severe ground shaking in comparison to the other parts of Silivri. The relative level of ground shaking should not be interpreted as absolute level and areas that are classified as *C* should not be regarded as areas that are unsuitable for urban settlements.

8.6. Conclusions

The major purpose of seismic microzonation is to offer urban planners and city officials some guidelines to mitigate earthquake risk. For this purpose, a seismic microzonation study was conducted for the Silivri Municipality in Istanbul. Assuming that the main factors affecting the earthquake characteristics on the ground surface are source and site factors, regional seismic hazard study was conducted at a regional scale adopting a single areal source model. The earthquake characteristics on the ground surface was determined based on the evaluation of site conditions with respect to equivalent shear wave velocity and with respect to site response analyses conducted for each representative soil profile. In determining the seismic microzonation for ground shaking, all available results were utilised by adopting a grid system to estimate the variation of spectral amplification in the investigated region, namely in the town of Silivri.

Acknowledgements

The authors would like to express their gratitude to Prof. Haluk Eyidoğan and Arif Cemal Seçkin, who made significant contributions to this study. It would not have been possible to conduct such comprehensive assessment without their input and support. The authors would also like to express their thanks and gratitude to the mayor of Silivri who has supported and financed the whole project.

CHAPTER 9
DYNAMIC ANALYSIS OF SOLID WASTE LANDFILLS AND LINING SYSTEMS

Pedro Simão Sêco e Pinto
National Laboratory of Civil Engineering (LNEC), University of Coimbra and New University of Lisbon, Portugal

9.1. Introduction

The need of construction of high solid waste landfill in order to protect human health and the environment has created new engineering challenges. In this framework the seismic behaviour of solid waste landfills has deserved considerable attention. This behaviour has been analysed by experimental methods or mathematical methods. The characterization of material properties for seismic design is a difficult task as due the heterogeneity of the material large samples are needed.

The seismic design of solid waste landfill follows the same procedures for the design of embankment dams and so both pseudo-static and deformational analysis methods are used. However it is important to assess the behaviour of the geosynthetic elements used in the cover systems and bottom lining systems to the seismic induced permanent displacements. The monitoring tasks and the safety control of landfills are analysed. Risk analysis and safety are addressed.

9.2. Performance of Solid Waste Landfills during Earthquakes

From the lessons learned from past earthquakes, such as Loma Prieta earthquake (Johnson et al., 1991; Buranek and Prasad, 1991; Sharma and Goyal, 1991) and Northridge earthquake (Matasovic et al., 1995; Stewart et al., 1994; Augello et al., 1995) it is important to stress that modern solid waste landfills withstand the design earthquake without damages to human health and environment.

From well documented case histories the following failure mechanisms can be selected:

- Sliding or shear distortion of landfill or foundation or both;
- Landfill settlement;
- Transverse and longitudinal cracks of cover soils;
- Cracking of the landfill slopes;
- Damage to the gas system header pipes;
- Tears in the geomembrane liners;
- Disruption of the landfill by major fault movement in foundation;
- Differential tectonic ground movements;
- Cracks about the contact between refuse landfill and canyon;
- Liquefaction of landfill or foundation.

The damage modes listed are not necessarily independent of each other.

A. Ansal (ed.), *Recent Advances in Earthquake Geotechnical Engineering and Microzonation*, 267–284.
© 2004 *Kluwer Academic Publishers*.

Experience has shown that well built waste landfills can withstand moderate shaking peak accelerations up to at least 0.2g with no harmful effects. Nevertheless this scenario, the integrity of solid waste landfills during strong earthquakes to achieve environmental and a public health objective deserves more consideration.

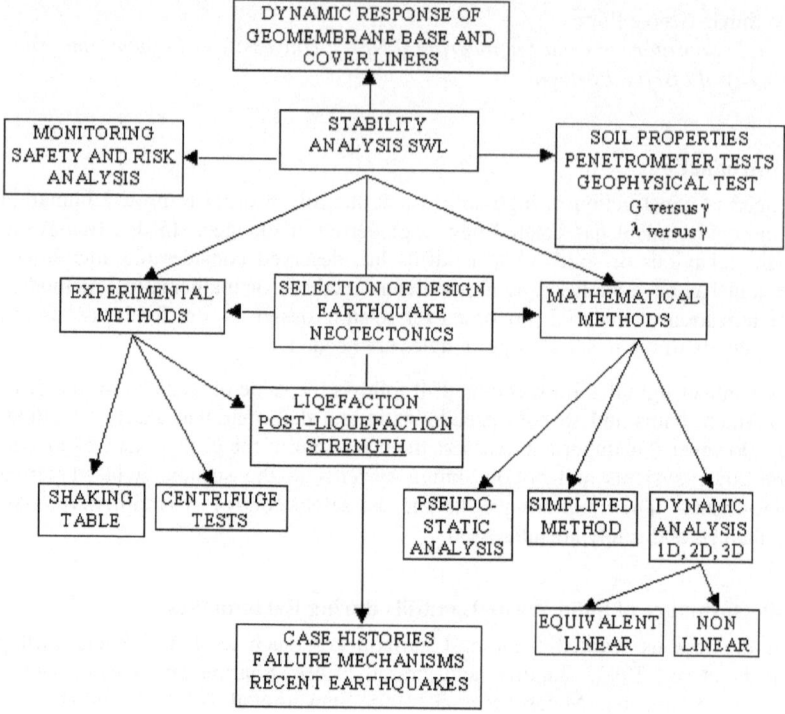

Fig. 9.1. Flowchart for solid waste landfills

9.3. Analysis of Solid Waste Landfills Stability during Earthquakes

9.3.1. INTRODUCTION

The stability analysis of solid waste landfills can be established by the flow chart presented in Figure 9.1.

The behaviour of solid waste landfills during the occurrence of earthquakes can be analysed by experimental methods or mathematical methods. Seismic design of solid waste landfills uses the same principles of seismic design of embankment dams (Sêco e Pinto, 1993). The capabilities and limitations of these methods are briefly summarised.

9.3.2. EXPERIMENTAL METHODS

Experimental methods are used to test predictive theories and to verify mathematical models. The most popular techniques for solid waste landfills are shaking table and centrifuge models.

Yegian et al. (1995) by conducting shaking table tests have concluded that: (i) the geosynthetic interface reduces the level of the acceleration pulses of the ground motion; and (ii) the geosynthetic interface acts as base isolator absorbing the wave energy through interface slip.

Centrifuge model tests have been carried out to understand the principle of waste-structure interaction and to investigate deformation induced stress redistribution within the waste body near a structure (Kockel et al., 1997)

9.3.3. MATHEMATICAL METHODS

The following dynamic analysis of embankment dams is used (Sêco e Pinto et al., 1995):
1. pseudo-static analyses;
2. simplified procedures to assess deformations;
3. dynamic analysis.

The slope stability of waste landfills is generally evaluated by limit equilibrium slope stability analyses. For the pseudostatic analyses a seismic coefficient value equivalent to the peak ground acceleration divided by 1.5 can be considered (Sêco e Pinto et al., 1998). For solid waste landfills an acceptable seismic behaviour is anticipated if the calculated pseudo-static factor of safety ranges from 1.3 to 1.5.

Simplified procedures to assess landfills deformations were proposed by Newmark (1965), Sarma (1975) and Makdisi and Seed (1977) and have given reasonable answers in areas of low to medium seismicity.

Newmark's original sliding block model considering only the longitudinal component was extended to include the lateral and vertical components of earthquake motion by Elms (2000).

The use of dynamic pore pressure coefficients along with limit equilibrium and sliding block approaches for assessment of stability of earth structures during earthquakes was demonstrated by Sarma and Chowdhury (1996).

Several finite element computer programs assuming an equivalent linear model in total stress have been developed for 1D (Schnabel et al., 1972; Idriss and Sun, 1992), 2D (Lysmer et al., 1974) and pseudo 3D (Lysmer et al., 1975). Since these models are essentially elastic the permanent deformations cannot be computed by this type of analysis and are estimated from static and seismic stresses with the aid of strain data from laboratory tests (cyclic triaxial tests or cyclic simple shear tests).

To overcome these limitations, nonlinear hysteretic models with pore water pressure generation and dissipation have been developed using incremental elastic or plasticity theory. The incremental elastic models have assumed a nonlinear and hysteretic behaviour for soil and the unloading-reloading has been modelled using the Masing criterion and incorporate the effect of both transient and residual pore-water pressures generated by seismic loading (Finn, 1987).

Sargent (1990) has introduced the concepts of verification and validation and the relations established between the three entities: the physical problem; the conceptual model; and the computer model and its numerical implementation are illustrated in Figure 9.2.

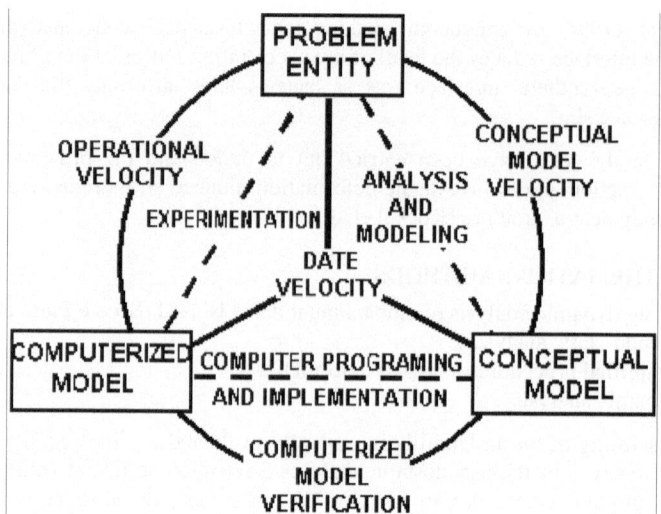

Fig. 9.2. Relations between physical problem, conceptual model and computer model
(after Sargent, 1990)

Verification intends to ensure that the computer program is correct and its represents
faithfully the conceptual model and validation applies essentially to the conceptual
model, and its ability to reproduce satisfactorily the physical phenomena.

A slightly different terminology is adopted by ICOLD (1993) that considers that the
numerical modelling process for dams should be checked in order to avoid unreliable
results considering the following aspects:

1. justification of the whole modelling method (the relevance to physical reality);
2. validation of the computer code;
3. quality assurance of the whole computation process.

9.3.4. SELECTION OF DESIGN EARTHQUAKES

The Code of Federal Regulations (US-CFR, 1991) requires for the new municipal solid
waste landfills to be designed for a maximum horizontal acceleration with a 10 percent
probability of exceedance (90 percent probability of non exceedance) in 50 year and
250 year of exposure periods. The related return periods are 475 years and 2375 years,
respectively.

It is important to refer that EC8 (1995) recommends for the seismic design of buildings
and bridges a return period of 475 years.

The selection of seismic design parameters for municipal solid waste landfill following
the dam projects depends on the geologic and tectonic conditions at and in the vicinity
of the site. The regional geologic study area should cover a 100km radius around the
site to include any major fault or specific attenuation laws.

The probabilistic approach quantifies numerically the contributions to seismic motion,
at the landfill site, of all sources and magnitudes larger than 4 or 5 Richter scale and
includes the maximum magnitude on each source.

The landfill should be designed for Operating Basis Earthquake (OBE) and Maximum Design Earthquake (MDE). Both depend on the level of seismic activity, which is displayed at each fault or tectonic province. For the OBE only minor damage is acceptable and is determined by using probabilistic procedures.

For the MDE only deterministic approach was used (ICOLD, 1983) but presently it is possible to use a deterministic and probabilistic approach. If the deterministic procedure is used, the return period of such an event is ignored, if the probabilistic approach is used a very long period is taken (ICOLD, 1989).

Neotectonics

The tectonic conditions should include tectonic mechanisms, location and description of faults (normal, strike and reverse), and estimation of fault activity (average slip rate, slip per event, time interval between large earthquake, length, directivity effects, etc), these factors are important to assess the involved risk.

Determination of neotectonic activity implies first the qualitative geomorphologic analysis of air photos and topographic maps. The GPS system is another powerful means of monitoring the crustal mobility. Cluff et al. (1982) have proposed the following classification for slip rates: extremely low to low for 0.001 mm/year to 0.01 mm/year, medium to high 0.1 mm/year to 1 mm/year and very high to extremely high 10 mm/year to 100 mm/year.

The most dangerous manifestation concerning the landfill stability and integrity is the surface fault breaking, intersecting the landfill site. The current practice is the deterministic approach in which the seismic evaluation parameters were ascertained by identifying the critical active faults, which show evidence of movements in Quaternary time.

Following (ICOLD, 1989) an active fault is a fault, reasonably identified and located, known to have produced historical fault movements or showing geologic evidence of Holocene (11000 years) displacements and which, because of its present tectonic sitting, can undergo movements during the anticipated life of man-made structures.

To assess if there is the potential for a significant amount of surface displacement beneath the dam several backhoe trenches are excavated with 3 to 4 meters deep and 30 to 50 meters long and should be inspected and log the exposures geologic features.

Recently a fault investigation method other than trenching has been developed, called the long Geo-slicer method in which long iron sheet piles with a flat U-shaped cross section are driven into an unconsolidated bed, iron plate shutters are inserted to face these iron sheet piles and the piles and shutters are pulled out to take undisturbed samples of strata of a certain width. This method is advantageous in regard to the ease of securing land for conducting investigations compared with trenching and the ease of bringing the strata samples back to the laboratory for detailed observations (Tamura et al., 2000).

When active faults are covered with alluvium geophysical explorations such as seismic reflection method, sonic prospecting, electric prospecting, electromagnetic prospecting, gravity prospecting and radioactive prospecting can be used (Takahashi et al., 1997). Of these the seismic reflection method can locate faults if geological conditions are favourable, and confirm the accumulation of fault displacements based on the amount of displacements in strata that increases with strata age.

Attenuation relations

Attenuation relations can be divided into 3 main tectonics classification shallow crustal earthquakes in active tectonics regions, regions subduction earthquakes and shallow crustal earthquakes in stable continental regions.

The following attenuation relations were proposed: Idriss model (1995) and Sadigh et al. (1997) model have only horizontal component and Abrahamson and Silva (1977) relation have been used for vertical component.

Sommerville et al. (1977) have shown that directivity has a significant effect on long-period ground motions for sites in the near-fault region.

9.3.5. SELECTION OF SOIL PROPERTIES FOR DYNAMIC ANALYSIS

The shear strength properties of waste landfills are not easily determined since the physical composition of the mixture makes it unsuitable for the conventional laboratory strength testing. The size of testing equipment is too small relative to the normal size of the refuse. Also the shear parameters of municipal solid waste show a broad variety and a differentiation between fresh and old wastes. To overcome this situation the waste properties are established based on the type of waste, the waste processing and the placement procedures.

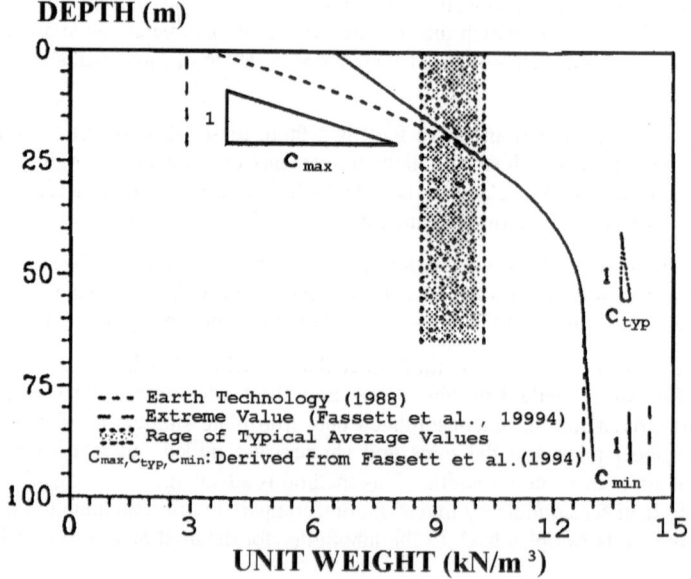

Fig. 9.3. Unit weight of MSW (after Kavazanjian, 1995; Reproduced with permission from Balkema)

Some properties are measured directly, such as dry density and water contents and other properties due the difficulties related with sampling are obtained from indirect methods combining with the existent knowledge of waste properties (Sêco e Pinto, 1997).

Total unit weights of the material are determined from in-place testing or laboratory compaction tests. Kavazanjian (1995) has proposed a unit weight profile with depth (Figure 9.3).

From literature survey the particle size distribution of municipal solid waste is shown in Figure 9.4 (Jessberg, 1994).

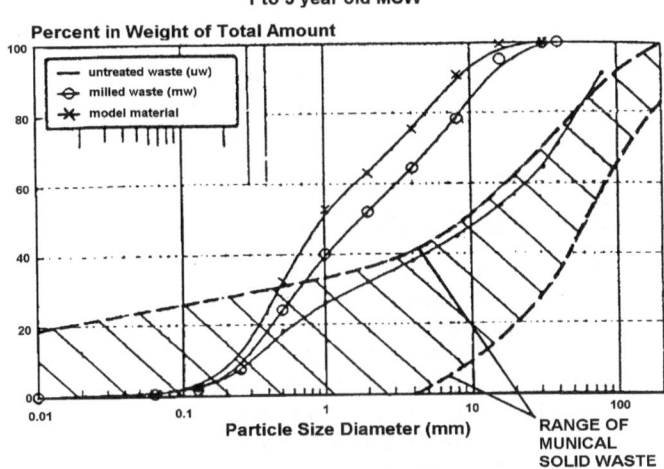

Fig. 9.4. Particle size distribution of waste for laboratory tests (after Jessberg, 1994; Reproduced with permission from Balkema)

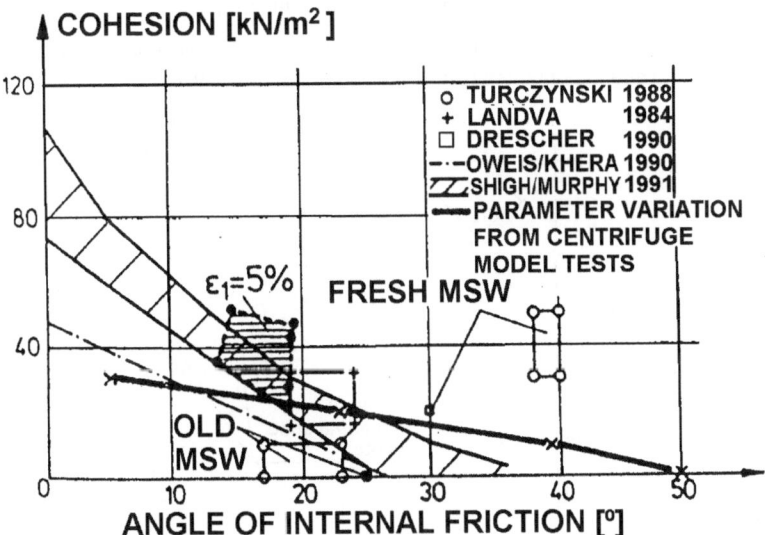

Fig. 9.5. Shear parameters of municipal solid waste (after Jessberg, 1996; Reproduced with permission from Balkema)

From results of laboratory and field tests the shear parameters of municipal waste exhibits a differentiation between fresh and old waste (Jessberg, 1996) (Figure 9.5). Also direct simple shear test laboratory tests on reconstituted large samples are used to determine large strain properties.

A wide range of reported V_s values for MSW compiled by Kavazanjian et al. (1996) is shown in Figure 9.6.

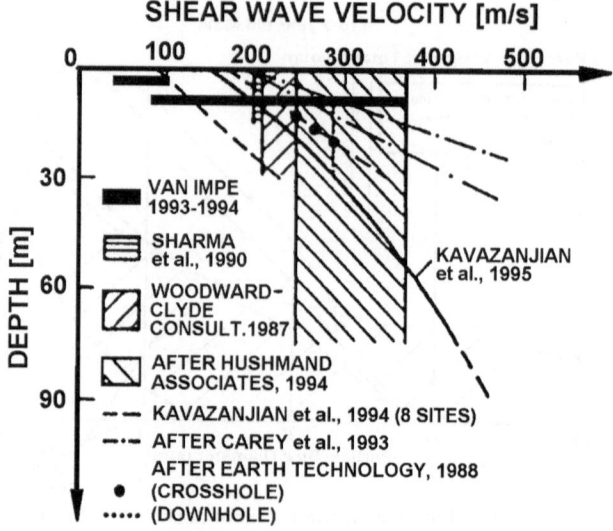

Fig. 9.6. Shear wave velocity of MSW (after Kavanzanjian et al., 1996; Reproduced with permission from Balkema)

To characterise the strength of solid waste Grândola landfill dynamic penetrometer tests were performed and the obtained results are shown in Figure 9.7.

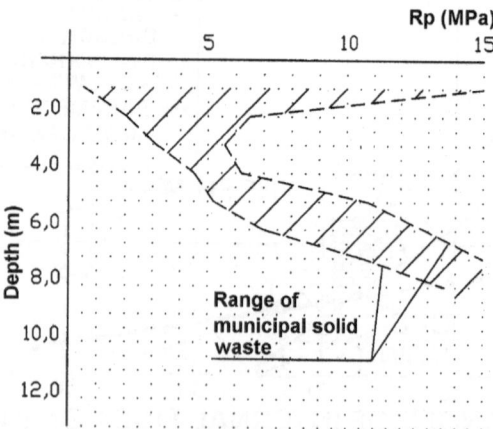

Fig. 9.7. Dynamic penetrometer tests (after Sêco e Pinto et al., 1999; Reproduced with permission from Balkema)

The measurement of the shear waves velocity by crosshole and downhole techniques need drilling boreholes in landfills. Spectral analysis of surface waves (SAWS) provide relatively accurate Vs profiles without the need for drilling and sampling the landfill material. Taking this into consideration geophysical measurements to estimate dynamic strain-dependent materials of solid wastes of Grandola landfill were implemented and a value of shear waves velocities between 330 - 350 m/s was obtained (Figure 9.8) (Sêco e Pinto et al., 1999).

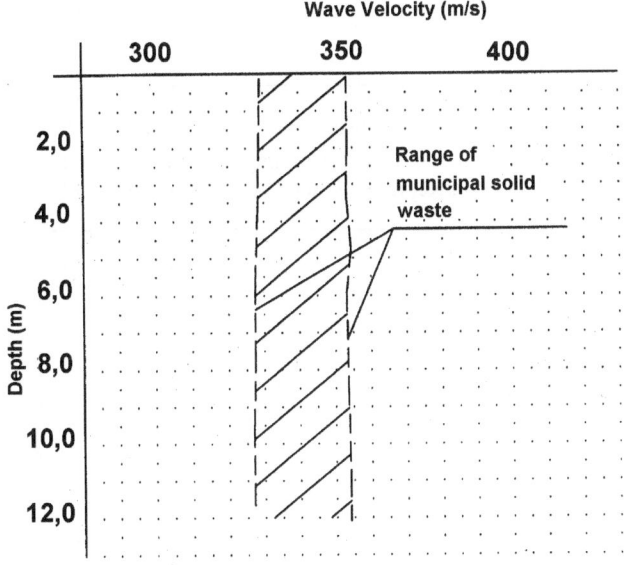

Fig. 9.8. Shear wave velocities of Grândola solid wastes (after Sêco e Pinto et al., 1999; Reproduced with permission from Balkema)

The obtained results have not shown a variation of shear wave velocities with depth, probably due the height of landfill being only 12m, and are in reasonable agreement with the results reported by Kavazanjian et al. (1995).

The variation of shear modulus G and damping ratio λ with shear strain can be derived by laboratory tests (Sêco e Pinto, 1990).

Limitations include health and safety constraints on sampling and testing of solid waste and the small size of test specimens relative to the size of the waste constituents.

For the variation of shear modulus and damping characteristics of waste materials, sandy silt material and silty material, with shear strain, the curves proposed by Singh and Murphy (1990) are presented in Figure 9.9.

The variation of shear modulus and damping ratio with shear strain for peat and clay material and the curves proposed by Kavazanjian and Matasovic (1995) for waste materials are presented in Figure 9.10.

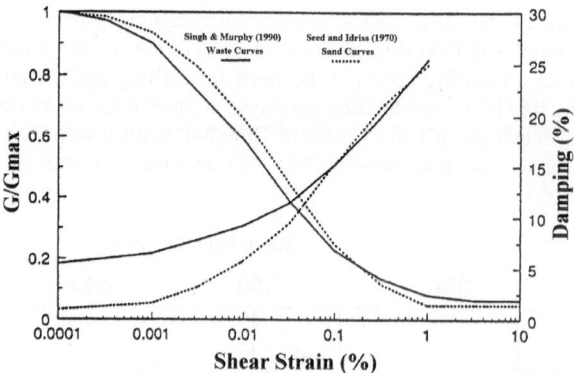

Fig. 9.9. Waste modulus degradation and damping curves used in study (after Singh and Murphy, 1990; Reproduced with permission from ASTM)

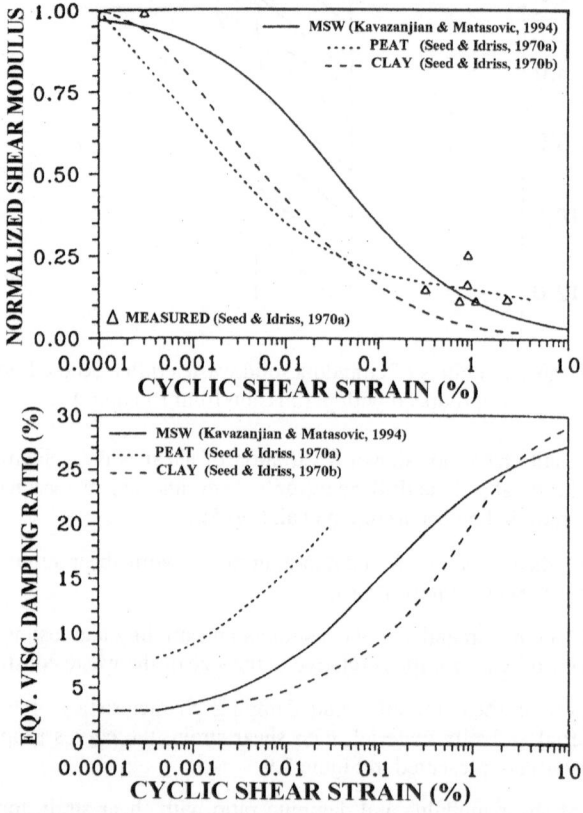

Fig. 9.10. Modulus and damping of MSW (after Kavazanjian and Matasovic, 1995, "Seismic analysis of solid waste landfills", *Proc. of Geoenvironment 2000*, ASCE Specialty Conference, New Orleans, Louisiana; Reproduced with permission from ASCE)

When solid waste landfills incorporate construction demolition debris the curves proposed for rockfill and gravel materials can be used. The shear interface resistance of liner and cover systems landfills has deserved increasing attention.

The dynamic properties of the geosynthetic liner can be replaced by the dynamic properties of the equivalent soil layer measured by shaking table tests.

9.3.6. SEISMIC RESPONSE ANALYSIS

The seismic responses obtained by computer finite element 1D programs are considered reasonable. These analyses are based on the solution of the equation of motion considering a homogenous and continuous soil deposit composed by horizontal soil layers and assuming a vertical propagation of shear waves.

Due the situation that slopes of landfills are flatter than slopes of earth dams and landfills decks are larger than dam crests, two dimensional response effects in landfills should be less significant than in earth dams. For the soil behaviour the equivalent linear methods is used and the shear modulus and damping ratio are adjusted in each iteration until convergence has occurred.

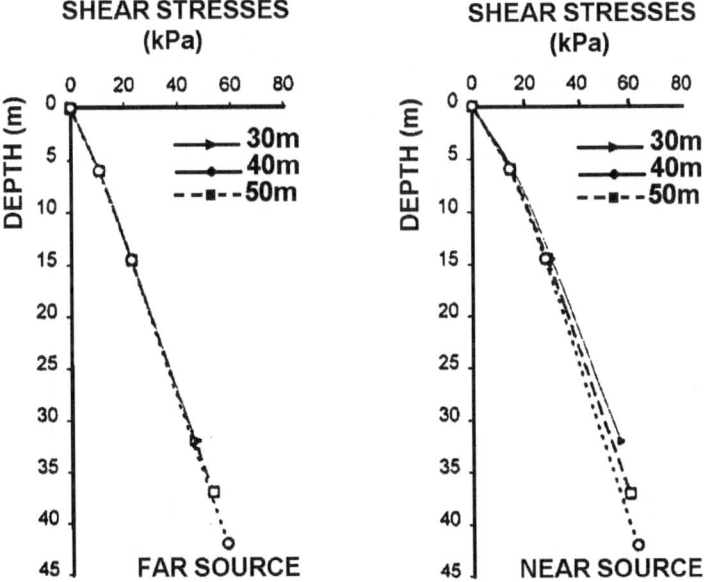

Fig. 9.11. Shear stresses distribution (after Sêco e Pinto et al., 1999; Reproduced with permission from Balkema)

Due to the uncertainties related with the materials properties, foundations geometry and also to check the influence of the seismic actions parametric and sensitivity studies are in general performed.

The shear stresses distribution and the acceleration distribution for solid waste Grandola landfill for three foundation geometries (30m, 40m and 50m depths) are presented in Figures 9.11 and 9.12 (Sêco e Pinto et al., 1999).

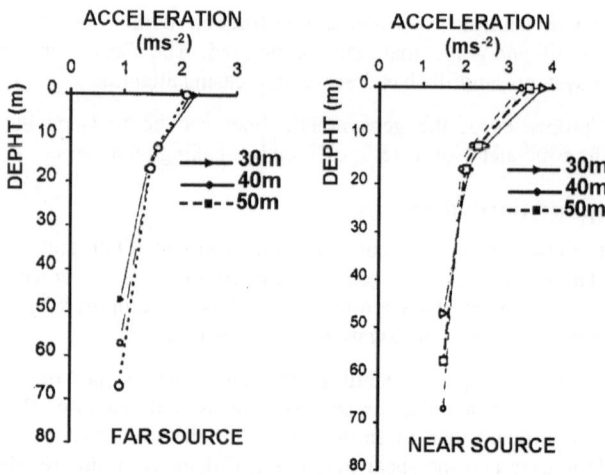

Fig. 9.12. Acceleration distribution (after Sêco e Pinto et al., 1999; Reproduced with permission from Balkema)

The seismic actions near source and far source were analysed. Due the geometry of the landfill (height and slopes) the effect of the HDPE geomembrane/geotextile liner was ignored, i.e. the dynamic properties of the geosynthetic liner was not replaced by the dynamic properties of the equivalent soil layer.

The Table 9.1 summarises for near source and far source the transference functions of acceleration (TFRA) between the bedrock and the ground level, the fundamental period of the layer (T_F), the maximum ground acceleration (Max A) and the amplification ratio (AR).

Table 9.1. Summary of the seismic analyses results

	Near Source	Near Source	Far Source	Far Source
Profile	P1	P2	P1	P2
TRFA	4.572	4.524	4.475	4.447
T_F (s)	0.409	0.489	0.414	0.499
MaxA (m/s²)	2.533	2. 321	1.656	1.607
A (m/s²)	0.945	0. 941	0.617	0.620
AR	2.68	2.467	2.683	2.592

It can be noticed that the amplification effects for the near source and for the far source are of the same order.

The values of TFRA decrease with the increasing of the thickness of foundation layer. The fundamental period values increase with the increasing of the thickness of foundation layer. The obtained results are in good agreement with the seismic performance of solid waste landfills.

Comparison the results of the analyses performed by SHAKE 91 and QUAD 4M codes

Rathje and Bray (1999) have concluded that: (i) the maximum seismic loading for base sliding within a landfill can be estimated conservatively with 1D analysis; (ii) the 1D analysis underpredicts the surface maximum horizontal acceleration (MHA) along the slope of a landfill by 10% on average, and by as much as 40 %; (iii) at the crest 1D analysis consistently underpredicts the MHA about 25%; (iv) along the deck the analysis is only moderately unconservative and the effect of base rock topography is not captured with 1D analysis.

It is important to stress that the dynamic characteristics of solid waste materials play an important role on the seismic response of landfill and this area deserves more consideration (Sêco e Pinto et al., 1998).

It is important to assess the dynamic shear strengths of liner materials due the effect of inertial forces in the refuse mass. Horizontal geosynthetic interfaces have a potential effect to modify the seismic response of overlying material.

Smooth HDPE geomembrane/geotextile liners reduce significantly the accelerations and shear stresses transmitted through the landfill profile, especially when the base acceleration exceeds 0.2g, as pointed by Yegian and Kadakal (1998). These effects should be taken into account to avoid unrealistic estimates of seismic acceleration, shear stresses and permanent deformations in a landfill.

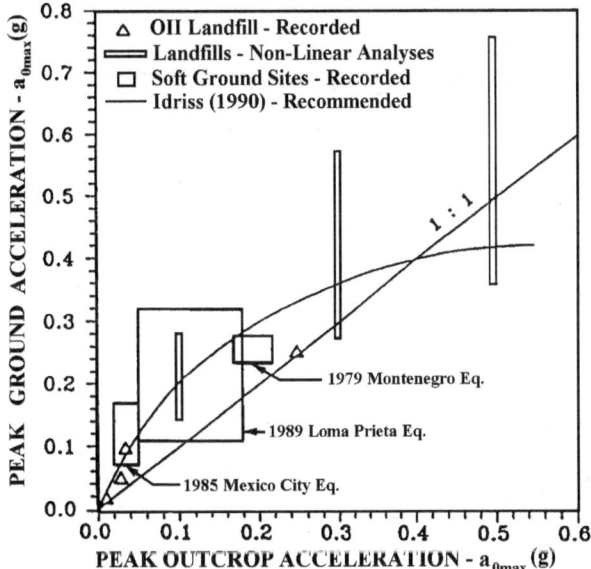

Fig. 9.13. Amplification of peak acceleration by landfills (after Kavazanjian and Matasovic, 1995, "Seismic analysis of solid waste landfills", *Proc. of Geoenvironment 2000*, ASCE Specialty Conference, New Orleans, Louisiana; Reproduced with permission from ASCE)

The effect of the wave propagation through the waste was analysed by Anderson and Kavazanjian (1995) and the results are presented in Figures 9.13 and 9.14.

Deformations ranging from 150 to 300mm are accepted in practice for design of geosynthetic liner systems. For cover systems large deformations can be accepted taking into consideration that most cover failures can be detected and repaired at reasonable costs.

During earthquakes inertial forces in the refuse mass may result in the mobilization of shear stress in excess of the dynamic shear strengths of liner materials.

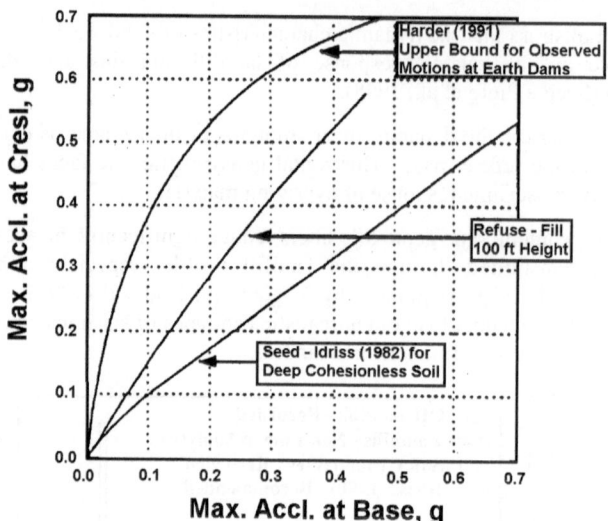

Fig. 9.14. Amplification of acceleration by earth dams and waste landfills (after Singh and Sun, 1995 "Seismic evaluation of municipal solid waste landfills", *Proc. of Geoenvironment 2000*, ASCE Specialty Conference, New Orleans, Louisiana, 22-24 February; Reproduced with permission from ASCE)

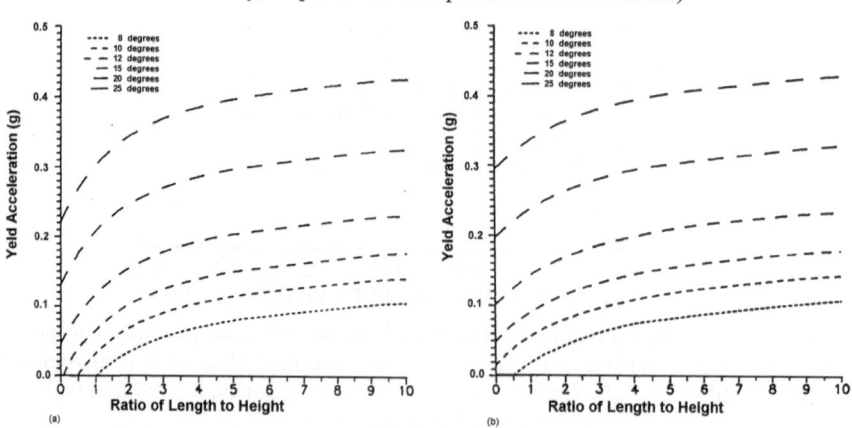

Fig. 9.15. Design charts for evaluating yield acceleration for lined landfills of (a) 2h:1v and (b) 4h:1v (after Shewbridge, 1996,"Yield acceleration of lined landfills", *J. of Geotechnical Eng.*, 122(2): 156-158; Reproduced with permission from ASCE)

Based on a simplified block analysis, Shewbridge (1996) presents in Figure 9.15 a chart to evaluate a yield acceleration of lined landfills.

The permanent displacement induced in the cover soil using four different earthquake records: Imperial Valley record of the 1940 El Centro Earthquake, Kern County record of the 1952 Taft earthquake, Capitola record of 1989 Loma Prieta earthquake and Newhall record of the 1994 Northridge earthquake computed by Ling and Leshchinsky (1997) is shown in Figure 9.16.

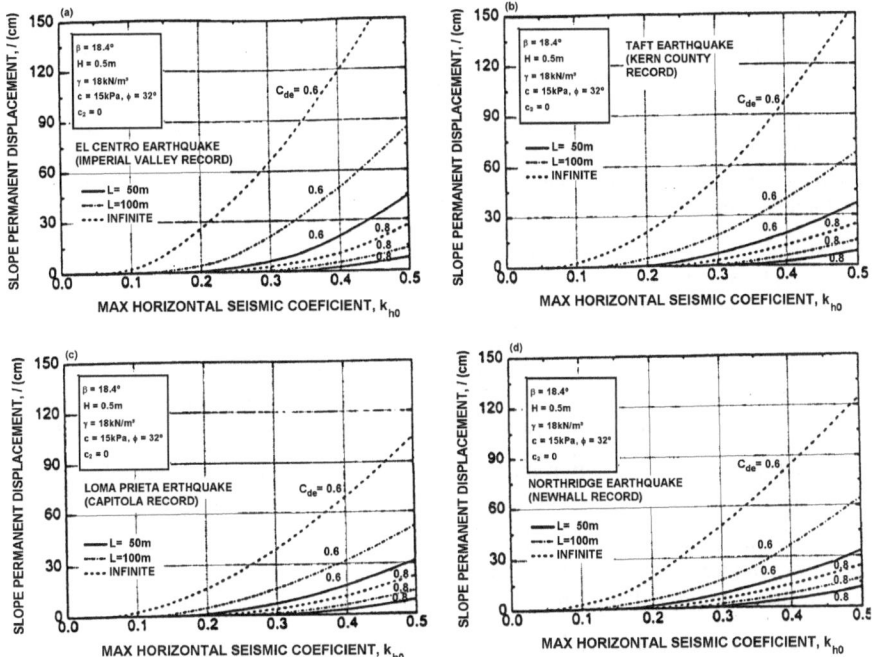

Fig. 9.16. Permanent displacements of finite and infinite slopes (after Ling and Leshchinsky, 1997. "Seismic stability and permanent displacement of landfill cover systems", *Journal of Geotechnical and Geoenvironmental Engineering*, 123(2): 113-122; Reproduced with permission from ASCE)

The allowable value for the calculated permanent seismic displacement of geosynthetic liner systems is 150 to 300mm. The upper value of 300mm is appropriate for simplified analyses which use upper bound displacement curves for generic Newmark displacement charts, residual shear strength and/or simplified seismic analyses (Kavazanjian, 1998). The lower value 150mm is more appropriate for more sophisticated analyses and formal Newmark displacement analyses.

The knowledge of interaction between waste and structures is still poor and mainly limited to field observations.

Centrifuge model tests have been carried out to understand the principle of waste-structure interaction and to investigate deformation induced stress redistributions within the waste body near a structure (Kockel et al., 1997).

Yegian and Kadakal (1998) have shown that smooth HDPE geomembrane/geotextile liners significantly reduce the accelerations and shear stresses transmitted through the landfill profile, especially when the base acceleration exceeds 0.2g. Ignoring these effects can result in unrealistic estimates of seismic acceleration, shear stresses and permanent deformations in a landfill. The dynamic properties of the geosynthetic liner were replaced by the dynamic properties of the equivalent soil layer measured by shaking table tests.

9.3.7. LIQUEFACTION ASSESSMENT

The methods available for evaluating the cyclic liquefaction potential of landfills or foundation are based on laboratory tests and field tests.

In general the following laboratory tests are used: (i) cyclic triaxial test, (ii) cyclic simple shear tests, (iii) torsional cyclic shear tests. Due the difficulties in obtaining high quality undisturbed samples field test such as *SPT* tests, *CPT* tests, seismic cone, flat dilatometer and methods based on electrical properties of soil are used.

To estimate liquefaction resistance from shear wave velocity there are two procedures: (i) methods based on a combination in situ shear wave velocity measurements and laboratory tests on undisturbed tube and in situ freezing samples from Tokimatsu et al. (1991); (ii) methods based on in situ shear wave velocity measurement and a correlation between liquefaction resistance and shear wave velocity deduced from liquefaction degree in the field from Stokoe et al. (1999). The assessment of liquefaction resistance from shear wave crosshole tomography was proposed by Furuta and Yamamoto (2000).

Liquefaction resistance of silty sands during seismic liquefaction conditions for various silt contents and confining pressures was investigated by Amini and Qi (2000).

The post-liquefaction strength of loose silty sediments is commonly less than that of sands, but moderately dense silts at shallow depths are generally dilative, making them more resistant to ground deformation than cleaner sands (Youd and Gilstrap, 1999).

A probabilistic method considering the uncertainty in the liquefaction criterion was proposed by Todorovska and Trifunac (1999).

9.4. Monitoring and Safety Control of Landfills

Landfill behaviour during construction and operation is monitored to check methods, results of analyses and model tests and to analyse it safety against deterioration of failure.

Seismic downhole-array data provide a unique source of information on actual soil behaviour over a wide range of loading conditions. Correlation and spectral analyses are performed to evaluate shear wave propagation characteristics, variation of shear wave velocity with depth, and site resonant frequencies and modal configurations (Elgamal et al., 1995).

In regard of seismic instrumentation of the response of the landfill to such seismic activity the type of instruments currently designated by accelerographs are strong - motion accelerographs, peak recording accelerographs and seismoscopes.

In comparison with manual readings the automatic system allows a rapid data processing of results a great number of instruments. Once in operation an automatic system allows a reduction of personal, both in the field and office. The automatic system and the central data processing allow a quicker updating of the information. An automatic system implies an increase of complexity, with the electronic equipment to be installed in unfavourable environment of temperature and humidity.

For data validation a preliminary check on the raw values (following the execution of function tests on measurement equipment) by comparing the actual values from the sensor readings with the established limits and data reduction (computation of engineering quantities) is performed For the interpretation of the measurements it is necessary to establish a procedure, a mathematic model that can be a statistical model a deterministic model or a hybrid model.

Safety control is the group of measures taken in order to have an up-to-date knowledge of the condition of the landfill and to detect in due time the occurrence of any anomalies to define actions to correct the situation or, at least, to avoid serious consequences.

9.5. Safety and Risk Analyses

Safety analysis for geotechnical structures, such as slopes, retaining walls, piles and shallow foundations implies the verification of limit states: ultimate limit states and serviceability limit states. For dams also two levels of safety are considered, depending of whether they correspond to normal conditions for use of the structures (current scenario) or are associated with an exceptional occurrence (failure scenarios).

From the above considerations it seems that for solid waste landfill a level of damage can be accepted provided there is no harmful discharge of contaminants to the environment.

For cover systems large displacements can be accepted taking into consideration that most cover failures can be detected and repaired at reasonable costs.

The allowable values of deformation of landfill systems, depends of several factors, namely of geosynthetic liner systems and gas recovery system.

Municipality waste landfills owners, regulatory authorities and consultants are interested in carrying out a risk analysis. Its purpose is to identify the main real risks associated with each type and height of landfill for all circumstances and can be conducted: (i) in extensive risk analysis of very large landfills, to substantiate reliably the probabilities chosen in event trees; (ii) in simplified risk analysis of smaller landfills, to focus low-cost risk analysis on a few main risks; (iii) and in identifying possibilities for reducing these risks through low-cost structural or non-structural measures.

Although the annual failure probability of landfills is lower than 10^{-6} in most cases, it may be higher for landfills in seismic areas.

Consideration of human behaviour is essential when assessing the consequence of failures: well organised emergency planning and early warning systems could decrease the number of victims and so the study of human behaviour plays an important role in assessment of risk analysis.

The results of a risk analysis can be used to guide future investigations and studies, and to supplement conventional analyses in making decisions on waste landfills safety improvements. With increasing confidence in the results of risk analyses, the level of risk could become the basis of safety decisions.

9.6. Final Remarks

In the precedent sections the different methods to analyse solid waste landfills and lining systems stability during earthquakes were presented. These tools are very important to assist the design engineer in incorporating the adequate design measures to prevent deleterious effects of earthquake shaking.

All the essential steps of good analyses, whatever the type of material are involved shall be performed with a sufficient degree of accuracy that the overall results can be extremely useful in guiding the engineer in the final assessment of seismic stability. This final assessment is not made by numerical results but shall be made by experienced engineers who are familiar with the difficulties in defining the design earthquake and the material characteristics, who are familiar with the strengths and limitations of analytical procedures, and who have the necessary experience gained from studies of past performance.

CHAPTER 10
EARTHQUAKE RESISTANT DESIGN OF SHALLOW FOUNDATIONS

Alain Pecker
Géodynamique & Structure, Bagneux, France

10.1. Introduction

Aseismic design of foundations still remains a challenging task for the earthquake geotechnical engineer. Leaving aside the seismic retrofit of existing foundations, which is a more difficult issue, even the design of new foundations raises issues which are far from being totally resolved. One of the main reasons stems from the complexity of the problem which requires skills in soil mechanics, foundation engineering, soil-structure interaction along with, at least some knowledge of structural dynamics.

A parallel between static design and seismic design reveals some similarity but also very marked differences. In the early days, static design of foundations put much emphasis on the so-called bearing capacity problem (failure behaviour); with the introduction of an appropriate safety factor, close to 3, the short term settlements were deemed to be acceptable for the structure. It is only with the increase in the understanding of soil behaviour and the development of reliable constitutive models that sound predictions of settlements could be achieved. Not surprisingly, earthquake geotechnical engineers have focused their attention on the non linear behaviour of soils and on the evaluation of the cyclic deformations of foundations. This was clearly dictated by the need for an accurate evaluation of the soil-structure interaction forces which govern the structural response. It is only during the last decade that seismic bearing capacity problems have been tackled. These studies have clearly been motivated by the foundation failures observed in the Mexico City (1985) and Kobe (1995) earthquakes.

These two aspects of foundation design have reached a state of development where they can be incorporated in seismic building codes; Eurocode 8 - Part 5 is certainly a pioneering code in that respect.

In this paper, which is an abridged version of the lecture presented at the GeoEng2000 Conference (Pecker and Pender, 2000) the fundamental aspects underlying the earthquake resistant design of shallow foundations will be reviewed and their implementation in seismic building codes will be discussed.

10.2. Aseismic Foundation Design Process

The aseismic design process for foundations is a "very broad activity requiring the synthesis of insight, creativity, technical knowledge and experience" (Pender, 1995). Information is required and decisions have to be made at various stages including:

1. the geological environment and geotechnical characterization of the soil profile;
2. the definition of the loads that will be applied to the foundation soil by the facility to be constructed;

A. Ansal (ed.), Recent Advances in Earthquake Geotechnical Engineering and Microzonation, 285–301.
© 2004 *Kluwer Academic Publishers.*

3. information about the required performance of the structure;
4. investigation of possible solutions with evaluation of load capacity, assessment of safety factors and estimates of deformations;
5. consideration of construction methods and constraints that need to be satisfied (finance and time);
6. exercise of judgment to assess potential risks.

Obviously the process described above is not a linear progression. Several iterations may be required, at least from step 1 to step 6, before arriving at a feasible, reliable and economic design. In the following we will focus on steps 2 and 4. We will assume that all the required information related to the soil characterization and structural performance is available. This in no way means that these two items are of secondary importance; the data listed under these items are probably the most difficult to assess and considerable experience is required as well as the exercise of judgment.

10.3. Evaluation of Seismic Demand

10.3.1. FUNDAMENTALS OF SOIL STRUCTURE INTERACTION

The earthquake design loads applied to the foundation arise from the inertia forces developed in the superstructure and from the soil deformations, caused by the passage of seismic waves, imposed on the foundations. These two phenomena are referred in the technical literature as inertial and kinematic loading. The relative importance of each factor depends on the foundation characteristics and nature of the incoming wave field.

The generic term encompassing both phenomena is Soil-Structure Interaction (SSI). However, more often, design engineers refer to inertial loading as SSI, ignoring the kinematic component. This situation stems from the fact that:

*   kinematic interaction may in some situations be neglected;
*   aseismic building codes, except for very few exceptions like Eurocode 8, do not even mention it;
*   kinematic interaction effects are far more difficult to evaluate rigorously than inertial interaction effects.

Figure 10.1 illustrates the key features of the problem under study (adapted from Gazetas and Mylonakis, 1998). It is presented in the general situation of an embedded foundation.

The soil layers away from the structure are subjected to seismic excitation consisting of numerous incident waves: shear waves (S waves), dilatational waves (P wave), surface waves (R or L waves). The nature of the incoming waves is dictated by seismological conditions but the geometry, stiffness and damping characteristics of the soil deposit modify this motion; this modified motion is the free field motion at the site of the foundation. Determination of the free field motion is in itself a challenging task because, as pointed out by Lysmer (1978), the design motion is usually specified at only one location, the ground surface, and the complete wave field cannot be back-calculated from this incomplete information; that is the problem is mathematically ill posed. Assumptions have to be made regarding the exact composition of the free field motion and it can be stated that no satisfactory solution is available to date.

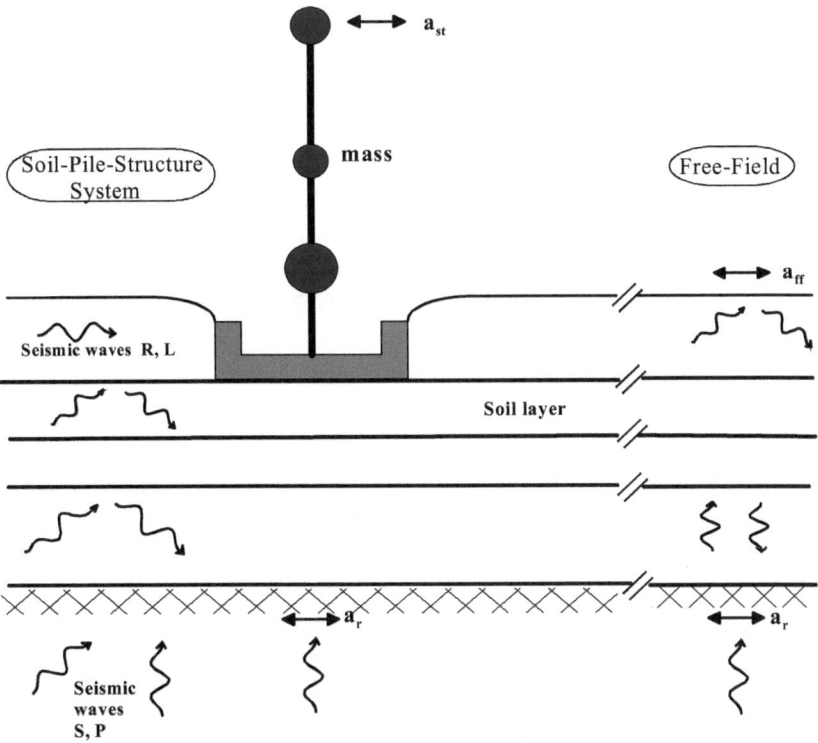

Fig. 10.1. Sketch of the soil-structure interaction problem

Let us now consider the motion around the structure and its foundation: the seismically deforming soil will force the embedded foundation to move, and subsequently the supported structure. Even without the superstructure, the motion of the foundation will be different from the free field motion because of the difference in rigidity between the soil on one hand, and the foundation on the other hand; the incident waves are reflected and scattered by the foundation. This is the phenomenon of kinematic interaction.

The motion induced at the foundation level generates oscillations in the superstructure which develop inertia forces and overturning moments at its base. Thus the foundation, the piles, and eventually the surrounding soil experience additional dynamic forces and displacements. This is the phenomenon of inertial interaction.

Obviously the foundation, in a broad sense, must be checked for the combined inertial and kinematic loading.

The above decomposition of the problem into three tasks (site response analysis, kinematic interaction, inertial interaction) is convenient for highlighting the various contributions of each to the final result. It does not necessarily imply that the steps must be performed separately as a complete interaction analysis (usually with the finite element method) is conceptually possible. However from a design and practical standpoint the computation of the foundation seismic loads usually follows the three step approach.

A direct (or complete) interaction analysis is very time demanding and not well suited for design, especially in 3D, which requires that the steps described above under item 2 to 6 be repeated several times. The multistep approach reduces the problem to more amenable stages and does not necessarily require that the whole solution be repeated again if changes occur, let's say, in the superstructure. In addition, it has the advantage that some of the intermediate steps can be neglected, as shown later.

The multistep approach is not only attractive for illustrating the fundamental aspects of soil structure interaction, it is also of great mathematical convenience. This convenience stems, in linear systems, from the superposition theorem (Kausel and Roesset, 1974). This theorem states that the seismic response of the complete system of Figure 10.2 can be computed in two steps:

- the kinematic interaction involving the response to base acceleration of a system which differs from the actual system in that the mass of the superstructure is equal to zero;
- the inertial interaction referring to the response of the complete soil-structure system to forces associated with accelerations equal to the sum of the base acceleration plus those accelerations arising from the kinematic interaction.

The later system is further divided into two consecutive steps:

- computation of the dynamic impedances at the foundation level;
- analysis of the dynamic response of the superstructure supported on the dynamic impedances and subjected to the kinematic motion, also called effective foundation input motion.

Provided each step described above is performed rigorously and under the restriction that the system remains linear, i.e. superposition is valid, the breakdown of the complete interaction analysis into consecutive steps is rigorous. For a mathematical description of the superposition theorem, the reader is referred to Kausel and Roesset (1974) or Gazetas and Mylonakis (1998).

With this, now classical, theoretical background on soil-structure interaction, in mind, one can proceed to examine the practical implementation of SSI in the state of practice. The extent to which this state of practice could be improved at a minimal cost will also be examined.

10.3.2. CODE APPROACH TO SOIL STRUCTURE INTERACTION ANALYSES

In almost every seismic building code, the structure response and foundation loads are computed neglecting the soil-structure interaction that is a fixed base analysis of the structure is performed. The belief is that SSI always plays a favorable role in decreasing the inertia forces; this is clearly related to the standard shape of code spectra which almost invariably possess a gently descending branch beyond a constant spectral acceleration plateau. Lengthening of the period, due to SSI, moves the response to a region of smaller spectral accelerations. However there is evidence that some structures founded on unusual soils are vulnerable to SSI. Examples are given by Gazetas and Mylonakis (1998) and Resendiz and Roesset (1985) for instance.

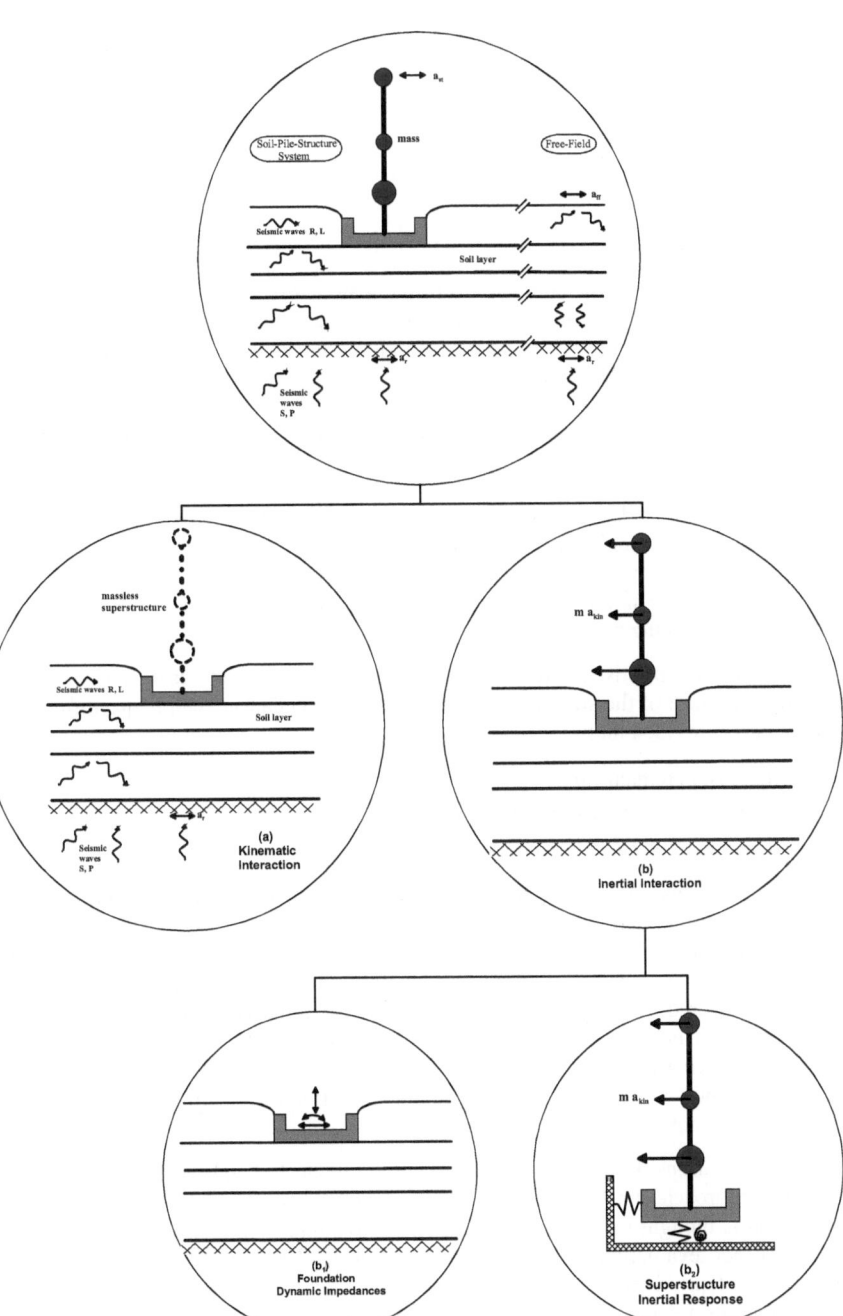

Fig. 10.2. Superposition theorem for soil-structure interaction problems

This has been recognized in some codes. Eurocode 8 states that "The effects of dynamic soil-structure interaction shall be taken into account in the case of:

- structures where P-δ effects play a significant role;
- structures with massive or deep seated foundations;
- slender tall structures;
- structures supported on very soft soils, with average shear wave velocity less than 100 m/s."

In addition, an annex to the code describes the general effect of SSI. To the best of our knowledge, Eurocode 8 is the only code which recognizes the importance of kinematic interaction; however, this is restricted to piled foundations and not mandatory for shallow or embedded foundations.

10.3.3. IMPROVED EVALUATION OF SEISMIC DEMAND

Linear Systems

With the tremendous development of computer facilities, there does not seem to be any rational reason for neglecting soil-structure interaction. Most building codes now require that the structural response be evaluated using a multimodal analysis, as opposed to a former monomodal analysis and this can be performed with most computer codes available on the market.

Referring to the multistep approach described previously, the last step of an SSI analysis (response of the structure connected to the impedances) can be performed on a routine basis provided that:

- the system remains linear;
- the kinematic interaction can be neglected;
- dynamic impedance functions are readily available.

Although the superposition theorem is exact for linear soil and structure, it can nevertheless be applied to moderately non linear systems. This can be achieved by choosing reduced soil characteristics which are compatible with the free field strains induced by the propagating seismic waves: this is the basis of the equivalent linear method, pioneered by Idriss and Seed (1968). This engineering approximation implies that all the soil non linearities arise from the passage of the seismic waves and that additional non linearities developed around the edges of a mat foundation, are negligible. Experience shows that it is a valid approximation in many situations where large soil instabilities do not occur.

For some situations, kinematic interaction can be neglected and the second step of the multistep approach can be bypassed. It must be realized however that, if kinematic interaction is thought to be significant, there is no simple means for evaluating it; as a matter of fact, evaluation of kinematic interaction is almost as difficult as solving the complete SSI problem. Obviously kinematic interaction is exactly zero for shallow foundations in a seismic environment consisting exclusively of vertically propagating shear waves or dilatational waves.

During the last decade, numerous solutions for the dynamic impedances of any shape foundations have been published (Gazetas, 1990). They are available for homogeneous soil deposits but also for moderately heterogeneous ones

Therefore provided all the aspects listed above are properly covered, seismic soil structure interaction can be covered at a minimal cost and reduces to the last step of the multistep approach: dynamic response of the structure connected to the impedance functions and subjected to the free field motion (equal to the kinematic interaction motion). However to be fully efficient, and to allow for the use of conventional dynamic computer codes, the impedance functions which are frequency dependent (Figure 10.4) must be represented by frequency independent values. The simplest versions of these frequency independent parameters are the so-called springs and dashpots. From the published results, it appears that only under very restrictive soil conditions (homogeneous halfspace, regular foundations) can these dynamic impedances be represented by constant springs and dashpots. Nevertheless, structural engineers still proceed using these values which, more than often, are evaluated as the static component (zero frequency) of the impedance functions.

However, fairly simple rheological models can be used to properly account for the frequency dependence of the impedance functions. These models can be developed using curve fitting techniques, or with physical insight, such as the series of cone models developed by Wolf (1994). Figure 10.3 shows examples of such models: Figure 10.3a is the model proposed by De Barros and Luco (1990) based on a curve fitting technique; Figure 10.3b is a class of cone models proposed by Wolf. With such models, which are most conveniently used in time history analyses, the actual dynamic action of the soil can be properly accounted for; even "negative stiffness", which are frequently encountered in layered soil profiles, can be apprehended with those models. As an illustrative example, Figure 10.4 presents the application of model 3a to an actual bridge pier foundation; the foundation is a large circular caisson, 90 m in diameter, resting on a highly heterogeneous soft soil profile. The "exact" impedances were computed using a frequency domain finite element analysis. Note the very good fit achieved by the model (square symbols) even for the negative stiffness of the rocking component. Clearly, implementation of such simple rheological models does not impose a heavy burden to the analyst and represents a significant improvement upon the lengthy and tedious iteration process in which springs and dashpots are updated to become compatible with the SSI frequencies.

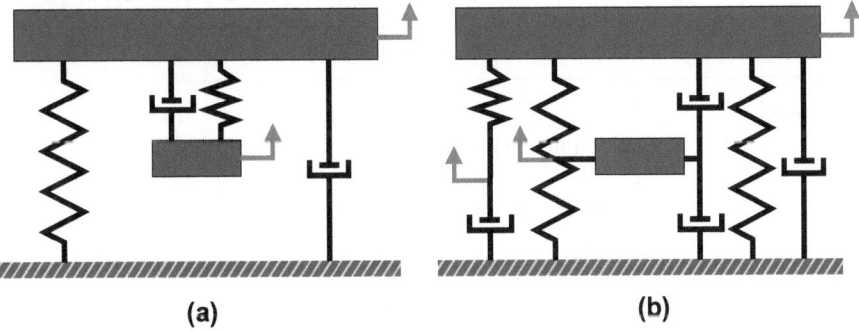

(a)   (b)

Fig. 10.3. Examples of cone models

However, care must be exercised when implementing the models sketched in Figure 10.3; due to the presence of the additional mass which induces inertia forces, the input motion, called the effective input motion, is different from the free field input motion. Although determination of the effective input motion is not difficult by itself, it might be more advantageous to make use of rheological models without masses; such models, made of an assembly of springs and dashpots, have recently been proposed and seem efficient for surficial foundations (Wu and Lee, 2002).

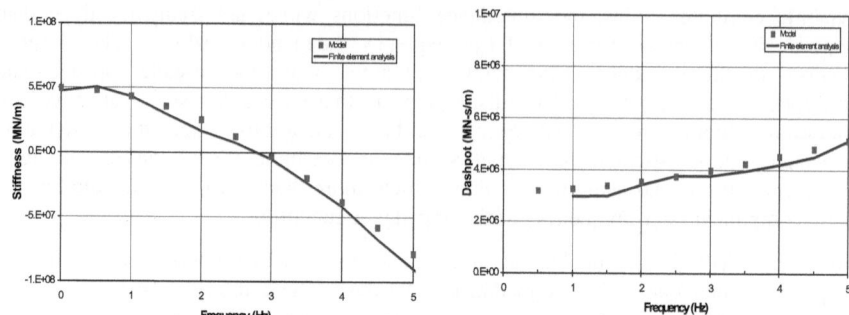

Fig. 10.4. Rocking dynamic impedances - example

Before moving onto consideration of nonlinear SSI, there in one section of Eurocode 8 which provides for a transition between the linear elastic approach discussed above and non linear methods discussed below. Table 10.1 (taken from Eurocode8) acknowledges that with increasing ground acceleration the soil adjacent to a shallow foundation will experience increasing shear strains and consequently the stiffness will decrease and the material damping increase. Table 10.1 suggests how the apparent average shear modulus and material damping of the soil adjacent will change with increasing peak ground acceleration and envisages that an elastic SSI calculation would be done with the modified values for the soil stiffness and damping. (Following this simplification there is, of course, no frequency dependence on the stiffness and damping parameters for the foundation).

Table 10.1. Average soil damping factors and average reduction factors (± one standard deviation) for shear wave velocity v_s and shear modulus G within 20m depth.($v_{s\,max}$ = average v_s value at small strain (< 10^{-5}), not exceeding 300m/s. G_{max} = average shear modulus at small strain.)

Ground acceleration ratio, α	Damping factor	$\dfrac{v_s}{v_{s,max}}$	$\dfrac{G}{G_{max}}$
0,10	0,03	0,9(±0,07)	0,80(± 0,10)
0,20	0,06	0,7(± 0,15)	0,50(+ 0,20)
0,30	0,10	0,6(∓0,15)	0,35(± 0,20)

Non Linear Systems

One of the main limitations of the multistep approach is the assumption of linearity of the system for the superposition theorem to be valid. As noted previously, some non linearities, such as those related to the propagation of the seismic waves, can be introduced but the non linearities specifically arising from soil-structure interaction are ignored. The generic term "non linearities" covers geometrical non linearities, such as foundation uplift, and material non linearities, such as soil yielding around the edges of shallow foundations, along the shafts of piles, and the formation of gaps adjacent to pile shafts. Those non linearities may be beneficial and tend to reduce the forces transmitted by the foundation to the soil and therefore decrease the seismic demand. This has long been recognized for foundation uplift for instance (see ATC 40).

Giving up the mathematical rigor of the superposition theorem, an engineering approximation to these aspects can be reached by substructuring the supporting medium into two sub-domains (Figure 10.5):

- a far field domain, which extends a sufficient distance from the foundation for the soil structure interaction non linearities to be negligible, non linearities in that domain are only governed by the propagation of the seismic waves,
- a near field domain, in the vicinity of the foundation where all the geometrical and material non linearities due to soil structure interaction are concentrated

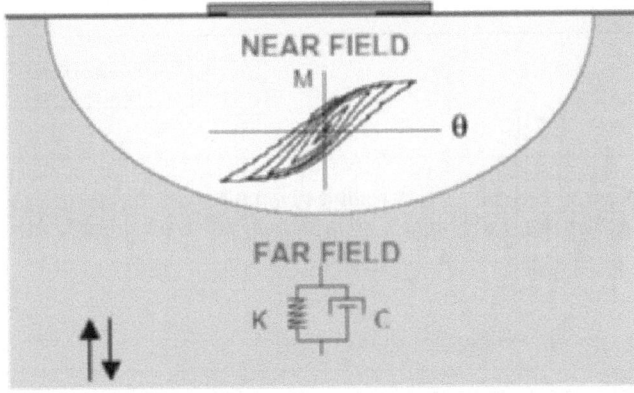

Fig. 10.5. Conceptual subdomains for dynamic soil structure analyses

The exact boundary between both domains is not precisely known but its location is irrelevant for practical purposes. This concept of far field and near field domains can be easily implemented if one assumes that the degrees of freedom of the foundation are uncoupled: the far field domain is modelled with the linear (or equivalent linear) impedance function whereas the near field domain is lumped into a non-linear macro-element. A simplified rheological representation of this sub-structuring is shown in Figure 10.6 (Pecker, 1998): the macro-element is composed of a finite number of springs and Coulomb sliders which are determined from curve fitting to the non-linear force-displacement (or moment-rotation) backbone curve, computed for instance with a static finite element analysis.

FAR FIELD   NEAR FIELD

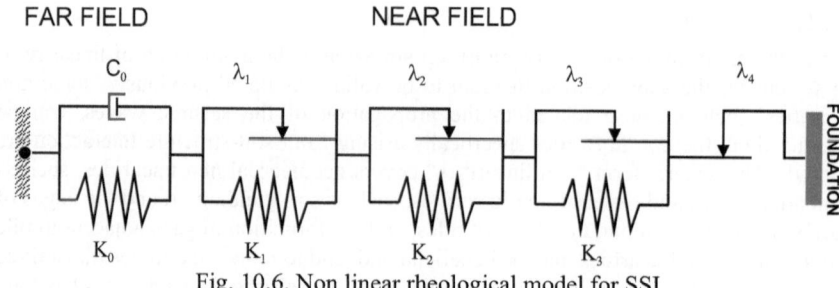

Fig. 10.6. Non linear rheological model for SSI

Damping in the near field domain arises only from material damping and obeys Masing's law; damping in the far field domain is of the viscous type. Calibration of this simplified rheological model against a rigorous 2D dynamic finite element analysis, including all the non linearities mentioned previously, shows very promising results. Figure 10.7 compares the overturning moments at the foundation level of a bridge pier foundation computed with both approaches: not only the amplitudes are correctly matched but also the phases are preserved.

This model has been extended in a more rigorous way to account for the coupling between the various degrees of freedom of the foundation, especially between the vertical and rotational one when uplift occurs (Cremer et al., 2001).

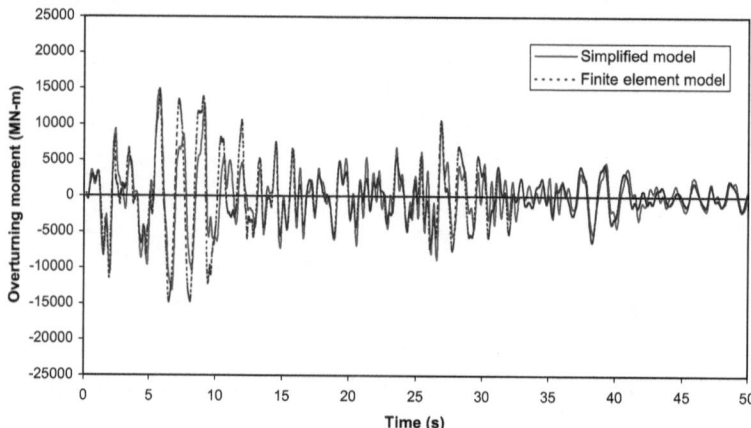

Fig. 10.7. Comparison between finite element analysis and the non linear rheological model for the foundation

10.4. Bearing Capacity for Shallow Foundations

Once the forces transmitted to the soil by the foundation are determined, the design engineer must check that these forces can be safely supported: the foundation must not experience a bearing capacity failure nor excessive permanent displacements. At this point a major difference appears between static, permanently acting loads, and seismic loads. In the first instance excessive loads generate a general foundation failure whereas seismic loads, which by nature vary in time, may induce only permanent irrecoverable displacements. Failure can therefore no longer be defined as a situation in which the safety factor becomes less than unity; it must rather be defined with reference to excessive permanent displacements which impede the proper functioning of the structure. Although this definition seems rather simple and the methodology has been successfully applied to dam engineering (Newmark, 1965), its implementation in a code format is far from an easy task. One of the difficulties is to define what are acceptable displacements of the structure in relation to the required performance. Another difficulty obviously lies in the uncertainty linked to the estimation of permanent displacements.

10.4.1. FUNDAMENTAL REQUIREMENT OF CODE APPROACHES

As an example of code documentation Eurocode 8 states that "The bearing capacity of the foundation shall be checked under the combination of the design action effects. To check the seismic bearing capacity of the foundation, the general expression and criteria provided in informative annex F may be used, which allow to take into account the load inclination and eccentricity arising from the inertia forces in the structure as well as the possible effects of the inertia forces in the supporting soil itself. The rise of pore water pressure under cyclic loading should be considered either in the form of undrained strength or as pore pressure in effective stress analysis. For important structures, non linear soil behaviour should be considered in determining possible permanent deformation during earthquakes."

More specifically, in most seismic codes the design engineer is required to check the following general inequality:

$$S_d \leq R_d \tag{10.1}$$

where S_d is the seismic design action (demand) and R_d the system design resistance (capacity). These two terms are explained below.

The design action represents the set of forces acting on the foundations. For the bearing capacity problem, they are composed of the normal force N_{sd}, shear force V_{sd}, overturning moment M_{sd} and soil inertia forces F developed in the soil. The actions N_{sd}, V_{sd} and M_{sd} arise from the inertial soil-structure interaction. The inertia force, $F = \rho a$ (ρ soil mass density, a acceleration), arises from the site response analysis and kinematic interaction. The term design action is used to reflect that these forces must take into account the actual forces transmitted to the foundation i.e. including any behaviour and over-strength factors used in inelastic design.

The design resistance represents the bearing capacity of the foundation; it is a function of the soil strength, soil-foundation interface strength and system geometry (for instance foundation width and length).

Obviously, inequality (10.1) must include some safety factors. One way is to introduce partial safety factors, as in Eurocode 8. This is not the only possibility and some other codes, like the New Zealand one, choose the Load and Resistance Factored Method (LRFD) and factor the loads and resistance (Pender, 1999). The Eurocode approach is preferred because it gives more insight in the philosophy of safety; on the other hand it requires more experimental data and numerical analyses to calibrate the partial safety factors.

With the introduction of partial safety factors inequality (10.1) is modified as follows:

$$S_d\,(\gamma_F.\ \text{actions}) \le \frac{1}{\gamma_{Rd}}\ R_d \left(\frac{\text{strength parameters}}{\gamma_m}, \text{geometry} \right) \qquad (10.2)$$

where "actions" represent the design action and "strength" the material strength (soil cohesion and /or friction angle, soil-foundation friction coefficient).

γ_F is the load factor applied to the design action: γ_F is larger than one for unfavorable actions and smaller than 1.0 for favorable ones.

γ_m is the material safety factor used to reflect the variability and uncertainty in the determination of the soil strength. In Eurocode 8, the following values are used: 1.4 on the undrained shear strength and cohesion and 1.25 on the tangent of the soil friction angle or interface friction coefficient.

γ_{Rd} is a model factor. It acts like the inverse of a strength reduction factor applied to the resistance in an LRFD code. This factor reflects the fact, that to evaluate the system resistance some approximations must be made: a theoretical framework must be developed to compute the resistance and like any model it involves simplifications, and assumptions which deviate from reality. It will be seen later on that the model factor is essential and can be used with benefit to differentiate a static problem from a seismic one.

10.4.2. THEORETICAL FRAMEWORK FOR THE PSEUDO-STATIC BEARING CAPACITY

Since the devastating foundation failures reported after the Mexico earthquake (Auvinet and Mendoza, 1986) a wealth of theoretical and experimental studies have been carried out to develop bearing capacity formulae which include the effect of the soil inertia forces (Sarma and Iossifelis 1990; Budhu and Al Karni, 1993; Richards et al., 1993; Zeng and Steedman, 1998).

The theoretical studies mentioned above are based on limit equilibrium methods (Chen 1975; Salençon, 1983); although they represent a significant improvement over the previous analyses which neglected the soil inertia forces, they suffer from limitations which restrict their use (Pecker, 1994):

- the horizontal accelerations of the soil and of the structure are assumed to have the same magnitude;
- the results are derived from an assumed unique failure mechanism which does not allow for foundation uplift;
- the methods only consider upper bound solutions without any indication on how close they are from the exact solution.

At the same time numerous studies have been initiated in France and Europe with the objective of providing more general solutions (Pecker and Salençon 1991, Dormieux and Pecker 1995, Salençon and Pecker 1994a, 1994b, Paolucci and Pecker 1997, PREC8, 1996). The solutions were developed within the framework of the yield design theory (Salençon 1983, 1990): the loading parameters N, V, M and F are considered as independent loading parameters thereby allowing for any combination of actions to be analyzed; many different kinematic mechanisms are investigated and lower bound solutions are also derived to (i) obtain the best possible approximation to the bearing capacity, (ii) bracket the true value to obtain a quantitative measure of the goodness of the solution. It is interesting to note that the results have been later completed by additional lower bound solutions which confirm the merit of the upper bound solutions and help to narrow the gap between upper and lower bound solutions (Ukritchon et al, 1998). Finally the results, mainly based on the upper bound solutions are cast in the general format (Pecker, 1997):

$$\phi \, (N, V, M, F) \leq 0 \qquad\qquad (10.3)$$

where $\phi \, ( ) = 0$ (Figure 10.8) defines in the loading parameter space the equation of a bounding surface.

Inequality (10.3) expresses the fact that any combination of the loading parameters lying outside the surface corresponds to an unstable situation; any combination lying inside the bounding surface corresponds to a potentially stable situation. The word potentially is used to point out that no assurance can be given since the solutions were derived from upper bound solutions. Indications on the merit of the solutions are obtained by comparison with the lower bound solutions and the model safety factor of Equation (10.2) is introduced to account for that uncertainty. The uncertainty is twofold: the solution is obtained from an upper bound approach and, although various kinematic mechanisms were investigated, their number remains necessarily limited when a comprehensive implementation of the upper bound theorem would require that all the conceivable mechanisms be investigated.

These results are put in a simple mathematical expression and implemented in the current version of Eurocode 8 (Annex F) and are applicable to cohesive and purely frictional materials. The equation, (Pecker, 1997), is:

$$\frac{\left(1-e\overline{F}\right)^{c_T} \left(\beta\overline{V}\right)^{c_T}}{\left(\overline{N}\right)^a \left[\left(1-m\overline{F}^k\right)^{k'} - \overline{N}\right]^b} + \frac{\left(1-f\overline{F}\right)^{c'_M} \left(\gamma\overline{M}\right)^{c_M}}{\left(\overline{N}\right)^c \left[\left(1-m\overline{F}^k\right)^{k'} - \overline{N}\right]^d} - 1 \leq 0 \qquad (10.4)$$

where : $\overline{N}=\dfrac{\gamma_{RD}N_{sd}}{N_{max}}, \quad \overline{V}=\dfrac{\gamma_{RD}V_{sd}}{N_{max}}, \quad \overline{M}=\dfrac{\gamma_{RD}M_{sd}}{B\,N_{max}}$.

N_{max} is the ultimate bearing capacity of the foundation under a vertical centered load, N_{sd}, V_{sd}, and M_{sd} are the design action effects at the foundation level, B the foundation width, and γ_{Rd} the material safety factor. The soil inertia forces are accounted for by the

normalized parameter $\overline{F}$ equal to $\rho a B/c_u$ for cohesive soils and to $a/g \, tan\phi$ for frictional soils. The other parameters entering Equation (10.4) are numerical parameters derived by curve fitting to the "exact" bearing capacity, the values of which can be found in Pecker (1997).

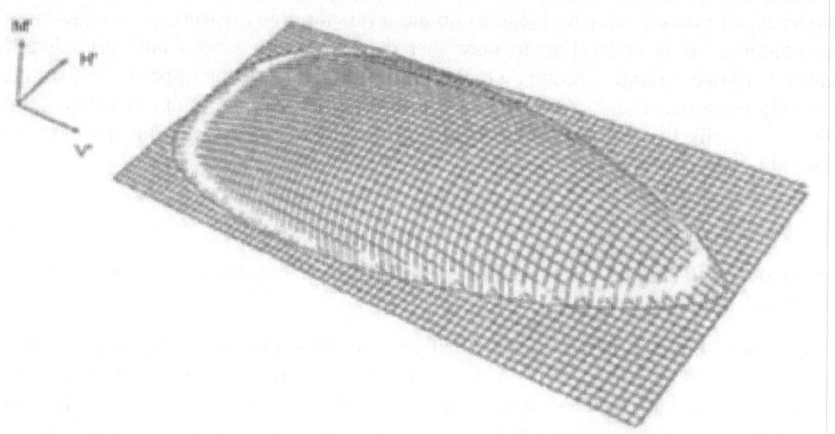

Fig. 10.8. Bounding surface for cohesive soils

10.5. Evaluation of Permanent Displacements

As noted previously and as recommended in Eurocode 8, in seismic situations, the permanent displacements should be evaluated. However such an evaluation is anything but an easy task. Probably the most rigorous approach would be to use a global model (finite element model) including both the soil and the structure. Obviously, the results depend on the non linear constitutive relationship used to model the soil behaviour and are only meaningful if a realistic model is used. Owing to this constraint, to computer limitations, and to the required skill from the analyst in geotechnical engineering, structural engineering, soil-structure interaction and numerical analysis, such an approach is seldom used in everyday practice.

The alternative approach, once the seismic forces are known, is to rely on a Newmark type of approach (Newmark, 1965). The bounding surface defined by Equation (10.3) is used as the surface defining the onset of permanent displacements. Sarma and Iossifelis (1990), Richards et al. (1993) used the Newmark's approach assuming that the soil moves together with the foundation in a rigid body motion. The method has been further extended by Pecker and Salençon (1991) considering a deformable soil body corresponding to the assumed kinematic failure mechanism. Using the kinetic energy theorem, these authors computed the foundation angular velocity, and by integration over time, the foundation permanent rotation. When applied to actual case histories the method proved to be reliable (Pecker et al., 1995).

A potential use of the method can be found for the development of a code like approach. Computed permanent displacements develop when the resultant of the design action lies outside the bounding surface: the larger the distance to the bounding surface, the greater the displacements.

This can be expressed mathematically by writing that for such situations:

$$S_d = \lambda.R_d \tag{10.5}$$

with $\lambda > 1$; $\lambda = 1$ corresponds to the onset of permanent displacements.

Comparing Equation (10.5) to Equation (10.2), it is readily apparent that allowing S_d to reach regions outside the bounding surface is equivalent to specifying a model safety factor γ_{Rd} smaller than 1.0. Therefore, γ_{Rd} can be used, in addition to reflecting the uncertainties in the model, to relax the constraint that at any time the capacity shall be larger than the demand, recognizing the fundamental difference between a static problem and a seismic one in which forces vary in time.

This approach has been implemented in Eurocode 8 and the tentative values proposed in its Annex F are intended to allow for the development of small permanent displacements in potentially non dangerous materials (medium to dense sand, non sensitive clay). These values range from 1.0 (medium dense to dense sands, non sensitive clays) to 1.5 (loose saturated sands) with intermediate values of 1.15 for loose dry sands. If this phenomenon were disregarded, γ_{Rd} values would always be larger than 1.0 (in the range 1.2 to 1.5).

In the case of non sensitive clays further justification for setting γ_{Rd} equal to 1.0 is the observation that shallow foundations in clay have generally been observed to perform well under seismic loading. As mentioned above, a reason for this may be the enhanced undrained shear strength available under rapid loading (Romo 1995, and Ahmed-Zeki et al 1999).

10.5.1. FURTHER DEVELOPMENTS: TOWARDS PERFORMANCE BASED DESIGN

One of the strong assumptions underlying the seismic bearing capacity checks is the independence between the computed design actions and soil yielding. Except for the sophisticated approaches involving the partition in near and far fields, the design actions are computed assuming quasi-linear foundation behaviour. However it is recognized that partial yielding of the foundation may affect the forces.

Attempts have been made by Nova and Montrasio (1991) for monotonic static loading based on the concept of a macro-element modelling the soil and foundation; the constitutive law for the macro-element is rigid plastic strain hardening with non associated flow rule. That concept of macro-element expressed in global variables at the foundation level has been extensively used in mechanics but seldom applied to soil-structure interaction. Paolucci (1997) and Pedretti (1998) have extended the method to seismic loading. These last two studies definitively prove that yielding of the foundation cannot be ignored in the evaluation of the design action.

A more general formulation has been proposed recently by Cremer et al. (2002). The developed macro-element taking advantage of the partition between near field and far field describes the cyclic behaviour of the foundation, reproduces the material non-linearities under the foundation (yielding) as well as the geometrical non-linearities (uplift), and accounts for the wave propagation in the soil. The strength criterion for the macro-element is represented by the bounding surface defined by the bearing capacity formula and a non associated flow rule with kinematic and isotropic hardening is used to compute the pre-failure displacements; the plastic model is coupled with an uplift model to integrate the influence of soil yielding on the uplift. Although presently restricted to strip shallow foundations the model shows some promising capabilities and should represent a step forward in the evaluation of permanent seismic displacements of shallow foundations.

10.6. Construction Detailing

Although the safety of a constructed facility does not rely only upon a blind application of seismic codes and standards which are used for its design and construction, those documents help significantly to minimize the most commonly encountered causes of deficiencies and failures.

Because all phenomena described previously cannot be analyzed with the necessary mathematical rigor and are not often relevant to even sophisticated calculations, construction detailing must always be enforced in seismic design of foundations. This is one of the major merits of seismic codes.

Many of these detailing practices, which are found in the most recent codes, are little more than common sense and by no means, the points raised herein, constitute an exhaustive list. However based on the author's experience, they represent the most common mistakes made in design by non-experienced designers:

- Foundations must not be located close to (or across) major active faults. Ground motions in the near field are far from being predictable and attempts to design buildings to accommodate such movements, especially the static co-seismic displacements associated with fault rupture, are almost hopeless.
- Liquefiable deposits and unstable slopes must always be treated before construction.
- The foundation system under a building must be as homogenous as possible unless construction joints are provided in the structure. In particular, for individual footings, the situation where some of them rest on a man-made fill and some on in-situ soils must always be avoided. It is also highly desirable that the foundations respect the symmetries of the building.
- The choice of the foundation system must always account for possible secondary effects such as settlements in medium-dense or loose dry sands, the post-earthquake consolidation settlements of clay layers, the settlements induced by the post-earthquake dissipation of pore pressures in a non liquefiable sand deposit. Raft foundations or end bearing piles are to be preferred whenever the anticipated magnitude of the settlements is high or when they can be highly variable across the building.
- Individual footings must always be linked with tie beams at the foundation level. These longitudinal beams must be designed to withstand the differential settlements between the footings.

10.7. Conclusions

The state of the art in the seismic design of shallow foundations is developing towards a rational, although still simplified, analysis. Sound, easy to use representations of the stiffness and damping of the soil foundation system can be provided to the structural engineer to perform the dynamic analysis of the structure when the system remains linear. The geotechnical engineer has at hand a theoretical framework to compute the seismic bearing capacity of foundations for which much progress has been achieved in the last decade leading to a rational, rigorous, approach. This bearing capacity is expressed in terms of a bounding surface defined by simple analytical formulae expressed in terms of dimensionless variables related to the inertial or kinematic loading parameters.

Recent developments in modeling the soil structure interface allow the designer to compute the earthquake induced permanent displacements of his foundation and therefore to rely on a, more rational, performance based approach for foundation design. It has been shown that performance based design can also be included in building codes as it has been tentatively done in the recent Eurocode 8 – Part 5 design code.

CHAPTER 11
BEHAVIOUR AND DESIGN OF DEEP FOUNDATION SUBJECTED TO EARTHQUAKES

Kohji Tokimatsu
Tokyo Institute of Technology, Tokyo, Japan

11.1. Introduction

Extensive soil liquefaction that occurred in the Hyogoken-Nambu earthquake (M=7.2) of January 17, 1995, damaged various structures and infrastructures in the reclaimed land areas along the coastline of Kobe. In particular, many of the quay walls in these areas moved up to several meters towards the sea due to liquefaction of their foundation soils and/or back-fills. This induced large horizontal ground movements as well as differential ground settlements near the waterfront. As a result, not only buildings with spread foundations but also those supported on piles settled and/or tilted without significant damage to their superstructures (Photo 11.1). Similar damage patterns were also observed at many buildings on liquefied level ground far away from the waterfront (Photo 11.2).

There was a serious concern that the piles of those buildings might have been damaged. Excavation surveys after the quake showed that pile heads in some tilted buildings actually failed (Photo 11.3) while others did not. This suggests that the piles might have failed at depths other than pile head (Photo 11.4) due to liquefaction-induced ground movement. However, little is known concerning the actual failure and deformation patterns of those piles, and their relation to ground displacements.

Photo 11.1. Tilted building in lateral spreading area

Photo 11.2. Tilted building in liquefied level ground

The object of this chapter is to summarize the failure and deformation modes of piles that experienced soil liquefaction and/or lateral spreading in the 1995 Kobe earthquake, together with cyclic and permanent ground displacements that might have developed during past earthquakes. Pseudo-static analyses using p-y curves are then presented and conducted for well-documented case histories of pile foundations from the Kobe earthquake.

A. Ansal (ed.), Recent Advances in Earthquake Geotechnical Engineering and Microzonation, 303–324.
© 2004 *Kluwer Academic Publishers.*

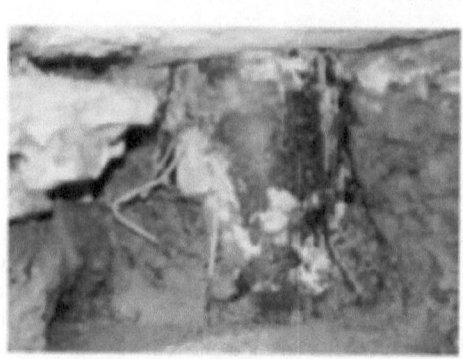

Photo 11.3. Damage to pile head

Photo 11.4. Damage to pile
near ground water table

11.2. Performance of Near-Surface Soils and Pile Foundations during the 1995 Hyogoken-Nambu Earthquake

11.2.1. SOIL LIQUEFACTION AND GROUND MOTION

Figure 11.1 is a map showing the area of heavy damage to buildings (Chuo-Kaihatsu, 1995) and the area where field manifestation of soil liquefaction was evident after the 1995 earthquake. Most of the buildings discussed in this paper were located in reclaimed land areas including Port and Rokko Islands, Fukaehama, and Mikagehama where soil liquefaction extensively occurred. Liquefaction developed to a lesser extent in central Port Island and southern Rokko Island where the fills had been treated or consisted of soils containing significant amounts of clay.

Also shown in Figure 11.1 are the particle orbits of ground displacements that have been double-integrated from the strong motion accelerograms obtained in those areas during the earthquake (CEORKA, 1995; Sekisui House, 1996). The peak cyclic ground displacements were 35cm at a non-liquefied site on Rokko Island, and 46cm and 55cm near liquefied areas on Port Island and in Fukaehama. It is estimated that about two thirds of these displacements resulted from shear strains induced in the reclaimed fills.

Figure 11.2 shows the acceleration response spectra with damping ratios of 10% for the ground motions at the same stations. The response acceleration in the period range less than 0.5s, which is equal to the natural period of buildings investigated in the study, is approximately equal to 0.3-0.5g.

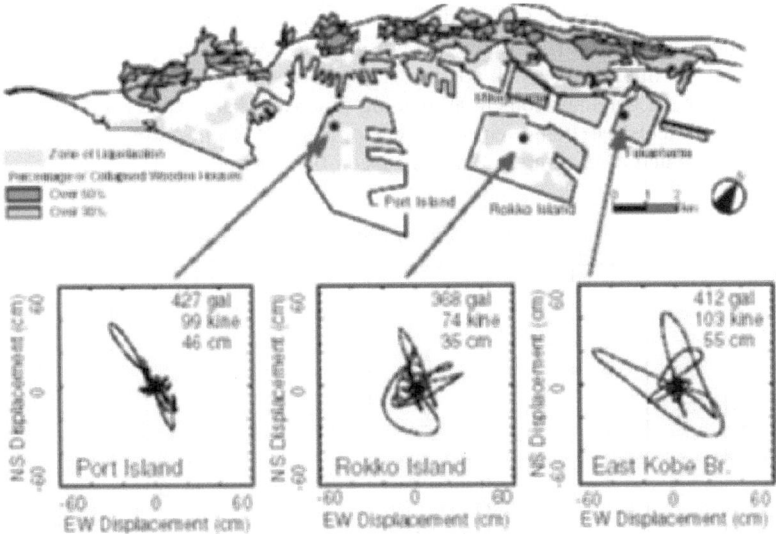

Fig. 11.1. Map showing zones of extensive building damage and liquefaction with displacement orbits of strong ground motions recorded in reclaimed lands

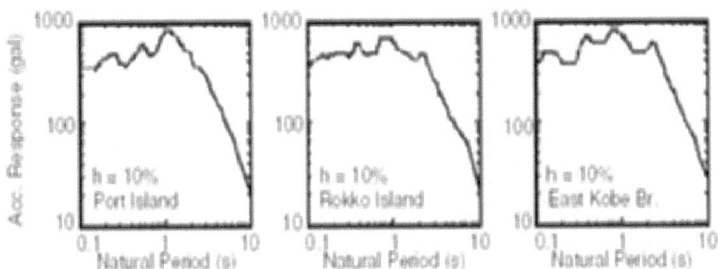

Fig. 11.2. Response acceleration spectra at three stations in reclaimed lands

11.2.2. CHARACTERISTICS OF PILE FOUNDATIONS OF BUILDINGS

The piles used in the Kobe area for building foundations are classified into precast concrete or steel pipe (S) piles and cast-in-place concrete piles. Most precast piles are hollow with outer diameters typically ranging from 35 to 60cm, which contrasts with solid cast-in-place concrete piles with diameters typically over 100cm. The precast concrete piles encountered in the Kobe area include prestressed concrete (PC) piles used before the 1980s and prestressed high strength concrete (PHC) piles used after the 1980s. To strengthen their capacity and ductility, steel pipe reinforced concrete (SC) piles and reinforced prestressed concrete (PRC) piles have been also used.

Table 11.1 summarizes the case histories to be analyzed in this study, with characteristics of piles. Figure 11.3 shows the relations of bending moment with curvature for those piles. Both PC and PHC piles have three different capacities for a given diameter, i.e., Types A, B, and C in ascending order of capacity. In the following, M_u is the bending moment at concrete crashing at extreme compression fibber of PC, PHC, and SC piles or at which compression fibber strain of S piles reaches a limiting value; M_y is the bending moment at yielding of tension bars of PC and PHC piles or at yielding of steel at extreme tension fibber of SC and S piles; and M_c is the bending moment at concrete cracking at extreme tension fibber of PC and PHC piles.

Table 11.1. Case histories analyzed in this study

Building or Site	Building Type	No. of Story	Depth of Pile head (m)	Pile Type	Pile Diameter (mm)	Pile Length (m)	Axial Force (MN)	Ground water Table (m)	Thickness of Reclaimed Fill (m)
A	RC	4	1.2	PHC-A	350	22	0.3	3	13
B	RC	5	1.7	SC+PHC-A	600	31	0.9	2.2	11.5
C	RC	4	1.73-2.13	SC+PHC-B	500	35-42	0.63	3	17
D	RC	6	1.78	S	500	30	0.66	3	13
E	NA	NA	1	SC+PHC-A	500	33	0	2	14
F	RC	3	1.65	PC-A	400	20	0.39	2	8
G	RC+S	2	2.5	S	406.4	27.55	0.63	3.5	15.4
H	SRC	NA	2.10-2.30	CC	1500	47-48	0.14	1.65	15

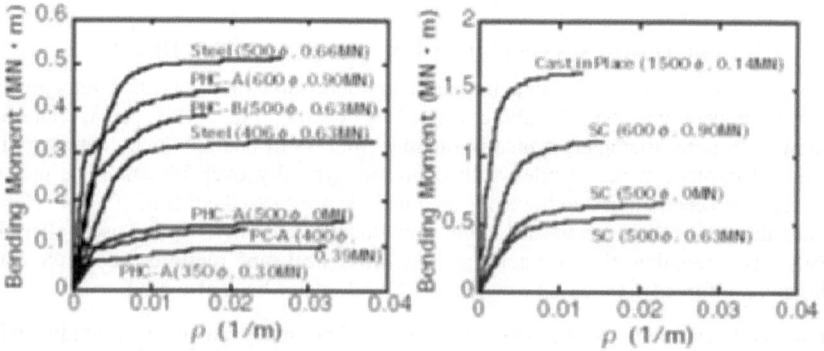

Fig. 11.3. Relation between bending moment and curvature of various piles

11.2.3. PILE DAMAGE FROM DETAILED FIELD INVESTIGATION

A large number of performances of pile foundations during the Kobe earthquake have been revealed based on field investigation including excavation of pile heads (Photo 11.5, e.g., Kansai Branch of Architectural Institute Japan (AIJ), 1996; AIJ et al., 1998). In addition to integrity tests for piles, several methods were used to detect pile damage below the ground surface. Borehole cameras (Photo 11.5, Oh-oka et al., 1996) have identified damage portions and severity of hollow or cored solid piles, and inclinometers (Photo 11.6, Shamoto et al., 1996) have provided data to estimate deformed shapes with depth of hollow piles.

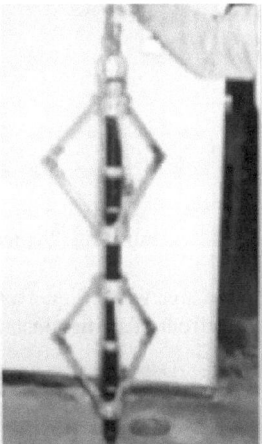

Photo 11.5. Borehole camera survey followed by   Photo 11.6. Inclinometer
excavation

The main findings from the field investigation are summarized below and illustrated in Figures 11.4 and 11.5 in the level ground without occurrence of lateral spread (Tokimatsu and Asaka, 1998, and Tokimatsu, 1999):

1. Damage concentrated near pile heads, and the top and the bottom of the liquefied layer (Figures 11.4(a)-(e)).
2. In some PC and PHC piles, damage occurred only near the top and/or bottom of liquefied layers, with no damage near pile heads (Figures 11.4(b)- (d));
3. Damage near pile heads often resulted in significant tilts of buildings with high aspect ratios (Figures 11.4(a), (b)).
4. Damage to piles through a thick non-liquefied crust did not necessarily lead to large tilts (Fig. 11.4(c); BTL Committee, 1998).
5. The failure and deformation modes of piles within a building were very similar to each other (Figures 11.4(a)-(e)).
6. Damage concentrated on PC and PHC piles, but no extensive damage to S and CC piles was reported.
7. PHC piles without any vertical load also suffered extensive damage near the bottom of the liquefied layer (Fig. 11.4(e); Horikoshi and Ohtsu, 1996).

8. In the treated level ground, buildings supported on piles bearing on firm soils
    beneath the fills often had no apparent settlement while the surrounding ground
    settled to some extent, creating vertical gaps around the bases of many
    buildings (Figure 11.4(f)).

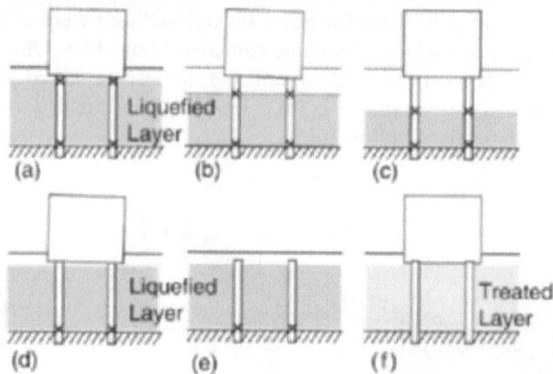

Fig. 11.4. Typical damage pattern of piles subjected in level ground

In the area where liquefaction-induced lateral spreading occurred, particularly along the
waterfront of artificial islands (Tokimatsu and Asaka, 1998, and Tokimatsu, 1999):

1. Damage also concentrated near the pile head, or the top and/or the bottom of
    the liquefied layer.
2. Damage was not limited to PC and PHC piles but extended to cast-in-place
    concrete piles (Tokimatsu et al., 1996) and some ductile S piles (JASPP, 1996).
3. The damage to PC and PHC piles often resulted in a large tilt of the
    superstructure, whereas the damage to S and CC piles rarely led to similar
    consequences.
4. A cast-in-place concrete pile foundation that carried only a small load was also
    damaged and displaced horizontally by as much as 1 m (Kuwabara and Yoneda,
    1998).
5. Damage to pile caps and foundation beams often preceded or accompanied the
    damage to S and CC piles.
6. The piles within a building near the waterfront showed different failure and
    deformation modes in the direction perpendicular to the shoreline as shown in
    Figs. 11.5(a) and 11.5(d) (Tokimatsu et al., 1997). In such a case, when facing
    the span side of the building with the sea on the left, the seaside pile cap rotated
    clockwise around its longitudinal axis, whereas the land-side pile cap rotated
    counter clockwise (Oh-oka et al., 1997a, 1997b).
7. Cast-in-place concrete piles surrounded by deep mixing walls as well as steel
    pipe piles driven in the ground treated by sand compaction piles did not suffer
    any serious damage (BTL Committee, 1998). Moderate to severe damage to
    their superstructures was, however, observed in these cases.
8. Cast-in-place concrete piles surrounded by cement column walls or continuous
    diaphragm walls did not suffer any serious damage (BTL Committee, 1998).
    The permanent horizontal displacements of bridge piers founded on diaphragm
    walls were negligibly small, while those of bridge piers founded on piles or

caissons were as large as a half of the permanent ground displacements nearby (Yokoyama et al., 1997).

The above findings confirm that, in addition to horizontal forces and overturning moments imposed on pile heads from superstructures, kinematic forces induced by dynamic and permanent ground displacements of liquefied and laterally spreading soils had significant impact on pile damage. In particular, the damage to piles without vertical loads confirms significant effects of ground movements.

The difference in failure and deformation modes of piles within a building near the waterfront as shown in Figures 11.5(a) and (d) probably reflects rapid changes in horizontal ground displacement. In addition, Figure 11.5 suggests that the lateral ground movement leads to more serious tilt in the span direction than that in the longitudinal direction. This indicates that to place the longitudinal direction of the building parallel to the direction of ground movement is effective for mitigating damage due to lateral ground movement.

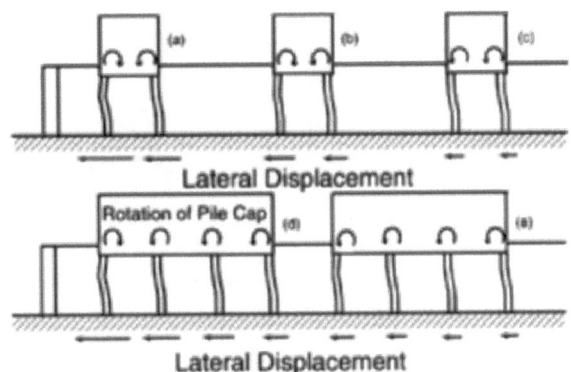

Fig. 11.5. Typical failure and deformation modes of pile foundations subjected to lateral spreading

The difference in damage of different pile type indicates that to use ductile piles or rigid foundations is also effective for mitigating damage resulting from lateral ground movement. Conversely, however, the inertial forces acting on their superstructures during shaking would increase with increasing rigidity of foundation. The earth pressure acting on the upstream side the building from laterally spreading soils would also increase with increasing rigidity of foundation.

11.3. Cyclic and Permanent Ground Displacements during Earthquakes

11.3.1. CYCLIC AND PERMANENT SHEAR STRAINS IN LIQUEFIED AND LATERALLY SPREADING GROUND

The field investigation described in the previous chapter has shown that the cyclic and permanent ground displacement to be developed in the liquefiable deposit had significant effects on pile performance, and thus these displacement patterns should be identified. Figure 11.6 summarizes the relation of cyclic and permanent shear strains with adjusted *SPT N*-values from the results of recent studies. In the figure, the cyclic shear strains in the level ground are plotted in Figure 11.7(a), while permanent shear strain near the waterfront are plotted in Figure 11.8(b). Also shown in Figure 11.8 in the solid curves are the limiting shear strains. Note that the maximum permanent shear strain is much larger than the cyclic shear strain for the same *N*-value.

Based on Figure 11.6, Tokimatsu and Asaka (1998) proposed preliminary charts for estimating maximum cyclic shear strain to be developed during and after earthquakes, as shown in Figure 11.7(a).

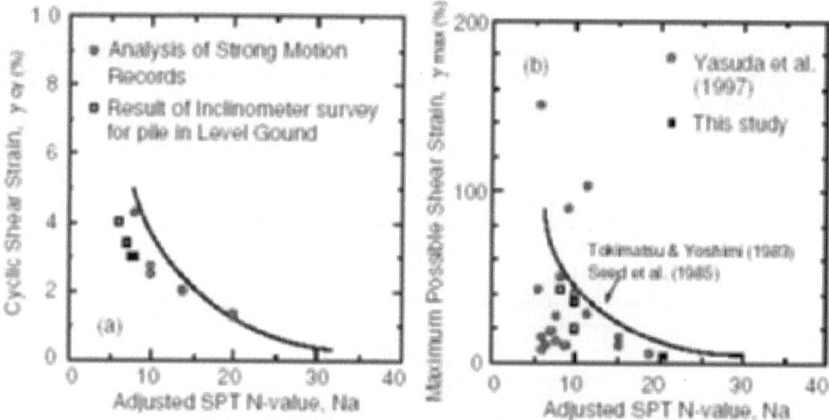

Fig. 11.6. Field correlation of cyclic and permanent shear strains with adjusted *SPT N*-value: (a) cyclic shear strain, (b) permanent shear strain

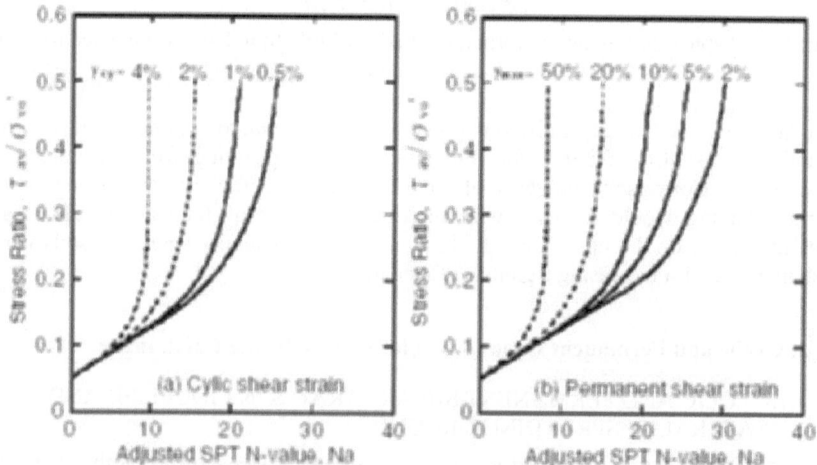

Fig. 11.7. Maximum cyclic and permanent shear strains due to soil liquefaction and lateral spreading

Figure 11.7(a) enables one to estimate cyclic ground displacement profile of a liquefied deposit in a simplified manner described below, as is the case in the liquefaction evaluations using *SPT N*-values (e.g., Tokimatsu and Yoshimi, 1983; Seed et al., 1985):

1. Determine adjusted *SPT N*-values, N_a, and equivalent cyclic stress ratios during earthquake, τ_{av}/σ_{vo}', with depth.
2. Estimate γ_{cy} from Figure 11.7(a), with depth.
3. Estimate a cyclic ground displacement profile, $f_{cy}(z)$, by integrating γ_{cy} upwards from the bottom of the liquefied layer, assuming γ_{cy} develops in the same horizontal direction.

For example, when an 8-m thick sand layer with $N_a=10$ liquefies extensively ($\gamma_{cy}=4\%$), the resulting ground surface displacement, $D_{cy}=f_{cy}(0)$, is estimated to be 32cm.

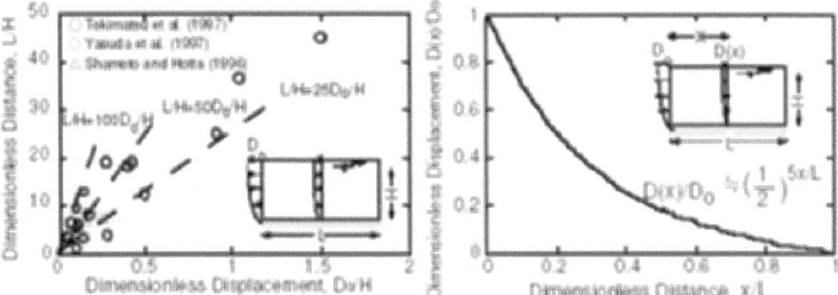

Fig. 11.8. Relation between horizontal displacement of waterfront and length of laterally spreading area

Fig. 11.9. Relation of horizontal ground displacement with distance from waterfront

11.3.2. PERMANENT GROUND DISPLACEMENT NEAR WATERFRONT

The permanent horizontal displacement induced by lateral spreading near the waterfront takes a maximum value at the waterfront and decreases with distance from the waterfront. Thus, the maximum value together with its attenuation characteristics should be identified. Figure 11.8 summarizes, based on recent studies, the relations between the horizontal ground surface displacement at the waterfront, D_o, and the length of the laterally spreading area, L, both normalized in terms of the thickness of the liquefied layer, H. The relation between the two may be expressed as:

$$L/H = (25\text{-}100)D_o/H \qquad (11.1)$$

This indicates that lateral spreading may extend horizontally inland about 50 times the horizontal ground surface displacement at the waterfront.

Figure 11.9 shows a typical relation of horizontal ground displacement, D, with distance from the waterfront, x, from the studies by Shamoto and Hotta (1996), and Ishihara et al. (1997) in which D and x are normalized in terms of D_o and L. The relation may be expressed as:

$$D(x)/D_o = (1/2)^{5x/L} \qquad (11.2)$$

Assuming that $L=50D_o$, the equation leads to

$$D(x) = (1/2)^{x/10Do} D_o \tag{11.3}$$

Figure 11.9 and Equations (11.2) and (11.3) show that the horizontal ground surface displacement in laterally spreading areas decreases to a half at $x=L/5$ and less than 1/5 at $x=L/2$.

The permanent ground displacement profile with depth z at distance x of a laterally spreading deposit, $f_{ls}(z,x)$, may be then approximated as:

for $z < z_w$

$$f_{ls}(z,x) = D(x) \tag{11.4}$$

for $z > z_w$

$$f_{ls}(z,x) = D(x)\cos(\pi(z - z_w)/2H) \tag{11.5}$$

$$= D(x)(1 - (z - z_w)/H) \tag{11.6}$$

in which z is depth below the ground surface, and z_w is depth of the groundwater table or the top of the liquefied layer.

D_o in Equation (11.2) depends significantly on the type and seismic design of quay walls as well as the strong motion characteristics and soil conditions behind and below the quay wall, and may be expressed as:

$$D_o = \min (D_{max}, D_w) \tag{11.7}$$

in which D_w is the displacement of the quay wall and D_{max} is the maximum possible ground surface displacement of the liquefied soil determined by integrating γ_{max} with depth. In this estimation, γ_{cy}, $f_{cy}(z)$, D_{cy}, and Figure 11.7(a) in the procedure described in Section 11.3.1 should be replaced by γ_{max}, $f_{max}(z)$, D_{max}, and Figure 11.7(b). If the quay wall moves seaward by 3 m with a 10-m thick liquefied sand layer having $N_a=10$ ($\gamma_{max}= 40\%$), the permanent ground displacement near the waterfront is expected to be 3m. In contrast, it is only 1 m for a layer with $N_a =20$ ($\gamma_{max}= 10\%$).

11.4. Pseudo-Static Analysis for Seismic Design of Pile Foundations

Seismic design of foundations may be made based on either dynamic response or pseudo-static analyses. In this section, a pseudo-static analysis based on Beam-on-Winkler-springs method is described, with emphasis placed on how the ground displacement determined in the previous section is incorporated into the design.

11.4.1. INERTIAL AND KINEMATIC FORCES ACTING ON FOUNDATION

Figure 11.10 schematically illustrates the soil-pile-structure interaction in liquefiable soils during and after an earthquake. Before liquefaction, the inertia force from the superstructure may dominate (Case I). After liquefaction during shaking, not only the inertial force but also the kinematic force induced by the cyclic ground displacement comes to play an important role (Cases II). Towards the end of shaking, residual shear strain may accumulate, resulting in permanent horizontal ground displacement. At this stage, the kinematic force due to permanent ground displacement may have a dominant effect on pile performance (Case III), particularly near quay walls which failed or moved seaward (Case III-b). Such permanent ground displacement may also occur even in a horizontally stratified soil deposit with level ground (Case III-a); however, it is generally less than and hardly ever exceeds the maximum cyclic ground displacement unless it is extremely loose.

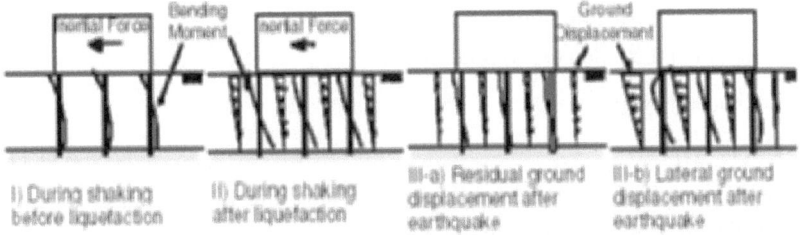

Fig. 11.10. Schematic figure showing soil-pile-structure interaction in liquefied and laterally spreading soil

The above discussions indicate that piles in the level ground could suffer the most severe loading conditions in either Case I or II, whereas piles in laterally-spreading soils also experience severe loading condition in Case III. These loading conditions should be properly considered in stress and deformation analysis of piles in liquefiable soils.

11.4.2. BEAM-ON-WINKLER-FOUNDATION METHOD

Simplified pseudo-static design methods using p-y curves for pile foundations (AIJ, 1988, 2001; JRA, 1997), i.e., a single pile supported on nonlinear Winkler springs as shown in Figure 11.11(a), are based on the following equation:

$$EI(d^4y/dz^4) = -kBy \qquad (11.8)$$

in which E and I are Young's modulus and moment of inertia of pile, y is horizontal displacement of pile, z is depth, k is coefficient of horizontal subgrade reaction, and B is pile diameter.

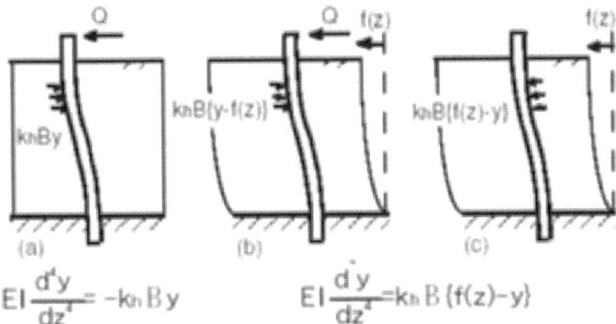

$$EI\frac{d^4y}{dz^4} = -k_h B y \qquad EI\frac{d^4y}{dz^4} = k_h B \{f(z) - y\}$$

Fig. 11.11. Schematic figure showing simplified pseudo-static analysis using p-y springs for pile foundations

If liquefaction and/or lateral spreading occur during and after shaking as shown in Cases II and III in Figure 11.10, the equation may be defined as:

$$EI(d^4y/dz^4)=kB\{f(z)-y\} \tag{11.9}$$

in which *f(z)* is either cyclic or permanent ground displacement profile described in Section 11.3, which is applied to the pile through the p-y springs as shown in Figures 11.11 (b) or (c).

11.4.3. NON-LINEAR P-Y SPRING

The value of k in Equation (11.8) is defined as:

$$k=2k_o/(1+|y/y_1|) \tag{11.10}$$

$$k=k_o(y/y_1)^{-0.5} \tag{11.11}$$

in which *y1* is reference displacement equal to either 1cm or 0.01*B* depending on the codes, and k_o may be defined as (JRA, 1980):

$$k_o = 56NB^{-3/4} \ (MN/m^3) \tag{11.12}$$

in which B is pile diameter in cm, *N* is *SPT N*-value. The ultimate lateral resistance or pressure, p_y, may be defined as:

$$p_y=3K_p\sigma_{vo}' \tag{11.13}$$

in which σ_{vo}' is the initial effective confining pressure, and K_p is the Rankine passive earth pressure coefficient.

When the ground displacement induced by soil liquefaction cannot be neglected in Equation (11.9), *y* in Equations (11.10) and (11.11) has to be replaced by $y_r =y-f(z)$. Both P_y and k_o of liquefied soils should be reduced, for example, according to the following equations, as illustrated in Figure 11.12 (AIJ, 2001).

$$k_{ol} = \beta_l\,k_o \tag{11.14}$$

$$p_{yl} = \alpha_l\,p_y \tag{11.15}$$

in which α_l and β_l are scaling factors in terms of *SPT N*-value as show in Figure 11.13 where α_l is tentatively equal to β_l (AIJ, 2001).

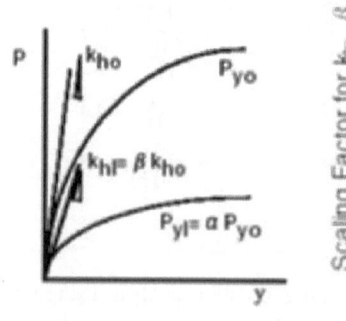

Fig. 11.12. Analytical model for
p-y spring

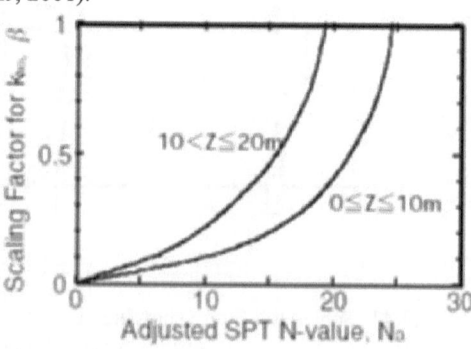

Fig. 11.13. Scaling factor for p-y spring

11.4.4. EARTH PRESSURE ACTING EMBEDDED FOUNDATION

The earth pressure acting on the embedded porting of foundation also depends on the relative displacement between soil and foundation. When the ground displacement is negligibly small, the soil always resists the foundation to deform, as shown in Figure 11.14(a). In contrast, when the ground displacement gets large, the soil tends to push the foundation, the extent of which depends on the rigidity of foundation, as show in Figures 11.14(b), (c). Such effects should be properly taken into account in the analysis.

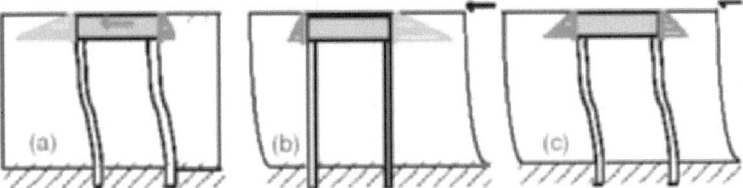

Fig. 11.14. Schematic figure showing earth pressure acting on foundations due to inertial force and ground displacement

11.5. Effects of Cyclic Ground Displacements on Pile Performance

While many PC and PHC piles in an extensively liquefied area suffered severe damage, most piles of the same types in the non-liquefied area in reclaimed lands survived without visible damage (Tokimatsu et al., 1996). This suggests that Case II shown in Figure 11.10 provides more severe loading conditions than Case I for pile foundations supporting buildings. In order to clarify this point, a case history of 35cm diam PHC piles supporting a 4-story building in Mikagehama (Shamoto et al., 1997; Building A in Table 11.1) is examined. Figure 11.15(c) shows a boring log of the site. A field survey showed that the piles cracked near the pile head as well as near the bottom of the fill as shown in Figure 11.15(d), causing a large tilt of the building.

Both Cases I and II are considered in the analysis. It is assumed that the piles are subjected to an inertial force with a base shear coefficient of 0.4 for both cases. In Case II, the cyclic ground displacement profile computed from the proposed method as shown in Figure 11.15(b) is also considered. The computed displacement on the ground surface is about 36cm, which appears consistent with the observed values described previously. A scaling factor for the horizontal subgrade reaction of 1/10 is used from Figure 11.13 for liquefied soils throughout the study, since a preliminary analysis indicates that the difference in scaling factor does not have a significant effect on estimated failure patterns of piles.

Figures 11.15(a) and (e) show the distribution of computed bending moments with depth and the relations between bending moment and curvature at critical depths for the two cases. The computed bending moments in Case II are close to M_u near the pile head and the bottom of the liquefied fill, while those in Case I are considerably smaller at all depths. The computed result of Case II is consistent with the pile damage in Figure 11.15(d), while that of Case I appears to reproduce the better performance of pile foundations in non-liquefied deposits. These results confirm that Case II creates more severe loading conditions than Case I and that the ground displacement in liquefied deposits has significant effects on pile performance during earthquakes. Thus, the analyses for Case I will be omitted hereafter.

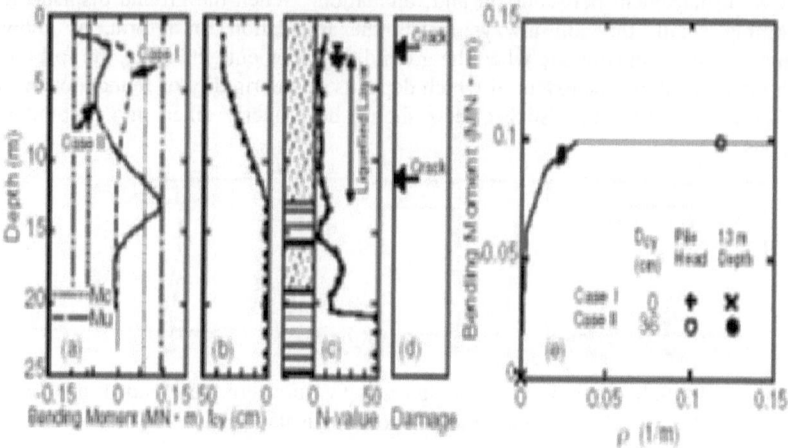

Fig. 11.15. Computed bending moment of PHC pile at Building A for Cases I and II, with ground displacement, boring log and pile damage

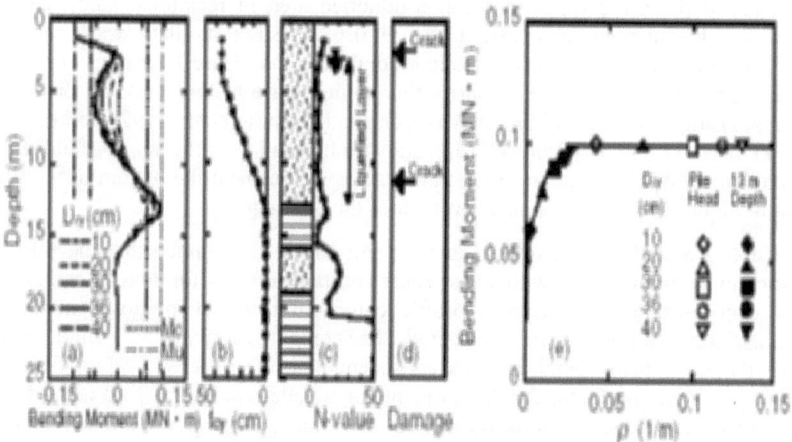

Fig. 11.16. Computed bending moment of PHC pile at Building A for Case II, with ground displacement, boring log and pile damage

In order to examine the effects of pile type and ground displacement on pile damage, similar computations are made for four buildings including the one described above and three supported on either SC+PHC-A piles, SC+PHC-B piles, or S piles (Buildings A-D in Table 11.1). The different field performance of two follower piles of different capacities and the critical behaviour of S piles may provide a good basis to identify the major factor influencing pile damage.

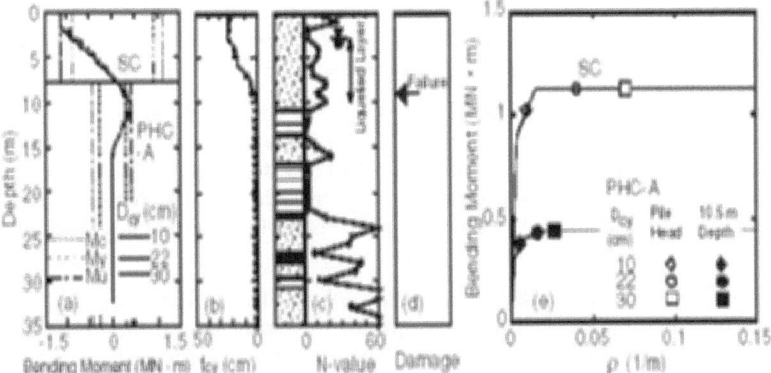

Fig. 11.17. Computed bending moment of pile at Building B for Case II, with ground displacement, boring log and pile damage

A five-story building supported on the SC+PHC-A piles in the center of Fukaehama (Building B) tilted by 1/29 to the northeast after the quake without any damage to superstructure (Nagai, 1997). 50-60cm diameter SC piles 6m long, and PHC-A piles 13 and 12m long were used as the upper, and middle and lower piles. The piles penetrated to a depth of about 33m through a reclaimed fill with a thickness of about 10 m (Figure 11.17(c)). An excavation survey suggested that the pile heads inclined to the southwest, the opposite direction of the tilting of the building, by 1/30-1/23. In addition, boring and inspection into a hollow space of a pile through the pile cap suggested that the pile was damaged and bent largely near the bottom of the liquefied layer (depths between 9.6 and 11.3m (Figure 11.17(d))).

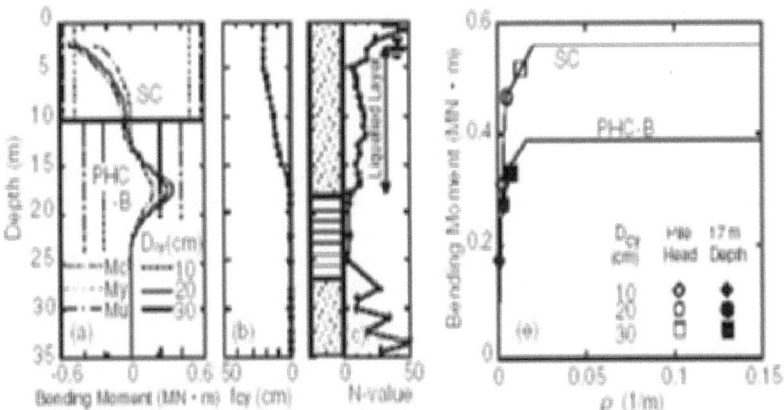

Fig. 11.18. Computed bending moment of pile at Building C for Case II, with ground displacement, boring log and pile damage

A four-story building supported on the SC+PHC-B piles in an untreated area on Port Island (Building C) suffered neither differential settlement nor structural damage (Fujii et al., 1996). 50 cm diameter SC piles 8m long, and PHC-B piles 13 and 14 m long were used as the upper, and middle and lower piles. The piles penetrated to a depth of about 37 m through a thick reclaimed fill (Figure 11.18(c)). An excavation survey with pile integrity tests on some exposed pile heads suggested that the piles had no damage.

Five-story school buildings supported on the 50 cm diameter S piles 30 m long in Fukaehama (Building D) experienced insignificant damage, though the ground surface around the building settled by 10 to 50 cm (JASPP, 1996). Figure 11.19(c) shows the boring log of the site. An excavation survey, however, showed that large residual deformation occurred at the horizontal steel plate connecting the pile head and the pile cap.

It is assumed in the analysis that the piles are subjected to inertial force with a base shear coefficient of 0.4 as well as kinematic forces resulting from the cyclic ground displacement profile shown in Figures 11.16(b)- 11.19(b). In addition, computations are also made, using the same base shear coefficient, for several ground displacement profiles scaled from the computed ones.

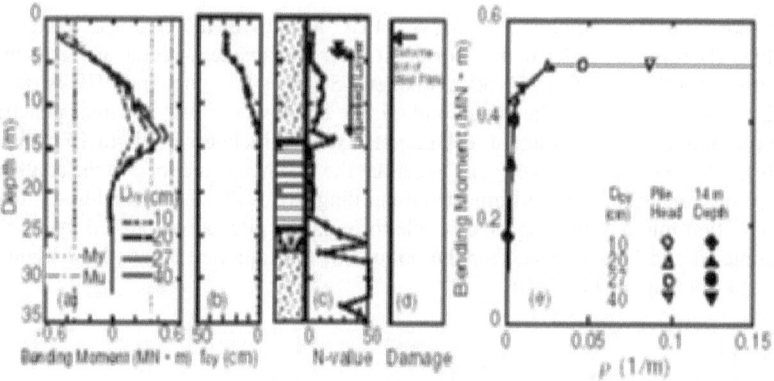

Fig. 11.19. Computed bending moment of pile at Building D for Case II, with ground displacement, boring log and pile damage

Figures 11.16(a)- 11.19(a) compare the distributions of the computed bending moment with depth for several ground surface displacements, and Figures 11.16(e)- 11.19(e) the relations between bending moment and curvature at critical depths, i.e., the pile head and/or the bottom of the liquefied layer. The computed results in Figures 11.16 to 11.19 show that the piles themselves may not suffer severe damage for a ground surface displacement less than 15 cm as the bending moment at any depth is below the critical value. In contrast, for ground surface displacements greater than 20-45 cm, pile foundations may suffer significant distress depending on their capacity and ductility, since the bending moments near the pile head and the bottom of the liquefied layer reach the critical value. For example, the bending moments at the critical depths of S+PHC-A pile in Figure 11.17 exceed M_u for a ground surface displacement of about 25 cm, while those of S+PHC-B pile in Figure 11.18 and of S pile in Figure 11.19 are still below M_y or M_u for the same displacement. In addition, most of the piles are estimated to experience severe stress conditions near the bottom of the liquefied layer as well as near the pile head. The above results appear consistent with the field performance and damage features of the various pile foundations in reclaimed lands summarized previously, suggesting the significance of ground displacement in pile damage.

A case history of a group of piles (Horikoshi and Ohtsu, 1996; Site E in Table 11.1) also provides a good basis to evaluate the effects of ground displacement, since neither inertia force nor vertical load from the superstructure was imposed on the piles during the earthquake. The piles 33 m long (SC pile 5 m long + PHC piles 13m and 15 m long) with diameters of either 40 or 50 cm had been driven through a reclaimed fill in the centre of Fukaehama about 350 m away from the shoreline. The pile heads were located at a depth in between 0.5 and 1.5 m with a groundwater table of 2 m. The piles experienced the earthquake before placement of any pile caps or foundation beams and failed at a depth of about 8 m, i.e., the interface between gravelly medium sand with *SPT N*-values of 5 and gravelly coarse sand with *SPT N*-values of 12 (Figure 11.20(c) and (d)).

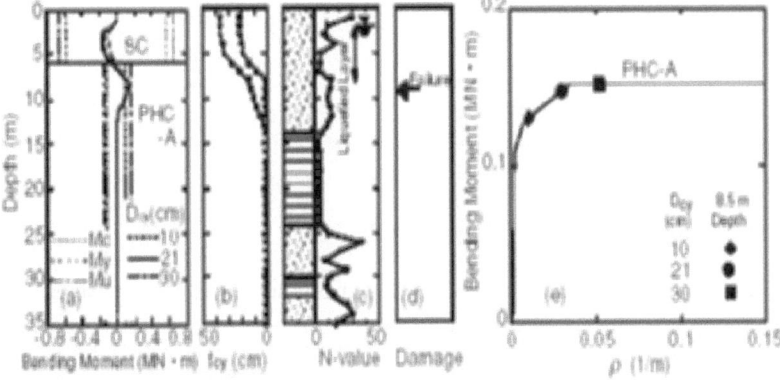

Fig. 11.20. Computed bending moment of pile at Site E for Case II, with ground displacement, boring log and pile damage

Figure 11.20 (b) shows the estimated cyclic displacement profiles for two cases where soil liquefaction is assumed to have developed only in a loose fill above 8 m depth or throughout the fill. The pseudo-static analysis has been conducted without any inertial forces using the estimated cyclic displacement profiles and their scaling values. Figures 11.20(a) and (e) summarize the computed results only for the former case since it has yielded more severe stress conditions in the pile. The computed moments near the interface between liquefied and non-liquefied layers reach the ultimate bending moment, M_u, for a ground surface displacement of about 25 cm. The computed result appears consistent with the observed damage pattern of the pile, confirming the significant effects of cyclic ground displacement in pile damage.

11.6. Effects of Permanent Ground Displacements on Pile Performance

It is conceivable that the difference in the failure mode near the waterfront such as shown in Figures 11.5(a), (d) might have been induced by the variation of horizontal ground displacements in the direction perpendicular to the shoreline such as shown in

Figure 11.9. The p-y analysis for a single pile cannot simulate such deformation patterns. Thus, an analytical model shown in Figure 11.21 is developed for Case III in which a group of piles connected with a foundation beam are subjected to permanent displacement profiles varying with distance from the waterfront.

First, case histories of two pile foundations in and around Fukaehama (Buildings F and G) are examined. Building F of three stories, supported on 40 cm diameter PC piles 20m long, was situated 6 m from a quay wall that displaced by about 2m towards the sea (Tokimatsu et al., 1997). Its span direction was perpendicular to the shoreline. Figure 11.22(d) shows a boring log of the site, which indicates that a loose fill 8m thick might have liquefied during the earthquake.

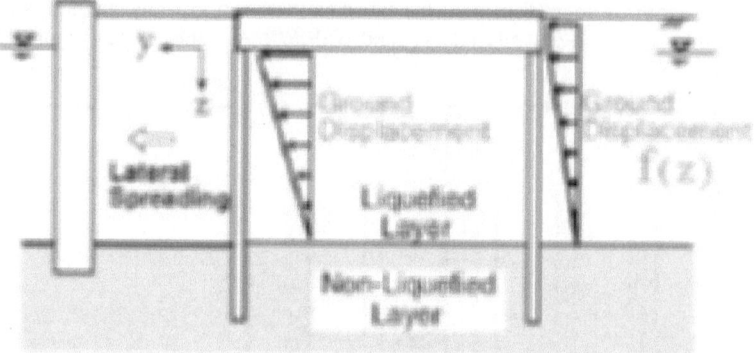

Fig. 11.21. Analytical model of pile foundation subjected to lateral spread

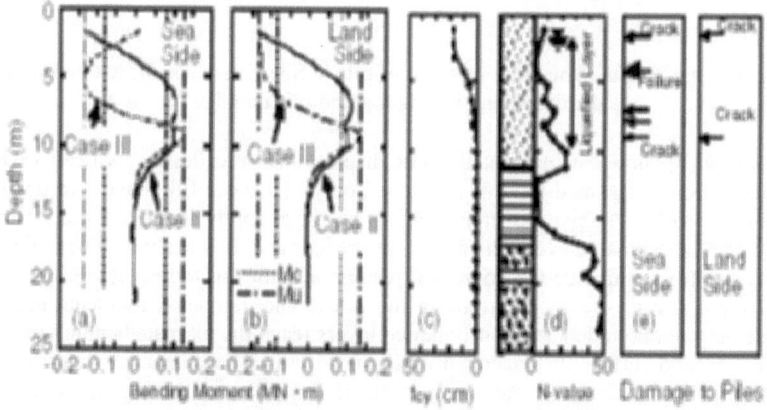

Fig. 11.22. Computed bending moment of pile at Building F for Cases II and III, with ground displacement, boring log and pile damage

A field survey conducted after the quake showed that the building had moved by about 80 cm and inclined by three degrees towards the sea, without any damage to its superstructure. An inclinometer survey also showed that the piles in the building failed in a different mode depending on their location. Namely, the piles on the sea side were bent towards the sea with failures at three depths, while the piles on the land side inclined simply towards the sea with failures at two depths, as shown in Figure 11.5(a).

Building G of two stories, supported on 40.6 cm diameter S piles, was situated about 100m away from a quay wall on the north side that displaced by about 4 m. Its span direction was perpendicular to the northern shoreline. Figure 11.23(d) shows a boring log of the site. Despite minor damage to the superstructure, an inclinometer survey showed that the pile heads displaced northern seawards by 20 to 60 cm and western seawards by 20 to 80 cm. The deformation modes of the piles within the building are very similar, in contrast to those near the waterfront.

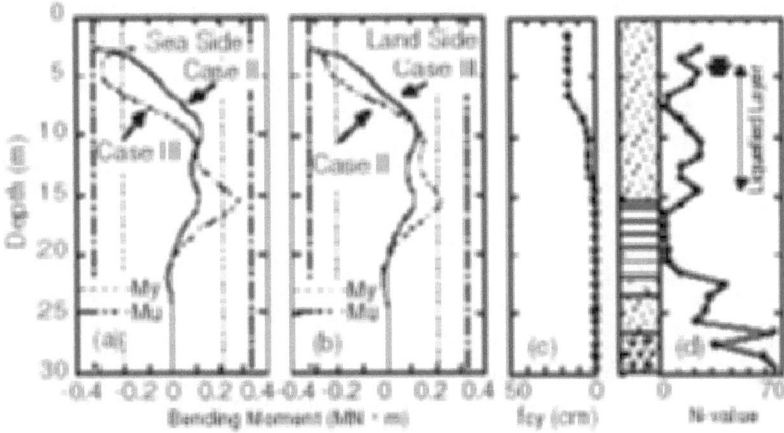

Fig. 11.23. Computed bending moment of pile at Building G for Cases II and III, with ground displacement, and boring log

The analysis is performed in the span directions of both buildings for both Cases II and III. In Case II, a base shear coefficient of 0.4 is assumed with a cyclic ground displacement profile as shown in Figure 11.22(c) or 11.23(c). In Case III, the assumed permanent ground surface displacements are 100cm and 60cm at both ends of the building near the waterfront, and 40 cm and 30cm at both ends of the building away from the waterfront. Figures 11.22(a), (b) and 11.23(a), (b) show the distributions of computed bending moments with depth for the two cases. In Figure 11.22, the computed bending moments in Case II are close to M_u near the pile heads and exceed M_c near the bottom of the liquefied fill, being consistent with the observed damage shown in Figure 11.22(e); however, the failure at the middle of the liquefied layer on the sea side pile can only be simulated in Case III. In addition, Case III appears to create more severe loading conditions than Case II for both cases, i.e., the bending moments at the bottom of the liquefied layers reach the ultimate values, which might have produced the observed large displacements of piles. Thus, the results from Case III are discussed hereafter.

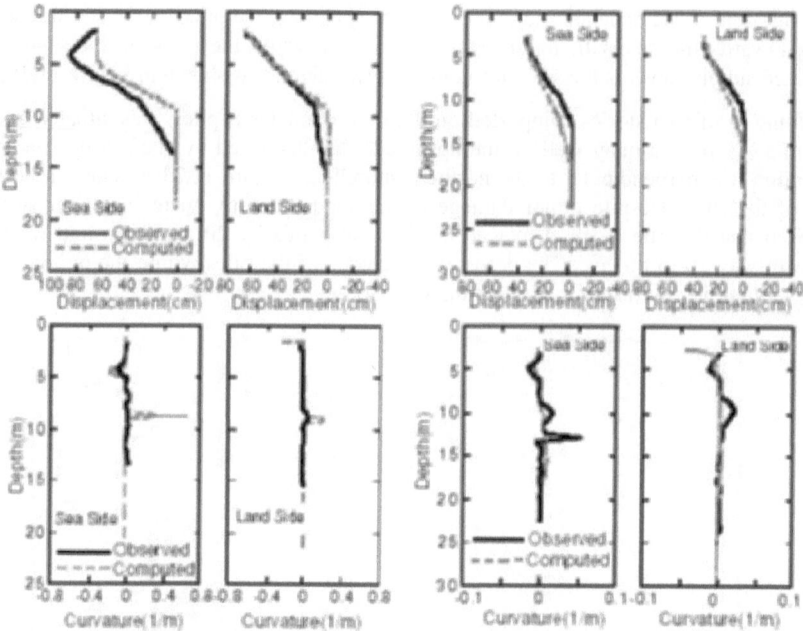

Fig. 11.24. Computed displacement and curvature of piles at building F, compared with the observed values

Fig. 11.25. Computed displacement and curvature of piles at Building G, compared with the observed values

The computed displacements of two pile foundations for Case III are shown in Figures 11.24 and 11.25 in broken lines. Also shown in the Figures in solid lines are observed displacement and curvature patterns from inclinometer surveys. It is found that in both cases the deformation and curvature patterns of the piles are in reasonably good agreement with the observed ones. The comparison of the assumed ground surface displacements with the computed displacements of the pile heads near the waterfront indicates that the soil pushes the pile on the sea side, while the pile pushes the soil on the land side. This difference could have induced the difference in failure patterns between the piles as shown in Figure 11.24, as well as the difference in the pile cap rotations observed on both sides of the buildings. A similar soil-pile interaction also occurred but to a lesser extent in the piles away from the waterfront, as shown in Figure 11.25. These findings indicate that the amount and the distribution of permanent ground displacements could have significant influence on the final deformation and failure patterns of pile foundation near the waterfront, although some of the damage might have occurred during shaking.

The failure of piles of a building under construction in a lateral spreading zone (Kuwabara and Yoneda, 1998; Building H in Table 11.1) also provides a good basis for analysis. Since these piles carried only 2% of the long-term design load, the kinematic force is considered to have dominated during and after shaking. The building 16.1 m wide in NS direction by 41.8 m long in EW direction was located about 25 m and 90 m inland from the east and south shorelines that displaced seawards by about 2.5 m and 3 m, respectively. Its span direction was perpendicular to the southern shoreline. This building was supported on 1.2-1.7 m diameter cast-in-place concrete piles that penetrated through a reclaimed fill about 15-m thick (Figure 11.26(c)) and down to a depth of about 48 m.

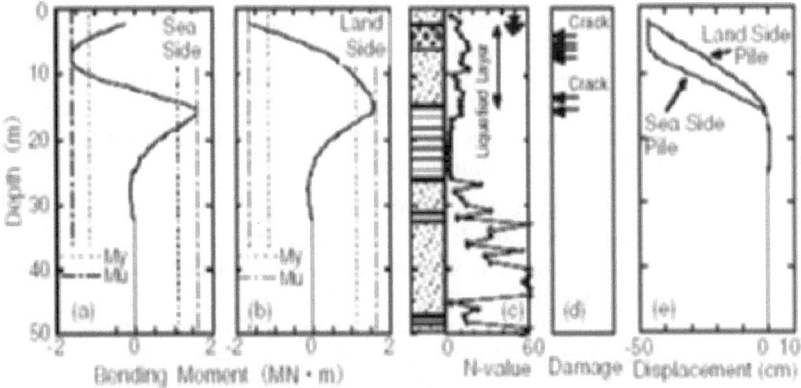

Fig. 11.26. Computed bending moment and displacement of piles at Building H, with ground displacement, boring log and pile damage

After the earthquake, the ground surface around the building settled by 30-50 cm. Vertical cracks were observed in almost all foundation beams, and the south side of the building settled by 2.8cm with respect to the north side. In addition to an excavation survey to examine damage to pile heads, a television camera was inserted into boreholes cored vertically through two east side piles with diameters of either 140 or 150cm. These surveys showed that horizontal and/or diagonal cracks occurred from the pile head to a depth of 5 m as well as at depths between 11 and 15.2m (Figure 11.26 (d)) and that the piles on the east end tilted towards southwest to south-southwest by 1/17-1/20 at least above a depth of 15.2m. Thus, the building might have been displaced towards that direction by at least 88cm, i.e., about 62-85cm southwards and 23-62cm westwards. The deformation of the foundation beams and the direction of the pile displacement suggest that the kinematic force might have been more crucial in the span direction than those in the longitudinal direction.

Since inertial forces during shaking were negligibly small, only Case III was considered with the maximum ground displacements of 75cm on the seaside and 50cm on the landside. Figures 11.26(a), (b) and (e) show the distribution of computed bending moments and displacements with depth. The computed bending moments indicate that failures might have occurred near the bottom of the liquefied layer as well as near the pile head, which shows fairly good agreement with the field observation shown in Figure 11.26(d). In addition, the computed displacement of 62cm of the pile head is consistent with the observed displacement. These findings confirm that the amount and the distribution of permanent ground displacements could have significant influence on the final deformation and failure patterns of pile foundations subjected to lateral spreading near the waterfront. Such effects therefore should be taken into account in the foundation design where lateral ground spreading is expected to occur.

11.7. Conclusions

Field performance of various pile foundations that experienced soil liquefaction and lateral spreading in the 1995 Hyogoken-Nambu earthquake have been compiled and summarized, together with their failure and deformation modes identified by surveys using borehole cameras and/or slope-indicators. Cyclic and residual shear strains in soil deposits during and after earthquakes have been estimated from field and aerial photographic surveys as well as analysis of strong motion records, and a simplified method for evaluating ground displacements has been developed. A simple p-y analysis using the ground displacement profile from the proposed method has been conducted to evaluate pile performance during the earthquake. The field observation together with the analytical results leads to the following conclusions:

1. The field investigation using a borehole camera and a slope-indicator has confirmed that failures of piles concentrated at the interface between liquefied and non-liquefied layers, as well as near the pile heads, indicating significant effects of ground displacements on pile damage.
2. The failure and deformation patterns of piles in the lateral spread zone vary with distance from the waterfront, probably due to the variation of lateral ground movements in the direction perpendicular to the waterfront.
3. Cyclic and permanent shear strains to de developed in sands during earthquakes are expressed in terms of normalized *SPT N*-values.
4. The simple p-y method combined with the proposed estimates for ground displacements can differentiate the damage from undamaged pile foundations subjected to cyclic and/or permanent ground displacements in the Kobe earthquake and can simulate reasonably well the failure and deformation modes of damaged piles. This indicates that the ground deformation could have played an important role on pile performance. Although preliminary and required further refinement and verification, the proposed model appears promising for evaluating the effects of ground displacements on piles.

Acknowledgements

The study described herein was made possible through the post-earthquake field investigation and their compilation conducted by various organizations including but not limited to the Committee on Building Foundation Technology against Liquefaction and Lateral Spreading, Japan Association for Building Research Promotion; the Committees on Damage to Building Foundations in Hyogoken-Nambu Earthquake, both Architectural Institute of Japan and Kansai Branch of AIJ; and Japanese Association for Steel Pile Piles. The strong motion accelerograms at the Higashi-Kobe Bridge station was provided by the Public Works Research Institute, Ministry of Construction. The authors express their sincere thanks to the above organizations and persons.

REFERENCES

Abrahamson, N. A. (2000) "State of the practice of seismic hazard evaluation", *Proceedings of GeoEng2000*, Invited papers, 1: 659-685

Abrahamson, N. A. and Shedlock, K. M. (1997) "Some comparisons between recent ground motion relations", *Seismological Research Letters*, 68(1): 9-23

Abrahamson, N. A. and Silva, W. J. (1997) "Empirical response spectral attenuation relations for shallow crustal earthquakes", *Seismological Research Letters*, 68(1): 94-127

Adams, J., Weichert, D. H., Halchuk, S. and Basham, P. W. (1996) Trial Seismic Hazard Maps of Canada – 1995: Final Values for Selected Canadian Cities, Geological Survey of Canada, Open File 3283, 97p

Ahmed-Zeki, A. S., Pender, M. J., and Fitch, N. R. (1999) "Strain rate effects on the undrained shear strength of Waitemata residual clay", *Proc. 8th Australia New Zealand Conference on Geomechanics*, Hobart, 2: 791-796

Aki, K. (1987) "Strong Motion Seismology", in *Strong Ground Motion Seismology*, Eds. M. Erdik and M. N. Toksoz, Reidel

Aki, K. and Richards, P. G. (1980) *Quantitative Seismology: Theory and Methods*, W.H. Freeman and Co., New York

Alarcon-Guzman, A., Chameau, J. L. and Leonards, G. A. (1986) "A New Apparatus for Investigating the Stress-Strain Characteristics of Sands", *Geotechnical Testing Journal*, 9(4): 204-212

Ambraseys, N. N, Simpson, K. A, and Bommer, J. J. (1996) "Prediction of horizontal response spectra in Europe", *Earthquake Engineering and Structural Dynamics*, 25(4): 371-400

Ambraseys, N. N and Finkel, C. F. (1995) *The Seismicity of Turkey and Adjacent Areas*, Muhittin Seren Publ. Istanbul, 240p

Ambraseys, N. N. (1995) "The prediction of earthquake peak ground acceleration in Europe", *Earthquake Engineering and Structural Dynamics*; 24(4): 467-490

Ambraseys, N. N. and Sarma, S. H. (1967) "The Response of Earth Dams to Strong Earthquakes", *Geotechnique*, 17: 181-283

Amini, F. and Qi, G. Z. (2000) "Liquefaction testing of stratified silty sands", *Journal of Geotechnical and Geoenvironmental Engineering*, 126(3): 208-217

Ampadu, S. K. and Tatsuoka, F. (1993) "Effect of Setting Method on the Behaviour of Clays in Triaxial Compressiom from Saturation to Undrained Shear", *Soils and Foundations*, 33(2): 14-34

Amato, A., Azzara, R., Chiarabba, C., Cimini, G. B., Cocco, M., Di Bona, M., Margheriti, L., Mazza, S., Mele, F., Selvaggi, G., Basili, A., Boschi, E., Courboulex, F., Deschamps, A., Gaffet, S., Bittarelli, G., Chiaraluce, L., Piccinini, D., and Ripepe, M. (1998). "The 1997 Umbria-Marche, Italy, earthquake sequence: a first look at the main shocks and aftershocks", *Geophys. Res. Lett.*, 25(15), 2861-2864

Anastasiadis, A., Apessou, M., and Pitilakis, K. (2002) "Earthquake hazard assessment in Thessaloniki, Greece: Level II – Site response analyses", *Proc. Int. Conf. on Earthquake Loss Estimation*, October 24-26, 2002, Bucharest, Romania

Anderson, J. G. and Quaas, R. (1988) "The Mexico Earthquake of September 19, 1985 - Effect of Magnitude on the Character of Strong Ground Motion: An Example from the Guerrero, Mexico Strong Motion Network", *Earthquake Spectra*, 4(3): 635-646

Anderson, J. G. and Hough, S. E. (1984) "A Model for the Shape of the Fourier Amplitude Spectrum of Acceleration at High Frequencies", *Bulletin of the Seismological Society of America*, 74(5): 1969-1993

Anderson, J. G. and Luco, J. E., (1983) "Parametric Study of Near-Field Ground Motion for a Strike-Slip Dislocation Model", *Bulletin of the Seismological Society of America*, 73(1): 23-43

Anderson, D.G. and Kavazanjian, E., Jr. (1995) "Performance of landfills under seismic loading". State of the Art (SOA 14), *Proc. of 3rd International Conference on Recent Advances in Geotechnical Earthquake Engineering and Soil Dynamics*, St. Louis, 3: 1557-1587

Andrews, D. J. (1986) "Objective determination of source parameters and similarity of earthquakes of different size", *Earthquake Source Mechanics* (eds. S. Das, J. Boatwright and C.H. Scholtz), American Geophys. Union, Washington D.C, 259-268

Ansal, A., Springman, S., Studer, J., Demirbaş, E., Önalp, A., Erdik, M., Giardini, D., Şeşetyan, K., Demircioğlu, M., Akman, H., Fäh, D., Christen, A., Laue, J., Buchheister, J., Çetin, Ö., Siyahi, B., Fahjan, Y., Gülkan, P., Bakir, S., Lestuzzi, P., Elmas, M., Köksal, D., and Gökçe, O. (2003) *Part 2C-Case Studies, Microzonation of Pilot Areas, Adapazarı, Gölcük, İhsaniye and Değimendere*, Research Task Group Report, World Institute for Disaster Risk Management project on Microzonation for Earthquake Risk Mitigation

Ansal, A. M. (2002) "Seismic Microzonation Methodology", *Proc. of 12th European Conf. on Earthquake Engineering*, Paper No.830, London, UK

Ansal, A. M. (1999) "Strong Motions and Site Amplification", Theme Lecture, *Second International Conf. on Earthquake Geotechnical Engineering*, Lisbon, Portugal, Balkema, Rotterdam, 3: 879-894

Ansal, A. M. and İyisan, R. (1998) "Uniform Risk in Site-Specific Seismic Hazard Analysis", *XI Danube European Conference on Soil Mechanics and Geotechnical Engineering*, Porec, Croatia, Balkema, Rotterdam, 317-324

Ansal, A. M. (1997) "Design Earthquake Characteristics for Istanbul" Proceedings of Prof.Dr.Rifat Yarar Symposium, Turkish Earthquake Foundation, 1: 233-244 (in Turkish)

Ansal, A. M. (1994) "Effects of Geotechnical Factors and Behaviour of Soil Layers during Earthquakes, State-of-the-Art Lecture", *Proc. of 10th European Con. on Earthquake Engineering*, Vienna, Austria, 1: 467-476

Ansal, A. M., Iyisan, R. Sengezer, B. S. and Gençoglu, S. (1993). "The damage distribution in March 13, 1992 Erzincan earthquake and effects of geotechnical factors", *Soil Dynamics and Geotechnical Earthquake Engineering*, Ed. P. Sêco e Pinto, Balkema, Rotterdam, 413-434

Applied Technology Council, ATC (1996) *ATC 40: The Seismic Evaluation and Retrofit of Concrete Buildings*, 2 volumes, Redwood City, CA

Applied Technology Council, ATC (1985) *ATC-13: Earthquake Damage Evaluation for California*, Applied Technology Council, Redwood City, CA

Architectural Institute of Japan, AIJ (1988, 2001) *Recommendations for design of building foundations*, 430p, (in Japanese)

Architectural Institute of Japan (1998) *Report on the Hanshin-Awaji Earthquake Disaster*, Building Series Volume 4, Wooden Structure and Building Foundations (in Japanese)

Architectural Institute of Japan (1990) *Ultimate Strength and Deformation Capacity of Buildings in Seismic Design*, 715p. (in Japanese)

Arias, A. (1970) "A Measure of Earthquake Intensity", *Seismic Design for Nuclear Power Plants*, Ed. R. Hanson, Massachusetts Institute of Technology Press, Cambridge, MA

Ashford, S.A. and Sitar, N. (1997) "Analysis of topographic amplification of inclined shear waves in a steep coastal bluff", *Bulletin of the Seismological Society of America*, 87(3): 692-700

Athanasopoulos, G. A., Pelekis, P. C. and Leonidou, E. A. (1999) "Effects of Surface Topography on Seismic Ground Response in the Egion (Greece) 15 June 1995 Earthquake", *Soil Dynamics and Earthquake Engineering*, (18):135-149

Atkinson, G. M. and Silva, W. (2000) "Stochastic modelling of California ground motions", *Bulletin of the Seismological Society of America*, 90: 255 –274

Atkinson, G. M. and Boore, D. M. (1998) "Evaluation of models for earthquake source spectra in eastern North America", *Bulletin of the Seismological Society of America*, 88: 917 –934

Atkinson, G. M. (1997) "Empirical ground motion relations for earthquakes in the Cascadia region", *Canadian Journal of Civil Engineering*, 24: 64-77

Augello, A. J., Bray, J. D., Seed., R, D., Matasovic, N., and Kavazanjian, Jr. E. (1995) "Solid waste landfill performance during the 1994 Northridge earthquake" *Proc. of 3rd International Conference on Recent Advances in Geotechnical Earthquake Engineering and Soil Dynamics*, St. Louis, 3: 1395-1401

Auld, B. (1977) "Cross-Hole and Down-Hole Vs by Mechanical Impulse", *Journal Geotechnial Engineering Division*, ASCE, 103(GT12): 1381-1398

Auvinet, G. and Mendoza, M. J. (1986) "Comportamiento de diversos tipos de cimentacion en la zone lascustre de la Ciudad de Mexico durante el sismo del 19 de septiembre de 1985", *Proceedings Symposium: Los Sismos de 1985*; Casos de Mecanica de Suelos, Mexico

Baguelin, F. and Jezequel, J. F. (1973) "Le Pressiometre Autoforeur" *Annales de l'ISTBTP*, 97: 135-159

Baldi, G., Bellotti, R., Ghionna, V.N., Jamiolkowski, M., and Lo Presti, D. C. F. (1989a) "Modulus of Sands from CPT's and DMT's", *Proc. XII International Conference on Soil Mechanics and Foundation Engineering*, Rio de Janeiro, 1: 165-170

Baldi, G., Jamiolkowski, M., Lo Presti, D. C. F., Manfredini, G., and Rix, G. J. (1989b) "Italian Experience in Assessing Shear Wave Velocity from CPT and SPT", *Earthquake Geotechnical Engineering, Proc. of Discussion Session on Influence of Local Conditions on Seismic Response*, XII ICSMFE, Rio de Janeiro, Brasil: 157-168

Baldi, G., Bellotti, R., Ghionna, V., Jamiolkowski, M., Pasqualini, E. (1986) "Drained penetration of sands. Part II", *IV Int. Geotechnical Seminar*, Singapore. NTI, Nanyang Technological Institute, 143-156

Bard, P.-Y., and Riepl-Thomas, J (1999) Wave Propagation in Complex Geological Structures and Local Effects On Strong Ground Motion, Chapter 2 of the book *"Wave Motion in Earthquake Engineering"*, Eds. E. Kausel and G.D. Manolis, WIT Press, ISBN 1-85312-744-2, 38-95

Bard, P.-Y. (1999) "Microtremor measurements: a tool for site effect estimation ?", State-of-the-art paper, *Second International Symposium on the Effects of Surface Geology on Seismic Motion*, Yokohama, December 1-3, 1998, Eds. Irikura, Kudo, Okada and Sasatani, Balkema, 3, 1251-1279

Bard, P.-Y. (1994) "Effects of Surface Geology on Ground Motion: Recent Results and Remaining Issues", *Proc. 10th ECEE*, Vienna, Austria, (1):305-323

Bard, P.-Y. and Meneroud, J.-P. (1987) "Modification du signal sismique par la topographie. Cas de la vallée de la Roya (Alpes-Maritimes)", *Bull. Liaison Laboratoires des Ponts-et-Chaussées*, Numéro spécial "Risques Naturels" 150-151, 140-151 (in French)

Bell, L. (1998) Building prototypes, seismic damage probability matrices for buildings located in southwestern BC, Report, University of British Columbia, BC

Bellotti, R., Jamiolkowski, M., Lo Presti, D. C. F., and O'Neill, D. A. (1996), "Anisotropy of Small Strain Stiffness in Ticino Sand", *Geotechnique*, 46(1): 115-131

Bellotti, R., Ghionna, V. N., Jamiolkowski, M., Robertson, P. K., and Peterson, R. W. (1989) "Interpretation of Moduli from Self-Boring Tests in Sand", *Géotechnique*, 39(2): 269-292

Bellotti, R., Ghionna, V. N., Jamiolkowski, M., Lancellotta, R., and Manfredini, G. (1986) "Deformation Characteristics of Cohesionless Soils from in Situ Tests", *Use of In Situ Tests in Geotechnical Engineering*, Geotechnical Special Pubblication, No. 6, Ed. S.P. Clemence, ASCE, 47-73

Bender, B. and Perkins, D. M. (1987) *SEISRISK III: A Computer Program for Seismic Hazard Estimation,* USGS Bulletin 1772, Denver

Beresnev, I. and Atkinson, G. (1997) "Modeling Finite Fault Radiation from the w^n Spectrum", *Bulletin of the Seismological Society of America*, 87: 67-84

Beyen, K and Erdik, M. (2002) "Two Dimensional Nonlinear Simulation of the Site Response of Adapazarı Plain under Aftershocks of the Kocaeli Earthquake", 17 August, 1999, *BSSA* Special Edition

Biot, M.A. (1956a) "Theory of Propagation of Elastic Waves in a Fluid-Saturated Porous Solid, I. Lower Frequency Range", *J. Acoust. Soc. Am.*, 28: 168-178

Biot, M.A. (1956b) "Theory of Propagation of Elastic Waves in a Fluid-Saturated Porous Solid, II. Higher Frequency Range", *J. Acoust. Soc. Am.*, 28: 179-191

Blanquera, A. (1999) Evaluation of structural earthquake damage to buildings in southwestern B.C., Masters Thesis, Department of Civil Engineering, University of British Columbia, Vancouver, B.C.

BTL Committee (1998) *Research Report on liquefaction and lateral spreading in the Hyogoken-Nambu earthquake* (in Japanese)

Bolt, B. (1969) "Duration of Strong Motion", *Proceedings of the Fourth World Conference on Earthquake Engineering*, Santiago, Chile, 1304-1315

Bommer J. J. and Martinez-Pereira, A. (2000) "Strong Motion Parameters: Definition, Usefulness and Predictability", *Proceedings of the Twelfth World Conference on Earthquake Engineering*, Auckland, New Zealand, 206

Bommer, J. J., Elnashai, A. S., Chlimintzas, G. O., and Lee, D. (1998) Review and development of response-spectra for displacement-based seismic design, ESEE Research Report 98-3, Civil Engineering Department, Imperial College, London

Boore, D. M. (2000) SMSIM – Fortran Programs for Simulating Ground Motions From Earthquakes: Version 2.0 – A Revision of OFR 96-80-A, USGS Open File Report OF 00-509

Boore D. M. and Joyner, W. B. (1997) "Site Amplification for Generic Soil Sites", *Bulletin of the Seismological Society of America*, 87(2): 327-341

Boore, D. M., Joyner, W. B., and Fumal, T. E. (1997) "Equations for Estimating Horizontal Response Spectra and Peak Acceleration from Western North American Earthquakes: A Summary of Recent Work", *Seismological Research Letters*, 68(1): 128-153

Boore, D. M., Joyner, W. B., and Fumal, T. E. (1993) *Estimation of response spectra and peak accelerations from western North American earthquakes: An interim report*, United States Geological Survey Open-File Report 93-509, Menlo Park, California, 72 p

Boore, D. M. (1983) "Stochastic simulation of high-frequency ground motions based on seismological models of the radiated spectra", *Bulletin of the Seismological Society of America*, 73:1865-1894

Boore, D. M.; Harmsen, S. C. and Harding, S. T. (1981) "Wave scattering from a step change in surface topography", *Bulletin of the Seismological Society of America*, 71(1): 117-125

Boore D. M. (1972) "A note on the effect of simple topography on seismic SH waves", *Bulletin of the Seismological Society of America*, 62: 275-284

Borcherdt, R.D. (1994) "Estimates of Site Dependent Response Spectra for Design (Methodology and Justification)", *Earthquake Spectra*, 10(4): 617-654

Borcherdt, R.D., Wentworth, C.M., Jannsen, A., Fumal, T., and Gibbs, J. (1991) "Methodology for Predictive GIS Mapping of Special Study Zones for Strong Ground Shaking in the San Francisco Bay Region, California", *Proceedings of the Fourth International Conference on Seismic Zonation*, Stanford, California, 3: 545-552

Borcherdt, R. D. (1970) "Effects of local geology on ground motion near San Francisco Bay", *Bulletin of the Seismological Society of America*, 60: 29-61

Bouchon, M. (1987) "Numerical Simulation of Earthquake Ground Motion", *Strong Ground Motion Seismology* (Eds. M.Erdik, M.N.Toksoz), NATO ASI Series, D. Reidel Publishing Company, 185-207

Brambati, A., Faccioli, E., Carulli, E. B., Culchi, F., Onofri, R., Stefanini, S. and Ulcigrai, F. (1980) "Studio de microzonizzazione sismica dell'area di Tarcento (Friuli)", *Edito da Regiona Autonoma Friuli-Venezia-Giulia* (in Italian)

Brune, J. N. (1976) "The Physics of Earthquake Strong Motion", *in Seismic Risk and Engineering Decisions*, Eds. C. Lomnitz and E. Rosenblueth, Elsevier

Brune, J. N. (1971) "Correction", *Journal of Geophysical Research*; 76: 5002

Brune, J. N. (1970) "Tectonic stress and the spectra of seismic shear waves from earthquakes", *Journal of Geophysical Research*, 75:4997-5009

Bruzzi, D., Ghionna, V. N., Jamiolkowski, M., Lancellotta, R. and Manfredini, G. (1986) "Self-Boring Pressuremeter in Po River sand", *Proc. 2nd Int. Symposium on Pressuremeter and Its Marine Applications*, ASTM STP 950: 57-74

Building Seismic Safety Council, BSSC (1994) *NEHRP Recommended Provisions Seismic Regulations for New Buildings, Part 2: Commentary*, FEMA 223A, 335p

Budhu, M. and Al-Karni, A. (1993) "Seismic bearing capacity of soils", *Geotechnique*, 33: 181-187

BUKOERI, Bosphorus University Kandilli Observatory and Earthquake Research Institute (2002) Seismic Data for Marmara Region, ftp://boun.edu.tr/kandilli/pub/

Buranek, D. and Prasad, S. (1991) "Sanitary landfill performance during the Loma Prieta earthquake", *Proc. of 2nd International Conference on Recent Advances in Geotechnical Earthquake Engineering and Soil Dynamics*, St. Louis, 2: 1655- 1660

Byrne, P. M., Salgado, F. M., and Howie, J. A. (1990) "Relationship between the Unload Shear Modulus from Pressuremeter Tests and the Maximum Shear Moduli of Sand", *Proc. 3rd Int. Symposium Pressuremeter*, Oxford, 231-242

Cabral, J. (1996) "Sismotectónica de Portugal", *Colóquio / Ciências*, Lisbon, (18) 39-58, (in Portuguese)

California Division of Mines and Geology, CDMG (1996) Probabilistic seismic hazard for the state of California, DMG Open File Report 96-08, 33p

Calosi, E., Ferrini, M., Cancelli, A., Foti, S., Lo Presti D.C., Pallara, O., D'Amato Avanzi, G., Pochini, A., Puccinelli, A., Luzi, L., Rainone, M., and Signanini, P. (2001) "Geological and Geotechnical investigations for the seismic response analysis at Castelnuovo Garfagnana in Central Italy", *Proc. XV ICSMGE, Earthquake Geotechnical Engineering Satellite Conference*, Istanbul, Turkey, *Lessons Learned from Recent Strong Earthquakes* (Ed. A. Ansal), 141-148

Campanella, R. G. and Stewart, W. P. (1990) "Seismic Cone Analysis Using Digital Signal Processing for Dynamic Site Characterization", *43rd Canadian Geotechnical Engineering Conference*, Quebec City

Campbell, K. W. (1997) "Empirical Near-source Attenuation Relationships for Horizontal and Vertical Components of Peak Ground Acceleration, Peak Ground Velocity and Pseudo-absolute Acceleration Response Spectra", *Seismological Research Letters*, 68(1): 154-179

Campos-Costa, A., Sousa, M. L., Carvalho, A., Bilé Serra, J., and Carvalho, E. C. (2002) "Regional seismic risk scenarios based on hazard deaggregation", *Proceedings, 12th European Conference on Earthquake Engineering*, paper 470, London, Elsevier Science Ltd

Campos-Costa, A., Sousa, M. L., and Oliveira, C. S. (1998) "Seismic Risk: Methods and application to Portugal", *Proceedings, 11th European Conference on Earthquake Engineering*, paper 518, Paris

Carvalho, E. C., Coelho, E., Campos-Costa, A., Sousa, M. L., and Candeias, P. (2002) "Vulnerability evaluation of residential buildings in Portugal", *Proceedings, 12th European Conference on Earthquake Engineering*, paper 696, London, Elsevier Science

Castelli, V. and Monachesi, G. (2001) "Seismic history and historical earthquake scenario for the town of Fabriano (Central Italy)", *Italian Geotechnical Journal*; XXXV (2)

Cavallaro, A., Lo Presti, D. C. F., Maugeri, M., and Pallara, O. (1998) "Strain Rate Effect on Stiffness and Damping of Clays", *Italian Geotechnical Review*, XXXII(4): 30-49.

Cazacliu, B. (1996) Comportment des sables en petites et moyennes déformations; réalisation d'un prototype d'essai de torsion compression confinement sur cylindre creux, Doctorat, ECP-ENTPE, Paris

Celebi, M. (1987) "Topographic and geological amplifications determined from strong-motion and aftershock records of the 3March 1985 Chile earthquake", *Bull. Seism. Soc. Am.* 77: 1147–1167

Chávez-García, F. J., Raptakis, D., Makra, K., and Pitilakis, K. (2000) "Site Effects at EUROSEISTEST – II, Results from 2D numerical modelling and comparison with observations", *Soil Dynamics and Earthquake Engineering*, 19(1): 23-39

Chavez-Garcia, F. J., Rodriguez, M., Field, E. H., and Hatzfeld, D. (1997) "Topographic site effects. A comparison of two non-reference methods", *Bul.of the Seismological Society of America*, 87: 1667-1673

Chávez-García, F. J., Sanchez, L.R., and Hatzfeld, D. (1996) "Topographic site effects and HVSR. A comparison between observations and theory", *Bull. of Seis. Society of America*, 86: 1559-1575

Chen, W. F. (1975) *Limit Analysis and Soil Plasticity*, Elsevier Science Publishing Company

Chen, X. (1993) "A systematic and efficient method of computing normal modes for multilayered half-space", *Geophys. J. Int.*, 115: 391-409

Chen, Y. (1988) *The Great Tangshan Earthquake*, Wheaton and Co. Ltd., Exeter, UK

Chin-Hsiung, L., Jeng-Yaw, H. and Tzay-Chyn, S. (1998) "Observed variation of earthquake motion across a basin-Taipei City", *Earthquake Spectra*, (14)1:115-134

Chung, R. M., Yokel, F. Y., and Wechsler, H. (1984) "Pore Pressure Buildup in Resonant Column Tests", *Journal of Geotechnical Engineering Division*, ASCE, 110(2): 247-261

Chuo-Kaihatsu Corporation (1995) *Reconnaissance report on the 1995 Hyogoken-Nambu earthquake - Great Hanshin Disaster*, (Hyogoken-Nambu earthquake)

Ciampoli, M. (1997) "Structural Dynamics and Response Spectra", *Proceedings of the Advanced Study Course on Seismic Risk "SERINA"*, Thessaloniki, Greece, 21-27 Sep. 1997, ITSAK, 405-461

Clayton, C. (1995) *The Standard Penetration Test: Methods and Use*, CIRIA Report No. 143

Cluff, L. S., Coppersmith, K. J., and Knuepfer, P. L. (1982) " Assessing degrees of fault activity for seismic microzonation" *Proc. of 3rd International Earthquake Microzonation Conf.*, Univ. of Washington, Seattle, 1: 113-118

Coburn, A. and Spence, R. (1992) *Earthquake Protection*, John Wiley and Sons, Ltd, Chichester, UK

Committee of Earthq. Obs. and Res. in the Kansai Area, CEORKA (1995) Digitized strong motion records in the affected area during the 1995 Hyogoken-Nambu earthquake

Constable, S. C., Parker, R. L., and Constable, C. G. (1987) "Occam's inversion: a pratical algorithm for generating smooth models from electromagnetic sounding data", *Geophysics*, 32 (3): 289-300

Cook, S. E. (1999) Evaluation of non-structural earthquake damage to buildings on southwestern B.C., Masters Thesis, Department of Civil Engineering, University of British Columbia, Vancouver, B.C.

Cornell, C. A. (1968) "Engineering Seismic Risk Analysis", *Bulletin of the Seismological Society of America*; 58(5): 1583-1606

Cremer C., Pecker A., and Davenne L. (2002) "Modelling of non linear dynamic behaviour of a shallow strip foundation with macro-element", *Journal of Earthquake Engineering*, 6(2): 175-211

Cremer, C., Pecker, A., and Davenne, L. (2001) "Cyclic macro-element for soil structure interaction - Material and geometrical non linearities", *International Journal for Numerical and Analytical Methods in Geomechanics*, 25: 1257-1284

Daminelli, R., Marcellini, A., Tento, A., Pignone, R., Martelli, L., Frassineti, G., and Pagani, M. (2000) "Seismic microzonation for land use planning of some municipalities on the Adriatic coast", *Proceedings of the Workshop "Mitigation of seismic risk; Support to recently affected European countries"*, Belgirate, Italy, 27-28 November 2000

Das, S. (1997) "Far field radiation from an earthquake source", *Proc. of the Advanced Study Course on Seismic Risk*, SERINA, ITSAK, 72-100

De Barros, F. C. and Luco, E. (1990) "Discrete models for vertical vibrations of surface and embedded foundations", *Earthquake Engineering and Structural Dynamics*, 19

Deodatis, G. (1996), "Non-stationary Stochastic Vector Processes: Seismic Ground Motion Applications", *Probabilistic Engineering Mechanics*, 11: 145-168

Di Benedetto, H., Cazacliu, B., Boutin, C., Donah, T., and Touret, J. P. (1997) "Comportement des sables avec rotation d'axes", *XIV Int. Conf. Soil Mechanics and Foundation Engineering*, Hamburg, Balkema, Rotterdam, 1: 279-282

Dickenson, S. E. and Seed, R. B. (1996) "Nonlinear dynamic response of soft and deep cohesive soil deposits", *Proc. of Int. Workshop on Site Response*, Yokosuka, Japan, 2: 67-81

Dobry, R. and Iai, S. (2000) "Recent Developments in Understanding of Earthquake Site Response and Associated Seismic Code Implementation", *Geoeng 2000*, Melbourne, 19-24 November, 1: 186-219.

Dobry, R., Ladd, R. S., Yokel, F. Y., Chung, R. M., and Powell, D., (1982) "Prediction of Pore Water Pressure Build-up and Liquefaction of Sands during Earthquake by the Cyclic Strain Method", *National Bureau of Standards Building, Science Series 138*, Washington, D.C.

Dobry, R., Idriss, I. M., and Ng, E. (1978) "Duration Characteristics of Horizontal Components of Strong-Motion Earthquake Records", *Bulletin of the Seismological Society of America*, 68(5): 1487-1520

Dobry, R., Oweis, I., and Urzua, A. (1976) "Simplified procedures for estimating the fundamental period of a soil profile", *Bulletin of the Seismological Society of America*, 66: 1293-1321

d'Onofrio, A., Silvestri, F., and Vinale, F. (1999) "Strain rate dependent behaviour of a natural stiff clay", *Soils and Foundations*, 39(2): 69-82

Dormieux, L. and Pecker, A. (1995) "Seismic bearing capacity of a foundation on a cohesionless soil", *Journal of Geotechnical Engineering*, ASCE, 121: 300 303

Dowrick, D. J. and Rhoades, D. A. (1997) "Vunerability of different classes of low-rise buildings in the 1987 Edgecumbe, New Zealand, Earthquake", *Bulletin of the New Zealand National Society for Earthquake Engineering*, 30(3): 227-241

Durukal, E. (2002) "Critical evaluation of strong motion in Kocaeli and Duzce (Turkey) earthquakes", *Soil Dynamics and Earthquake Engineering*, 22 (7): 589-609

EERI Committee on Seismic Risk (1989) "The basics of seismic risk analysis", *Earthquake Spectra*, 5(4): 675-702

Elgamal, A. W., Zeghal, M., and Parra, E. (1995) "Identification and modelling of earthquake ground response" *First International Conference on Earthquake Geotechnical Engineering,* Japan

Elms, D. (2000) "Refinements to the Newmark sliding block model", *Proc. of 12th World Conference on Earthquake Engineering,* Paper No. 2132, Auckland, New Zealand

Erdik, M., Şeşetyan, K., Demircioğlu, M., Siyahi, B., and Akman, H. (2003) "Assessment of the Seismic Hazards in Adapazarı, Gölcük, Değirmendere and İhsaniye Provinces in Northwestern Turkey", Ch. 3 in *Part 2C-Case Studies, Microzonation of Pilot Areas, Adapazarı, Gölcük, İhsaniye and Değirmendere,* Research Task Group Report, World Institute for Disaster Risk Management project on Microzonation for Earthquake Risk Mitigation

Erdik, M. and Durukal, E. (2003) "Simulation Modelling of Strong Ground Motion", in *Earthquake Engineering Handbook,* Chen and Scawthorn, Editors, CRC Press LLC

Erdik, M. (2001) Report on 1999 Kocaeli and Duzce (Turkey) Earthquakes, in *Structural Control for Civil and Infrastructure Engineering,* Eds. F. Casciati, G. Magonette, World Scientific, (http://www.iiasa.ac. at /Research/RMS/july2000/Papers/erdik.pdf)

Erdik, M. and Durukal, E. (2001) "Ahybrid Procedure for the Assessment of Design Basis Earthquake Ground Motion for Near-Fault Conditions", *Soil Dynamics and Earthquake Engineering,* 21: 431-443

Ergin, K, Güçlü, U. and Aksoy, G. (1971) *Earthquake Catalogue for Turkey and Its Vicinity, 1965-1970,* I.T.U. Mining Faculty, Earth Physics Institute Publication, No.28

Ergin, K., Güçlü, U., and Uz, Z. (1967) *Earthquake Catalogue for Turkey and Its Vicinity 11-1964,* I.T.U. Mining Faculty, Earth Physics Institute Publication, No.24

Eurocode 8, EC8 (2003) Draft prENV 1998-1 Jan 2003, Design of structures for earthquake resistance, Part 1: General rules, seismic actions and rules for buildings (Draft No 4), European Committee for Standarization, Brussels, Belgium

Eurocode 8, EC8 (2002) Draft No4 - prEN1998-5 St34 Nov 02, Design of structures for earthquake resistance Part 5: Foundations, retaining structures and geotechnical aspects

Eurocode 8, EC8 (2002) Draft No4 - prEN1998-2 St34 Apr 02, Design of structures for earthquake resistance Part 5: Foundations, retaining structures and geotechnical aspects

Eurocode 8, EC8 (2002) Draft No.5, prEN 1998-1 May 2002, Part 1: General Rules, Seismic Actions and Rules for Buildings, Doc CEN/TC250/SC8/N317

Eurocode 8, EC8 (1999) prEN 1998-5, Part 5: Foundations, retaining structures and geotechnical aspects

Eurocode 8, EC8 (1999) ENV 1998-5, Design Provisions for Earthquake Resistance of Structures- Part 5: Foundations, Retaining Structures and Geotechnical Aspects

Eurocode 8, EC8 (1999) ENV 1998-1-1, Design of structures for earthquake resistance of structures, Part 1-1: General rules – seismic actions and general requirements for structures, Brussels, CEN

Eurocode 7, EC7 (2001) Draft prENV 1997-1 Oct 2001, Geotechnical design, Part 1: General Rules

Faccioli, E. (1991) "Seismic amplification in the presence of geological and topographic irregularities", *Proc. of 2nd Intern. Conf. on Recent Advances in Geotechnical Earthq. Engrg.,* 1779-1797

Fahey, M. and Carter, J. P. (1993) "A Finite Element Study of the Pressuremeter Test in Sand Using a Non-Linear Elastic Plastic Model", *Canadian Geotechnical Journal,* 30: 348-362

Fahey, M. (1991) "Measuring Shear Modulus in Sand with the Self-Boring Pressumeter", *Proc. of 10th European Conference on Soil Mechanics and Foundation Engineering,* Florence, Italy, Balkema, Rotterdam, 1: 73-76

Fäh, D., Rüttener, E., Noack, T. and Kruspan, P. (1997) "Microzonation of the City of Basel", Journal of Seismology, (1):87-102

FEMA (1997) NEHRP Guidelines for the Seismic Rehabilitation of Buildings, FEMA Report 273: Federal Emergency Management Agency

Ferrini, M. (coord.), Foti, S., Lo Presti, D., Luzi, L., Pergalani, F. Petrini, V. Pochini, A., Puccinelli, A., Signanini, P., and Socco, V. (2000) "La riduzione del rischio sismico nella pianificazione del territorio: le indagini geologico tecniche e geofisiche per la valutazione degli effetti locali", *CISM*, Lucca 3-6 Maggio 2000

Field, E. H. and Jacob, K. (1995) "A comparison and test of various site response estimation techniques, including three that are non reference-site dependent", *Bulletin of the Seismological Society of America*, 85: 1127-1143

Field, E. H., Jacob, K. H., and Hough, S. E. (1992) "Earthquake weak motion estimation: a weak motion case study", *Bulletin of the Seismological Society of America*, 82: 2283-2307

Finn, W. D. L. (1991) "Geotechnical engineering aspects of seismic microzonation", *Proc. 4th Intern. Conf. Seismic Zonation*, August 25-29, Stanford, California, E.E.R.I. (editor), Oakland CA, 1: 199-250.

Finn, W. D. L. (1987) Chapter 3 – Geomechanics, in *Finite Element Handbook*, McGraw-Hill, Editor H. Hardestuncer

Fioravante, V., Jamiolkowski, M., and Lo Presti, D. C. F. (1994) "Stiffness of Carbonatic Quiou Sand", *Proc. of 13th Int. Conf. on Soil Mechanics and Foundation Engineering*, New Dehli, India, 1: 163-167

Fioravante, V., Jamiolkowski, M., Lo Presti, D. C. F., Manfredini, G. and Pedroni, S. (1998) "Assessment of coefficient of earth pressure at rest from shear wave velocity measurements", *Géotechnique*, 48(4): 1-10

Foti, S. (2000) Multistation methods for geotechnical characterization using surface waves, PhD Diss, Politecnico di Torino, Torino, Italy

Foti, S., Lancellotta, R., Sambuelli, L., and Socco, L.V. (2000) "Notes on fk analysis of surface waves" *Annali di Geofisica*, 43(6): 1199-1210

Franceschina, G., Marcellini, A., and Pagani, M. (2001) "Expected reference seismic motion at Fabriano" *Italian Geotechnical Journal*, XXXV (2): 48-58

Frankel, A. D., Mueller, C. S., Barnhard, T. P., Leyendecker, E. V., Wesson, R. L., Harmsen, S. C., Klein, F. W., Perkins, D. M., Dickman, N. C., Hanson, S. L. and Hopper, M. G. (2000) "USGS National Seismic Hazard Maps", *Earthquake Spectra*, (16)1:1-15

Fuhriman, M. D. (1993) Cross-Hole Seismic Tests at two Northern California Sites Affected by the 1989 Loma Prieta Earthquake, M. Sc. Thesis, University of Texas at Austin

Fujii, S., Cubrinvski, M., Hayashi, T, Shimazu, S., and Tokimatsu, K. (1996) "Response analysis of buildings with a pile foundation on a liquefied ground", *Proc., 31nd Japan National Conf. on Geotechnical Engineering*, 1: 1139-1140 (in Japanese)

Furuta and Yamoto (2000) "Liquefaction assessment by shear wave crosshole tomography tests", *Proc. of 12th World Conference on Earthquake Engineering*, Auckland, New Zealand, Paper No. 831

Ganji, V., Gucunski, N., and Nazarian, S. (1998) "Automated Inversion Procedure for Spectral Analysis of Surface Waves," *Journal of Geotechnical and Geoenvironmental Engineering*, 124(8): 757-770

Gazetas, G. and Mylonakis, G. (1998) "Seismic soil structure interaction: new evidence and emerging issues". *Geotechnical Earthquake Engineering and Soil Dynamics*, ASCE, 2: 1119-1174

Gazetas, G., Dakoulas, P. and Papageorgiou, A. (1990) "Local Soil and Source-Mechanism Effects in the 1986 Kalamata (Greece) Earthquake", *Earthquake Engineering and Structural Dynamics*, (19): 431-453

Gazetas, G. (1990) Foundation Vibration, *Foundation Engineering Handbook*, 2nd Edition, Ed. Hsai-Yan Fang, Chap. 15

Geli, L., Bard, P.-Y, and Jullien, B. (1988) "The effect of topography on earthquake ground motion, A review and new results", *Bulletin of the Seismological Society of America*, 78: 42-63

Gençoğlu, S., İnan, E. and Güler, H. (1990) *Earthquake Hazard for Turkey*, Publication of the Chamber of Geophysical Engineers, Ankara, (in Turkish)

Ghionna, V., Karim, M., and Pedroni, S. (1994) "Interpretation of Unload-Reload Modulus from Pressuremeter Tests in Sand", *Proc. 13th Int.Conf. on Soil Mechanics and Foundation Engineering*, New Delhi, India, 115-120

Ghazi, A., and Yeroyanni, M. (1997) Communication to the Commission on Earthquakes, European Commission, Director-General, Science, Research and Development, EUR 16993 EN, Brussels

Giovinazzi, S. and Lagomarsino, S. (2003) "Seismic risk analysis: a method for the vulnerability assessment of built-up areas", *Proceedings, European Safety and Reliability Conference*, ESREL, Maastricht

Goto, S., Tatsuoka, F., Shibuya, S., Kim, Y. S., and Sato, T. (1991) "A simple gauge for local small strain measurements in the laboratory", *Soils and Foundations*, 31(1): 169-180

Grunthal, G. (1998) *European Macroseismic Scale 1998*, Luxembourg: Cahiers du Centre Europeén de Geodynamique et de Séismologie

Gucunski, N. and Woods, R. D. (1991) "Use of Rayleigh modes in interpretation of SASW test", *Proc. 2th Int. Conf. Recent Advances in Geotechnical Earthq. Eng. and Soil Dyn*, St. Louis, 1399-1408

Gueguen, P., Chatelain, J. - L., Guillier, B., Yepes, H. and Egred, J. (1998) "Site Effect and Damage Distribution in Pujili (Ecuador) after the 28 March 1996 Earthquake", *Soil Dynamics and Earthquake Engineering*, (17), 329-334

Gutenberg, B. and Richter, C. F. (1954) *Seismicity of the Earth and Associated Phenomena*, 2.nd Ed., Princeton University Press, 440p.

Güçlü, U., Altınbaş, G. and Eyidoğan, H. (1986) *Earthquake Catalogue for Turkey and its Vicinity (1971-1975)*, I.T.U. Mining Faculty, Earthscience Research Center, Publication, No. 30 (in Turkish)

Hanks, T. C. (1982) "f_{MAX}", *Bulletin of the Seismological Society of America*, 72: 1867-1879

Hanks, T. C. and McGuire R. K. (1981) "The character of high-frequency strong ground motion", *Bulletin of the Seismological Society of America*, 2071-2095

Hartzell, S., Carver, D. and Williams, R. A. (2001) "Site Response, Shallow Shear-Wave Velocity and Damage in Los Gatos, California, from the 1989 Loma Prieta Earthquake", *BSSA*, (91)3:468-478

Hartzell, S. H. (1978) "Earthquake aftershocks as Green's functions", *Geophysical Research Letters*, 53: 1425-1436

Haskell, N. A. (1953) "The dispersion of surface waves on multilayered media", *Bullettin of the Seismological Society of America*, 43 (1): 17-34

HAZUS 99 (1999) *Natural Hazard Estimation Methodology – Earthquakes*, Federal Emergency Management Agency, USA

Heaton, T. H., Tajima, F., and Mori, A.W. (1986) "Estimating ground motions using recorded accelerograms", *Surv. Geophys.*, 8: 25-83

Hensolt, W.J.and Brabb, E. E. (1990) Maps showing elevation of bedrock and implicationsfor design of engineered structures to withstand earthquake shaking in San Mateo County, California, USGS Open File Report 90-496

Hight, D. W., Gens A., and Symes, M. S. (1983) "The Development of a New Hollow Cylinder Apparatus for Investigating the Effects of Principal Stress Rotation in Soils", *Geotechnique*, 33(4): 355-383

Hisada, Y. (1995) "An efficient method for compunting Green's functions for a layered half-space with sources and receivers at close depths (part 2)", *Bulletin of the Seismological Society of America*, vol. 85

(4), pp. 1080-1093

Hisada, Y. (1994) "An efficient method for compunting Green's functions for a layered half-space with sources and receivers at close depths", *Bul. of the Seismological Society of America*, 84(5): 1456-1472

Horike, M. (1985) "Inversion of phase velocity of long-period microtremors to the S-wave-velocity structure down to the basement in urbanized areas", *J. Phys. Earth*, 33: 59-96

Horikoshi, K. and Ohtsu, H. (1996) "Investigation of PC piles damaged by the Hyogoken-Nambu earthquake", *Proc., 31st Japan National Conf. on Geotechnical Engineering*, 1: 1227-1228 (in Japanese)

Housner, G. W. (1989) *Competing Against Time*, Report to Governor Deukmejian of California, Governor's Board of Inquiry on the (1989) Loma Prieta Earthquake, George W. Housner, Chairman

Housner, G. W. and Jennings, P. C. (1964) "Generation of Artificial Earthquakes", *ASCE Journal of Engineering Mechanics Div.*, 90: 113-150

Husid, R. L. (1969) "Analisis de Terremotos: Analisis General", Revista del IDIEM, Universidad de Chile, Santiago, Chile, 8(1, May), 21-42

Iai, S., Ichi, K., Liu, H. and Morita, T. (1998) "Effective stress analyses of port structures", *Soils and Foundations*, Special Issue on Geotechnical Aspects of the January 17, 1995 Hyogoken-Nambu Earthquake, (2): 97-114

ICOLD (1993) Computer Software for Dam, Validation, Bulletin No. 94, International Conference on Large Dams or International Committee on Large Dams

ICOLD (1989) Selection Seismic Parameters for Large Dams – Guidelines, Bulletin No. 72, International Conference on Large Dams or International Committee on Large Dams.

ICOLD (1983) Seismicity and Dam Design, Bulletin No. 46, International Conference on Large Dams or International Committee on Large Dams

Idriss, I. M. (1995) "An overview of earthquake ground motions pertinent to seismic zonation", *Proc. of 5th Int. Conf. Seismic Zonation*, Nice, 2111-2126

Idriss, I. M. and Sun, J. I., (1992) *Users Manual for SHAKE91*, A Computer Program for Conducting Equivalent Linear Seismic Response Analyses of Horizontally Layered Soil Deposits, Program Modified Based on the Original SHAKE Program Published in December 1972, by Schnabel, Lysmer and Seed", Centre for Geotechnical Modelling, Department of Civil and Environmental Engineering, University of California, Davis, California, USA 12p

Idriss, I. M. (1991) "Earthquake Ground Motions at Soft Soil Sites", *Proceedings of the Second International Conference on Recent Advances in Geotechnical Earthquake Engineering and Soil Dynamics*, March 11-15, 1991, St. Louis, MO, Ed., S. Prakesh, University of Missouri-Rolla, 3

Idriss, I. M., (1990) "Response of soft soil sites during earthquakes", *Proceeding of the H. Bolton Seed Memorial Symposium* (Ed. J. M. Duncan), BiTech Publishers, Vancouver, BC, 273-289

Idriss, I. M., Dobry, R. and Singh, R. D. (1978) "Nonlinear behaviour of soft clays during cyclic loading", *Journal of Geotechnical Engineering*, ASCE, (GT12): 1427-1447

Idriss, I. M. and Seed, H. B. (1968) "Seismic response of horizontal soil layers", *Journal of Soil Mechanics and Foundation Division*, ASCE, 96(SM4)

Imai, T. and Tonouchi, K. (1982) "Correlations of N-Values with S-Wave Velocity", *Proc. ESOPT II*, Amsterdam, 2: 67-72

INE (1994) *Censos 91, Resultados definitivos*, Lisbon: Instituto Nacional de Estatística

Ionescu, F. (1999) Comportamento sforzi-deformazioni della sabbia di Toyoura da prove di taglio torsionale in condizioni di carico monotono e ciclico, Ph. D. Thesis, Politecnico di Torino Department of Structural and Geotechnical Engineering, (in Italian)

Irikura K. (1983) "Semiempirical Estimation of Strong Ground Motions During Large Earthquakes", *Bull. Disast. Res. Inst*, Kyoto University, 33: 63-104

Isenhower, W. M., Stokoe, K. H., and Allen, J. C. (1987) "Instrumentation for Torsional Shear Resonant Column Measurements Under Anisotropic Stresses", *Geotechnical Testing Journal*, 10(4): 183-191

Isenhower, W. M. and Stokoe, K. H. (1981) "Strain Rate Dependent Shear Modulus of San Francisco Bay Mud", *International Conference on Recent Advances in Geotechnical Earthquake Engineering and Soil Dynamics*, St. Louis, Missouri

Ishihara, K. and Cubrinovski, M. (1998) "Soil-pile interaction in liquefied deposits undergoing lateral spreading", *Proc. of 11th Danube-European Conf. on Soil Mechanics and Geotechnical Eng.*, 51-64

Ishihara, K., Yoshida, K., and Kato, M. (1997) "Lateral spreading of liquefied deposits during the 1995 Kobe earthquake", *Proc. 3rd Kansai Int. Geotechnical Forum on Comparative Geotechnical Eng.*, 31-50

Ishihara, K. (1997) "Terzaghi oration: Geotechnical aspects of the 1995 Kobe earthquake", *Proc. of the 14th International Conference on Soil mechanics and Foundation Engineering,* 4: 2047-2073

Ishihara, K. (1996) *Soil Behaviour in Earthquake Geotechnics*, Oxford Science Publications, Oxford, UK

Ishihara, K., Yasuda, S., and Nagase, H. (1996) "Soil characteristic and ground damage", *Soils and Foundations*, Special Issue on Geotechnical Aspects of the January 17, 1995 Hyogoken-Nambu Earthquake, 109-118

Isoyama, R., Ishida, E., Yune, K., and Shirozu, T. (1998) "Seismic damage estimation procedure for water supply pipelines", *Proceedings, IWSA International Workshop "Anti-Seismic Measures on Water Supply"* (Water and Earthquake'98 Tokyo), 26-36

Iyisan, R. (1996) "Correlations between Shear Wave Velocity and In-situ Penetration Test Results", *Technical Journal of Turkish Chamber of Civil Engineers*, 7(2): 1187-1199 (in Turkish)

Jamiolkowski, M., Lo Presti, D. C. F. and Froio, F. (1998) "Design Parameters of Granular Soils from In-Situ Tests", *XI Europe-Danube Conference on SMGE*, Porec Croatia, Balkema, 65-94

Jamiolkowski, M., Lancellotta, R., and Lo Presti, D. C. F. (1995) "Remarks on the Stiffness at Small Strains of Six Italian Clays" Keynote Lecture 3, *IS Hokkaido*, Balkema, Rotterdam, 2: 817-836

Jamiolkowski M. and Lo Presti D. C. F. (1991) "Anisotropy of Soil Stiffness at Small Strain", Panel Discussion to Session 1, *9th ARC on SMFE*, Bangkok, Thailand, 2: 195-197

Jamiolkowski, M., Ghionna, V., Lancellotta, R., Pasqualini, E. (1988) "New correlations for penetration tests in design practice", *1st Int. Symp. on Penetration testing*, Orlando, Florida, USA,Balkema, 1: 263-296

Japanese Road Association, JRA (1980, 1997) *Specifications for Road Bridges*, Vol. IV (in Japanese)

Japanese Association for Steel Pile Piles, JASPP (1996) *Investigation report on steel pipe pile foundations in the Hyogoken-Nambu earthquake- Part II*, 156p.

Jardine, R. J. (1995) "One Perspective of the Pre-Failure Deformation Characteristics of Some Geomaterial", Keynote Lecture 5, *IS Hokkaido*, Balkema, Rotterdam, 2: 855-886

Jardine, R. J. (1992) "Some Observations on the Kinematic Nature of Soil Stiffness", *Soils and Foundations*, 32(2): 111-124

Jardine, R. J. (1985) Investigations of Pile-Soil Behaviour with Special Reference to the Foundations of Offshore Structures, Ph.D. Thesis, University of London, UK

Jarpe, S., Hutchings, L., Hauk, T., and Shakal, A. (1989) "Selected Strong- and Weak-Motion Data from the Loma Prieta Earthquake Sequence", *Seismological Research Letters*, (60): 167-176

Jessberg, H. L. (1996) "ISSMFE/TC5 activities - Technical Committee on Environmental Geotechnics", *Proc. 2nd International Congress on Environmental Geotechnics*, Osaka, 1- 28

Jessberg, H. L. (1994) "Emerging problems and practices in environmental geotechnology", *Proc. of 13th*

International Conference on Soil Mechanics and Foundation Engineering, (1): 271- 281

JGS, Japanese Geotechnical Society (2000) Report on the Investigations of the 1999 Kocaeli Earthquake, Turkey (in Japanese)

JGS, Japanese Geotechnical Society (1998) *Remedial measures against soil liquefaction; from investigation and design to implementation,* Balkema, 433p.

JGS, Japanese Geotechnical Society (1991) Proc. of the Symposium on Counter Measurement against Liquefaction, (in Japanese)

Jibson, R. (1987) Summary of research on the effects of topographic amplification of earthquake shaking on slope stability, Open-File Report 87-268, U.S. Geological Survey, Menlo Park, Cal.

Johnson, M., Lew, M., Lundy, J., and Ray, M., E. (1991) "Investigation of sanitary slope performance during strong ground motion from the Loma Prieta Earthquake of October 17. 1989", *Proc. of 2nd Int. Conf. on Recent Advances in Geotechnical Earthquake Engng. and Soil Dynamics,* St. Louis, (2): 1701-1708

Jones, R. B. (1958) "In-situ measurement of the dynamic properties of soil by vibration methods", *Geotechnique,* 8(1): 1-21

Joyner, W. and Fumal, T. E. (1984) "Use of measured S-wave velocity for predicting geologic site effects on strong ground motion", *Proceedings of the Eight World Conference on Earthquake Engineering,* San Francisco, Calif., 2: 777-783

Justo, J. L. and Salwa, C (1998) "The 1531 Lisbon Earthquake", *Bulletin of Seismological Society of America,* 88(2): 319-328

Kafka, A.L. and Walcott, J.R. (1998) "How Well Does the Spatial Distribution of Smaller Earthquakes Forecast the Location of Larger Earthquakes in the Northeastern United States", *SRL,* 69(5): 428-440

Kanai, K. (1957) "The requisite conditions for predominant vibration of ground motion", *Bull. Earthquake Res. Inst. Tokyo University*; 31: 457

Kanamori, H. (1977) "The Energy Release in Great Earthquakes", *Journal of Geophysical Research,* 82: 2921-2987

Kanatani, M., Okada, S. and Yasuda, S. (2000) "Countermeasures against liquefaction-induced flow", *Tsuchi-toKiso,* 48(506): 43-48, (in Japanese)

Kansai Branch of Architectural Institute of Japan (1996) Report on case histories of damage to building foundations in Hyogoken-Nambu earthquake, Report presented by Committee on Damage to Building Foundations, 400p. (in Japanese)

Kausel, E. and Roesset, J. M. (1981) "Stiffness matrices for layered soils", *Bullettin of the Seismological Society of America,* 71(6): 1743-1761

Kausel, E. and Roesset, J. M. (1974) "Soil Structure Interaction for Nuclear Containment Structures". *Proc. ASCE, Power Division Specialty Conference,* Boulder, Colorado

Kavazanjian, E. Jr. (1998) "Current issues in seismic design of geosynthetic cover systems". *Proc. of 6th International Conference on Geosynthetics,* 1: 219-226

Kavazanjian, E. Jr., Matasovic, N., Stokoe, K. and Bray, J. (1996) "In situ shear wave velocity of solid waste from surface wave measurements", *Proc. of 2nd International Congress on Environmental Geotechnics,* Ed. M. Kamon, (1): 97-102

Kavazanjian, E., Jr. (1995) "Evaluation of MSW properties for seismic analysis", *Geoonvironmental Geotechnics,* Balkema, Rotterdam, 1126-1141

Kavazanjian, E. Jr. and Matasovic, N. (1995) "Seismic analysis of solid waste landfills", *Proc. of Geoenvironment 2000,* ASCE Specialty Conference, New Orleans, Louisiana, 22-24, February

Kavazanjian, E. Jr., Matasovic, N., Bonaparte, R., and Schmertmann, G. R. (1995) "Evaluation of MSW properties for seismic analysis", *Proc. of Geoenvironment 2000*, ASCE Specialty Conference, New Orleans, Louisiana, 22-24, February

Kawase, H. (1988) "Time Domain Response of a Semicircular Crayon for Incident SV, P and Rayleigh Waves Calculated by Discrete Wave Number Boundary Element Method", *BSSA*, (78):1415-1437

Kawase, J. and Aki, K. (1990) "Topography Effect at the Critical SV-Wave Incidence: Possible Explanation of Damage Pattern by the Whittier Narrows,California, Earthquake of 1 October 1987." *Bulletin of the Seismological Society of America*, 80(1): 1-22

Keilis- Borok (1957) "Investigation of the Mechanisms of Earthquakes", *Sov.Res.Geophys.*, 4

Khawaja, A. S. (1993) Damping Ratios from Compression and Shear Wave Measurements in the Large Scale Triaxial Chamber, M. Sc Thesis, University of Texas at Austin

Kiku, H., Yoshida, N., Yasuda, S., Irisawa, T., Nakazawa, H., Shimizu, Y., Ansal, A., and Erken, A. (2001) "In-situ penetration tests and soil profiling in Adapazarı, Turkey", *Proc. of the Satellite Conference on Lessons Learned from Recent Strong Earthquakes*, 15th ICSMGE, 259-265

Kim, D-S. and Stokoe, K. H. II (1994) "Torsional Motion Monitoring System for Small Strain (10-5 to 10-3) Soil Testing", *Geotechnical Testing Journal*, 17(1): 17-26

King, S. A. and Kiremidjian, A. S. (1994) *Regional Seismic Hazard and Risk Analysis Through Geographic Information Systems*, The John A. Blume Earthquake Engineering Center, Report No.111, Stanford, CA

Kockel, R., Konig, D., and Syllwasschy, O. (1997) "Three Basic Topics Mechanics on Waste Mechanics" *Proc. of 14th International Conference on Soil Mechanics and Foundation Engineering*, 3: 1831- 1837

Ktenidou, O-J. (2003) Numerical and experimental analysis of strong ground motion - Study of the effect of local soil conditions at CORSEIS array and the city of Aegion, Greece, Diploma Thesis. Civil Engineering Department, Aristotle University of Thessaloniki (in Greek)

Kuwabara, F. and Yoneda, K. (1998) "An investigation on the pile foundations damaged by liquefaction at the Hyogoken Nambu earthquake", *Journal of Struct. Constr. Engrg.*, AIJ, No. 507 (in Japanese)

Lachet, C., Hatzfeld, D., Bard, P. Y., Theodulidis, N., Papaioannou, C., and Savvaidis, A. (1996) "Site Effects and Microzonation in the City of Thessaloniki-Comparison of Different Approaches", *BSSA*, 86(6):1692-1703

Lachet, C. and Bard, P.Y. (1994) "Numerical and Theoretical Investigations on the Possibilities and Limitations of Nakamura's Technique", *Journal of Phys. Earth*, (42):377-397.

Lai, C. G., Rix, G. J., Foti, S., and Roma, V. (2002) "Simultaneous Measurement and Inversion of Surface Wave Dispersion and Attenuation Curves", *Soil Dynamics and Earthquake Eng.*, 22(9-12): 923-930

Lai, C. G., Pallara O., Lo Presti, D. C. F., and Turco, E. (2001) *Low-Strain Stiffness and Material Damping Ratio Coupling in Soils, Advanced Laboratory Stress-Strain Testing of Geomaterials*, TC29 Summary Book, 15th ICSMGE, Istanbul, Turkey. August 27-31, 2001. Eds. F.Tatsuoka, S.Shibuya and Kuwano.

Lai, C. G. and Rix, G. J. (1998) Simultaneous inversion of Rayleigh phase velocity and attenuation for near-surface site characterization, Technical Report GIT-CEE/GEO-98-2, Georgia Institute of Technology

LeBrun, B., Hatzfel, D., and Bard, P.-Y. (1999) "Experimental study of ground motion on a large scale topography". *J. of Seismology*, 3(1): 1-15

Lermo, L. and Chavez-Garcia, F. J. (1993) "Site effect evaluation using spectral ratios with only one station", *Bulletin of the Seismological Society of America*, 83: 1574-1594

Ling, H. I. and Leshchinsky, D. (1997) "Seismic stability and permanent displacement of landfill cover systems", *Journal of Geotechnical and Geoenvironmental Engineering*, 123(2): 113-122

Lo Presti, D., Pallara, O., Jamiolkowski, M. and Cavallaro, A. (1999) "Anisotropy of small strain stiffness of undisturbed and reconstituted clays", *Proc. IS Torino 99*, Balkema, 1: 3-10

Lo Presti, D. C. F., Pallara, O., and Cavallaro, A. (1997) "Damping Ratio of Soils from Laboratory and In-Situ Tests", *Proc. XIV ICSMFE, Seismic Behaviour of Ground and Geotechnical Structures*, Balkema, Rotterdam, 391-400.

Lo Presti, D. C. F., Jamiolkowski, M., Pallara, O., and Cavallaro, A. (1996) "Rate and Creep Effect on the Stiffness of Soils", *GSP No. 61*, ASCE, 166-180

Lo Presti, D. C. F., Pallara, O., Lancellotta, R., Armandi, M., and Maniscalco, R. (1993) "Monotonic and Cyclic Loading Behaviour of Two Sands at Small Strains", *Geotechnical Testing Jour.*, 16(4): 409-424

Lo Presti, D. C. F. and O'Neill, D. A. (1991) "Laboratory Investigation of Small Strain Modulus Anisotropy in Sand", *Proc. of ISOCCT1*, Postdam, NY

Lo Presti, D. C. F. (1989) "Proprieta' dinamiche dei terreni", *Proc. XIV CGT*, Department of Structural Engineering, Politecnico di Torino (in Italian)

Lo Presti, D. C. F. and Lai, C. (1989) Shear Wave Velocity in Soils from Penetration Tests, R.R. Dip. di In. Strutturale, Politecnico di Torino, (21)

Luco, J. E. and Anderson, J. G., (1983) "Steady State Response of an Elastic Half-Space to a Moving Dislocation of Finite Width", *Bulletin of the Seismological Society of America*, 73(1): 1-22

Lysmer, J. (1978) "Analytical procedures in soil dynamics", *Earthquake Engineering and Soil Dynamics,* Pasadena - Ca, 3: 1267-1316

Lysmer, J., Udaka, T., Tsai, C., and Seed, H. B. (1975) FLUSH-A computer program for approximate 3D analysis of soil-structure interaction problems, Report No. UCB/EERC 74-4, University of California, Berkeley

Lysmer, J., Udaka, T., Seed, H. B., and Hwang, R. (1974) LUSH 2-A computer program for complex response analysis of soil-structure systems, Report No. UCB/EERC 75-30, University of California, Berkeley

Makdisi, F. I. and Seed, H. B. (1977) A Simplified procedure for estimating earthquake-induced deformations in dams and embankments, Report No EERC 79-19, University of California, Berkeley

Makra, K., Raptakis, D., Chavez-Garcia, F. J., and Pitilakis, K. (2001) "Site effects and Design Provisions: The case of Euroseistest", *J. Pure and Applied Geophysics.* 158(12): 2349-2367

Makra, K. (2000) Contribution to the evaluation of site response for complex soil structure (Euroseis-test valley) using experimental and theoretical approaches, Ph.D. Thesis, Dept. of Civil Engineering, Aristotle University of Thessaloniki, (in Greek)

Marcellini, A., Daminelli, R., Tento, A., Franceschina, G., and Pagani, M. (2001a) 'The Umbria Marche Microzonation Project: Outline of the project and the example of Fabriano results" Italian Geotechnical Journal, XXXV (2): 28-35

Marcellini, A., Daminelli, R., Franceschina, G., and Pagani, M. (2001b) "Regional and local seismic hazard assessment", *Soil Dynamics and Earthquake Engineering*, 21, 415-429

Marcellini, A., Daminelli, R., Pagani, M., Riva, F., Tento, A., Crespellani, T., Madiai, C., Vannucchi, G., Frassineti, G., Martelli, L., Palumbo, A., and Viel, G. (1999) "Seismic Microzonation of some municipalities of the Rubicone area (Emilia-Romagna Region)", *Proceedings of the Eleventh European Conference on Earthquake Engineering* - "Invited lectures" volume, Paris, September 1998; 339-350

Marcellini, A., Iannaccone, G., Romeo, R.W., Silvestri, F., Bard, P.Y., Improta, L., Meneroud, J.P., Mouroux, P., Mancuso, C., Rippa, F., Simonelli, A.L., Soddu, P., Tento, A. and Vinale, F. (1995a) "The Benevento Seismic Risk Project. I- Seismotectonic and Geotechnical Background", *Proc. 5th International Conference on Seismic Zonation*, Nice, France, (1):802-809

Marcellini, A., Bard, P.Y., Iannaccone, G., Meneroud, J.P., Mouroux, P., Romeo, R.W., Silvestri, F., Duval, A.M., Martin, C. and Tento, A. (1995b) "The Benevento Seismic Risk Project. II- The microzonation", *Proc. 5th International Conference on Seismic Zonation*, Nice, France, (1):810-817

Marcellini, A., Pagani, M., and Riva, F. (1994) "Dependence of expected response spectrum shape on Gutenberg-Richter's value and standard deviation of attenuation law", *Proceedings of the XXIV General Assembly ESC*, Athens, September 19-25 1994, 3: 1619-1627

Marchetti, S. (1997) "The Flat Dilatometer: Design Applications", *Proc. III Int. Geotechnical Engineering Conference*, Cairo, 421-428

Marchetti, S. (1980) "In Situ Tests by Flat Dilatometer", *Journal of Geotechnical Engineering Division*, ASCE, 106(GT3): 299-321

Masing, G. (1926) "Eigenspannungen und verfestigung beim messing", *Proc. 2nd Int. Cong. of Applied Mechanics*, Zurich, Switzerland, (in German)

Matasovic, N., Kavazanjian, E. Jr., Augello, A. J., Bray, J. D., and Seed, R. B. (1995) "Solid Waste Landfill Damage Caused by 17 January 1994 Northridge Earthquake", *The Northridge, California, Earthquake of 17 January 1994, California Department of Conservation, Division of Mines and Geology Special Publication 116*, Eds. M. C. Woods and R.W. Seiple, Sacramento, California, 43-51

Mayne, P. W. and Martin, G. K. (1998) "Commentary on Marchetti Flat Dilatometer Correlations in Soils", *Geotechnical Testing Journal*, 21(3): 222-239.

Mayne P. W. and Rix, G. J. (1993) "G_{max} - q_c Relationships for Clays", *Geotechnical Testing Journal*, 16(1): 54-60

McGuire, K. R. (1995) "Probabilistic Seismic Hazard Analysis and Design Earthquakes: Closing the Loop", *BSSA*, (85)5:1275-1284

McGuire, K. R. and Hanks, T. C. (1980) "RMS Acceleration and Spectral Amplitudes of Strong Ground Motion during the San Fernando, California, Earthquake", *Bulletin of the Seismological Society of America*, 70:1907-1919

Medvedev, S. V. (1977) *Seismiceskoe Mikrorayonirovanie*, Isdatelusto Nauka, Moscow, 248 p., (in Russian)

Medvedev, J. (1962) *Engineering Seismology*, Academia Nauk Press, Moscow

Mendes-Victor, L. A., Oliveira, C. S., Pais, I., and Teves-Costa, P. (1993) "Earthquake damage scenarios in Lisbon for disaster preparedness", *NATO Advanced Research Workshop on An Evaluation of Guidelines for Developing Earthquake Damage Scenarios for Urban Areas*, Edition M. Erdik and B. Tucker, Istanbul, October 1993, 195-218

Midorikawa, S. (1987) "Prediction of Isoseismal Map in Kanto Plain due to Hypothetical Earthquake" *Journal of Structural Dynamics*, 33B: 43-48

Mitchell, J. K., Lodge, A. L., Coutinho, R. Q., Kayen, R. E., Seed, R. B., Nishio, S., and Stokoe, K.H. II (1994) In Situ Test Results from four Loma Prieta Earthquake Liquefaction Sites: SPT, CPT, DMT and Shear Wave Velocity, EERC Report No. UCB/EERC - 94/04

Miura, K., Miura, S., and Toki, S. (1986) "Deformation Behaviour of Anisotropic Dense Sand Under Principal Stress Axes Rotation", *Soils and Foundations*, 26(1) 36-52

Mok, Y. J. (1987) Analytical and Experimental Studies of Borehole Seismic Methods, Ph.D. Thesis, University of Texas at Austin

Monahan, P. A., Levson, V. M., Henderson, P., and Sy, A. (2000) Relative Liquefaction and Amplification of Ground Motion Hazard Maps of Greater Victoria, Report and Expanded Legend to Accompany British Columbia Geological Survey, Geoscience Map 2000-3, BC Geological Survey Branch, 25 p.

Nagai, K. (1997) "Lessons concerning foundation design in Hyogoken-Nambu earthquake", *Kenchiku Gijyutsu*, 564: 84-93 (in Japanese)

Nakamura, Y. (2000) "Clear identification of fundamental idea of Nakamura's technique and its applications", *Proceedings of 12th World Conference on Earthquake Engineering*, CD-Rom, 2656

Nakamura, Y. (1989) "A method for dynamic characteristics estimation of subsurface using microtremor on the ground surface", *Quarterly Report of Railway Technical Research Institute (RTRI)*, 30 (1): 25-33

Narula, P. L., Chaubey, S. K., and Shashank, S. (2002) "Bhuj, India Earthquake Reconnaissance Report", *Earthquake Spectra*, Supplement A to Vol 18, 45-50

National Building Code of Canada, NBCC (1995) Associate Committee on the National Building Code, National Research Council of Canada, Ottawa, Ontario

Nazarian, S. (1984) In situ determination of elastic moduli of soil deposits and pavement systems by Spectral-Analysis-of-Surface waves method, PhD Thesis, University of Texas at Austin

Nechtschein S., Bard, P.Y., Gariel, J.C., Ménéroud, J.P., Dervin, P., Cushing, M., Gaubert, C., Vidal, S. and Duval, A.M.(1995) "A topographic effect study in the Nice region", *Proc. Fifth International Conference on Seismic Zonation*, Nice, 2:1067–1074

NEHRP (2000) *Recommended provisions for seismic regulations for new buildings and other structures, Part 1: Provisions*, FEMA 368, Building seismic safety council of the National Institue of Building Sciences, USA

Neumann, F. (1954) *Earthquake Intensity and Related Ground Motion*, University of Washington Press, Seattle, WA

Newmark, N. M. (1965) "Effects of earthquakes on dams and embankments", *Geotechnique*, 15(2): 139-160

Nigbor, R. L. and Imai, T. (1994) "The Suspension P-S Velocity Logging Method", *Proc. XIII ICSMFE*, New Delhi, India TC # 10, 57-61

Nova, R. and Montrasio, L. (1991) "Settlements of shallow foundations on sand", *Geotechnique*, 41(2): 243-256.

Oh-oka, H., Onishi, K., Nanba, S., Mori, T., Ishikawa, K., Koyama, S., and Shimazu, S. (1997a) "Liquefaction-induced failure of piles in 1995 Kobe earthquake", *Proc., 3rd Kansai Int. Geotechnical Forum on Comparative Geotechnical Engineering*, 265-274

Oh-oka, H., Katoh, F., and Hirose, T. (1997b), "An investigation about damage to steel pipe pile foundations due to lateral spreading", *Proc., 32nd Japan National Conf. on Geotechnical Engineering*, 1: 929-930 (in Japanese)

Oh-oka, H., Iiba, M., Abe, A., and Tokimatsu, K. (1996) "Investigation of earthquake-induced damage to pile foundation using televiewer observation and integrity sonic tests", *Tsuchi-to-kiso, JSG*, 44(3): 28-30 (in Japanese)

Ohta, Y. and Goto, N. (1978) "Empirical shear wave velocity equations in terms of characteristic soil indexes", *Earthquake Engineering and Structural Dynamics*, 6

Ohta, Y. and Goto, N. (1976) "Estimation of s-wave velocity in terms of characteristic indices of soil", *Butsuri-Tanko*, 29(4): 34-41

Ohtsuki, A. and Harumi, K. (1983) "Effect of topography and subsurface inhomogeneities on seismic SV waves", *Earthquake Engineering and Structural Dynamics*, 11: 441-462

Oliveira, C. S. and Sánchez-Cabañero, J. (2002) "Evolución histórica del análisis de la peligrosidad sísmica. Tendencias para nuevos desarrollos", *Primero Congreso Nacional de Ingeniería Sísmica*, Murcia, 1999, Tomo II, Ministerio do Fomento de España, Serie Monografías, 57-89 (in Spanish)

Oliveira, C.S., Mota de Sá, F., and Pais, I. (2000) "Towards on-line damage assessment caused by an earthquake", *Proceedings, 6th International Conf. on Seismic Microzoning*, Palm Springs, CD Rom

Oliveira, C. S. and Pais, I (1993) "Technical approaches for earthquake emergency planning. Recent applications to the City of Lisbon", *Proceedings, International Conference on Natural Risk and Civil Protection*, Belgirate, CEC, DGXII-R&D. Ed. Chapman and Hall, 57-72

Onur, T. (2001) Seismic Risk Assessment in Southwestern British Columbia, Ph.D. Thesis, Department of Civil Engineering, University of British Columbia, Vancouver, BC

Pais, I., Mota de Sá, F., and Oliveira, C. S. (1999) "Construção de um Indicador de Vulnerabilidade Sísmica das Redes de Abastecimento. Aplicação à Rede de Gás de uma Área-Piloto de Lisboa", *Proc., 4th Portuguese Meeting on Seismology and Earthquake Engineering*, EST/Universidade do Algarve, Faro, (in Portuguese)

Pais, I., Teves-Costa, P., and Cabral, J. (1996) "Emergency management of urban systems under earthquake damage scenarios", *Proc., 11th World Conference on Earthquake Engineering, Acapulco*, CD-Rom

Panza, G. F., Vaccari, F., Costa, G., Suhadolc, P., and Faeh, D. (1996) "Seismic input modelling for zoning and microzoning", *Earthquake Spectra*, 12: 529-566

Paolucci, R. (1997) "Simplified evaluation of earthquake induced permanent displacement of shallow foundations", *Journal of Earthquake Engineering*, 1(3): 563-579

Paolucci, R. and Pecker, A. (1997) "Seismic bearing capacity of shallow strip foundations on dry soils", *Soils and Foundation*, 37(3): 95-105

Papa, V., Silvestri, F., and Vinale, F. (1988) "Analisi delle Proprietà di un Tipico Terreno Piroclastico mediante Prove di taglio Semplice", *Proc. Gruppo Nazionale di Coordinamento per gli Studi di Ingegneria Geotecnica*, Monselice, Italy, 1: 265-286

Papageorgiou, A., Halldorsson, B., and Dong, G. (2000) "Target Acceleration Spectra Compatible Time Histories", University of Buffalo, Dept. of Civil, Structural and Environmental Engrg., NY, http://civil.eng.buffalo.edu/engseislab/

Papageorgiou, A. S. and Aki, K. (1983) "A specific barrier for the quantitative description of inhomogeneous faulting and the prediction of strong motion: II. Application of the model", *Bulletin of the Seismological Society of America*; 73: 953-978

Papazachos, B. C., Karakostas, B. E., Papaioannou, Ch., Papazachos, C. B. and Scordilis, E. (1997) A Catalogue for the Aegean and Surrounding Area for the Period 550BC-1995.

Patel, N. S. (1981) Generation and Attenuation of Seismic Waves in Downhole Testing, M. Sc. Thesis, GT81-1, Department of Civil Engineering, University of Texas at Austin

Pecker, A. and Pender, M. (2000) "Earthquake Resistant Design of Foundations: New Construction". Invited lecture, *GeoEng2000*, Melbourne, 1: 313-332

Pecker, A. (1998) Rion Antirion Bridge - Lumped parameter model for seismic soil structure interaction analyses - Principles and validation, Geodynamique et Structure Report FIN-P-CLC-MG-FOU-X-GDS00060 - Prepared for Gefyra Kinopraxia

Pecker, A. (1997) "Analytical formulae for the seismic bearing capacity of shallow strip foundation", *Seismic Behaviour of Ground and Geotechnical Structures*, Seco e Pinto Ed., Balkema, 261-268

Pecker, A., Salençon, J., Auvinet, G., Romo, M.P. and Verzurra, L. (1995) Seismic bearing capacity of foundations on soft soils, Final Report to European Commission - Contract CI1 - CT92-0069

Pecker, A. (1994) "Seismic design of shallow foundations", State of the Art, *10th European Conference on Earthquake Engineering*, Duma Ed., Balkema, 1001-1010

Pecker, A. and Salençon, J. (1991) "Seismic bearing capacity of shallow strip foundations on clay soils", CENAPRED, *Proc. of the Int. Workshop on Seismology and Earthquake Engineering*, Mexico, 287-304

Pedretti, S. (1998) Nonlinear seismic soil-foundation interaction: analysis and modelling method, PhD Thesis Dpt Ing Structurale, Politecnico di Milano

Pender, M. J. (1995) "Earthquake Resistant design of Foundations", Keynote address, *Pacific Conference on Earthquake Engineering*, PCEE95, Melbourne

Pender, M. J. (1999) "Geotechnical Earthquake Engineering design practice in New-Zealand". *Proceedings of 2nd International Conference on Earthquake Geotechnical Engineering.* Ed. P.Sêco e Pinto, Balkema

Pereira de Sousa, F. L. (1932) *O terramoto do 1° de Novembro de 1755 em Portugal e um estudo demográfico*, Vol I-IV, Serviços Geológicos, Lisbon, (in Portuguese)

Peruzza, L. (2000) "Pericolosità sismica in termini probabilistici", Eds. A.Marcellini and P.Tiberi, *La Microzonazione sismica di Fabriano*, Biemmegraf, 28-29, (in Italian)

Phillips, W.S. and Aki, K. (1986) "Site amplification of coda waves from local earthquakes in central California" *Bulletin of the Seismological Society of America,* 79: 627-648

Pinto, L. D. (1998) Personal Communication

Pitilakis, K. et al. (2003) Experimental and theoretical analyses of site effects toward the improvement of soil classification and design spectra in EC8 and Greek Seismic Code, Final Report Organization of Seismic Planning and Protection

Pitilakis, K. et al. (2003) Microzonation study of Thessaloniki, Final report

Pitilakis, D. (2002) Numerical and experimental analysis of strong soil motion, Study of site effects in Corseis array and in the town of Aigion.Diploma Thesis, Civil Engineering Department, Aristotle University of Thessaloniki (in Greek)

Pitilakis, K. D., Makra, K. A., and Raptakis, D. G. (2001) "2D vs. 1D site effects with potential applications to seismic norms: The cases of EUROSEISTEST and Thessaloniki", Invited Lecture, *Proc. XV ICSMGE Satellite Conference on Lessons Learned from Recent Strong Earthquakes*, Istanbul, 123-133

Pitilakis, K., Raptakis, D., Lontzetidis, K., Tika-Vassilikou, Th., and Jongmans, D. (1999) "Geotechnical and geophysical description of EURO-SEISTEST, using field, laboratory tests and moderate strong motion recordings", *J. of Earthquake Engineering*, 3(3): 381-409

Pradhan, T. B. S., Tatsuoka, F., and Horii, N. (1988) "Strenght and Deformation Characteristics of Sand in Torsional Simple Shear", *Soils and Foundations*, 28(3): 131-148

PREC 8 (1996) (Prenormative Research in Support of Eurocode 8) "Seismic behaviour and design of foundation and retaining structures", Facioli-Paolucci Eds., Report No.2

Priolo, E. (2001) "Deterministic computation of the reference ground motion in Fabriano (Marche, Italy)", *Italian Geotechnical Journal*, XXXV (2)

Puci I and Lo Presti D.C.F. (1998) "Damping measurement of reconstituted granular soils in Calibration Chamber by means of seismic tests" *Proceedings of the 11th European Conference on Earthquake Engineering*, 6-11 September 1998, Paris, Balkema, Paper No. 206

Pugliese, A. and Sabetta, F. (1989) "Stima di spettri di risposta da registrazioni di forti terremoti Italiani", *Ingegneria Sismica*, 6: 3-14, (in Italian)

Ramberg, W. and Osgood, W. R. (1943) Description of Stress-Strain Curves by Three Parameters, Tech. Note 906, Nat. Advisory Commitee for Aeronautics, Washington DC.

Raptakis et al. (2003a) "Site effect due to lateral propagation in Thessaloniki, Greece – I. Soil structure and Observational evidences" (submitted to *Bulletin Earthquake Engineering*)

Raptakis et al. (2003b) "Site effect due to lateral propagation in Thessaloniki, Greece – II. 2D theoretical approaches" (submitted to *Bulletin Earthquake Engineering*)

Raptakis, D. G., Chavez-Garcia F., Makra, K. A. and Pitilakis, K. D. (2000) "Site Effects at Euroseistest-I. 2D Determination of the Valley Structure and Confrontation of the Observations with 1D Analysis", *Soil Dynamics and Earthquake Engineering*, 19(1): 1-22

Raptakis, D., Theodulidis, N., and Pitilakis, K. (1998) "Data analysis of the Euroseistest strong motion array in Volvi (Greece): standard and horizontal to vertical ratio techniques", *Earthquake Spectra*, 14: 203-224

Rathje, E. M. and Bray, J. D. (1999) "Two dimensional seismic response of solid waste landfills", *Proc. of 2ⁿᵈ International Conference on Earthquake Geotechnical Engineering*, Lisbon, Ed. P. S. Sêco e Pinto, Balkema, Rotterdam, 2: 655-660

Rathje, E. M., Abrahamson, N. A., and Bray, J. D. (1998) "Simplified Frequency Content Estimation of Earthquake Ground Motions", *J.Geotech. and Geoenv.Enrg.*, ASCE, 124(2): 150-159

Redpath, B. B. and Lee, R. C. (1986) In-Situ Measurements of Shear-Wave Attenuation at a Strong-Motion Recording Site, Prepared for USGS Contract No. 14-08-001-21823

Redpath, B. B., Edwards, R. B., Hale, R. J., and Kintzer, F. C. (1982) Development of Field Techniques to Measure Damping Values for Near Surface Rocks and Soils, Prepared for NSF Grant No. PFR-7900192

Rey, J., Faccioli, E., and Bommer, J. J. (2002) "Derivation of design soil coefficients (S) and response spectral shapes for Eurocode 8 using the European Strong-Motion Database", *J. of Seismology*, 6 (4):547-555

Reynolds, J. M. (1997) *An Introduction to Applied and Environmental Geophysics*, J.Wiley and Sons, New York

Richards, R., Elms, D.G. and Budhu, M. (1993) "Seismic bearing capacity and settlements of shallow foundations", *Journal of Geotechnical Engineering*, ASCE, 119(7): 662-674

Riepl, J., Zahradnik, J., Plicka, V., and Bard, P.-Y. (2000) "About the efficiency of Numerical 1-D and 2-D Modelling of Site Effects in Basin Structures", *Pure and Applied Geophysics*, 157: 319-342

Riepl, J. (1997) Effets de site: évaluation expérimental et modélisations multidimensionnelles: application ou site test EURO-SEISTEST (Grèce), Thése Université Joseph Fourier – Grenoble I

Risk Engineering, Inc. (1997). EZ-Frisk™ (Version 4.0 User's Manual), Boulder, Colorado, 92p.

Rix G.J., Lai C.G., and Foti S. (2001) "Simultaneous measurement of surface wave dispersion and attenuation curves", *Geotechn. Testing J.*, ASTM, 350-358

Rix, G. J., Lai, C. G., and Spang, A. W. Jr. (2000) "In situ measurement of damping ratio using surface waves", *J. Geotechechnical and Geoenvironmental Engineering*, ASCE, 126: 472-480

Rix, G. J. and Stokoe, K. H. II (1991) "Correlation of Initial Tangent Modulus and Cone Penetration Resistance", *Proc. I ISOCCT*, Postdam, New York

Rix, G. J. (1988) Experimental Study of Factors affecting the Spectral Analysis of Surface Waves Method, Ph. D. Thesis, University of Texas at Austin, 315 p.

Rix, G. J. (1984) Correlation of Elastic Moduli and Cone Penetration Resistance, M. Sc. Thesis, University of Texas at Austin

Robertson, P. K. and Campanella, R. G. (1983) "Interpretation of Cone Penetration Tests - Part I: Sands", *Canadian Geotechnical Journal*, 20(4): 718-733

Rocha, F., Serrano, S., Anderson, M., and Oliveira, C. S. (2003) "Estudos de risco sísmico no âmbito do SNPC", *Proc. 3rd Simpósio da Assembleia Portuguesa de Meteorologia e Geofísica*, Aveiro. (in Portuguese)

Rocha, F., Serrano, S., Anderson, M., and Oliveira, C. S. (2002) "Emergency planning for the Metropolitan

Area of Lisbon", Abstract, *XXIII General Assembly European Geophysical Society*, Nice

Roësset, J. M., Chang, D. W., and Stokoe, K. H. II, (1991) "Comparison of 2-D and 3-D Models for Analysis of Surface Wave Tests", *5th International Conference on Soil Dynamics and Earthquake Engineering*, Karlsruhe, Germany, 111-126.

Rojahn, C., King, S. A., Scholl, R. E., Kiremidjian, A. S., Reaveley, L. D., and Wilson, R. R. (1997) "Earthquake Damage and Loss Estimation Methodology and Data for Salt Lake Country", Utah (ATC-36), *Earthquake Spectra*, 13(4): 623-642

Rollins, K. M., Evans, M. D., Diehl, N. B., and Daily, W. D. III (1998) "Shear Modulus and Damping Relationship for Gravels", *J. of Geotechnical and Geoenvironmental Eng.*, ASCE, 124(5): 396-405

Romo, M. (1995) "Clay behaviour ground response and soil-structure interaction studies in Mexico City", *Proc. 3rd. Conf. on Recent Advances in Geotech. Earthq. Engng. and Soil Dynamics*, 2: 1039-1051

Rovelli, A., Bonamassa, O., Cocco, M., Di Bona, M., and Mazza, S. (1988) "Scaling laws and spectral parameters of the ground motion in active extensional areas in Italy", *Bulletin of the Seismological Society of America*; 78: 530-560

RSA (1983) Regulamento de segurança e acções para estruturas de edifícios e pontes, Decreto-Lei no: 235/83. Lisbon: Imprensa Nacional - Casa da Moeda, (in Portuguese)

RSCCS (1958) Regulamento de segurança das construções contra os sismos, Decreto no: 41 658", Lisbon: Imprensa Nacional - Casa da Moeda, (in Portuguese)

Sabetta, F. and Bommer, J. J. (2002) "Modification of the spectral shapes and subsoil conditions in Eurocode 8", *Proc., 12th European Conference on Earthquake Engineering*, paper 518, London

Sabetta, F. and Pugliese, A. (1987) "Attenuation of peak horizontal acceleration and velocity from Italian strong-motion records", *Bulletin of the Seismological Society of America*; 77(2):1491-1513

Sadigh, K., Chang, C.-Y., Egan, J. A., Makdisi, F. I., and Youngs, R. R., (1997) "Attenuation Relationships for Shallow Crustal Earthquakes Based on California Strong Motion Data," *Seismological Research Letters*, 68(1): 180-189

Safak, E. (1988) "Analytical Approach to Calculation of response Spectra from Seismological Models of Ground Motion", *Earthquake Engineering and Structural Dynamics*, 16:121-134

Salençon, J. and Pecker, A. (1994a) "Ultimate bearing capacity of shallow foundations under inclined and eccentric loads, Part 1: Purely cohesive soil". *European Journal of Mechanics A/Solids*, 14(3): 349-375

Salençon, J. and Pecker, A. (1994b) "Ultimate bearing capacity of shallow foundations under inclined and eccentric loads, Part II: Purely cohesive soil without tensile strength". *European Journal of Mechanics A/Solids*, 14(3): 377-396

Salençon, J. (1990) "An introduction to the yield design theory and its application to soil mechanics", *European Journal of Mechanics A/Solids*, 9(5): 477-500

Salençon, J. (1983) *Calcul à la rupture et analyse limite*, Presses de l'Ecole Nationale des Ponts et Chaussées, Paris

Sànchez-Salinero, I. (1987) Analytical investigation of seismic methods used for engineering applications, PhD. Diss., University of Texas at Austin

Sanchez-Sesma F. J., Herrera I., and Aviles J. (1982) "A boundary method for elastic wave diffraction: application to scattering of SH waves by topographic irregularities", *Bulletin of the Seismological Society of America*, 72: 473-490

Saragoni, G. R. and Hart, G. C. (1974) "Simulation of artificial earthquakes", *Earthquake Engineering and Structural Dynamics*, 2:249-267

Sargent, R. C. (1990) "Validation of mathematical models". *Proc. of Geoval.-90 Conf.*, Stockholm

Sarma, S. H. and Chowdhury, R. (1996) "Simulation of pore pressure in earth structures during earthquakes", *Proc. of 11th World Conference on Earthquake Engineering,* Acapulco, Mexico

Sarma, S. K. and Iossifelis, I. S. (1990) "Seismic bearing capacity factors of shallow strip footings", *Geotechnique,* 40: 265-273

Sarma, S. H. (1975) "Seismic stability of earth dams and Embankments", *Geotechnique,* 25(4): 743-76

Scandone, P., Patacca, E., Meletti, C., Bellatalla, M., Perilli, N., and Santini, U. (1990) "Struttura geologica e schema sismotettonico della Penisola Italiana", *Atti del convegno Gruppo Nazionale per la Difesa dai Terremoti 1990,* 1: 119-135, (in Italian).

Schnabel, P. B., Lysmer, J. and Seed, H. B. (1972) Shake: A computer program for earthquake response analysis of horizontally layered sites, Report No. UCB/EERC 72-12, University of California, Berkeley

Schwartz, D. P. and Coppersmith, K. J. (1984) "Fault behavior and characteristic earthquakes: Examples from the Wasatch and San Andreas faults", *J. Geophysics. Research,* 89: 5681- 5698

SEAOC Blue Book (1998) PBSE Guidelines: Part 2, Appendix G, Guidelines for Performance Based Seismic Engineering, Structural Engineers Association of California, Sacramento, CA

Sêco e Pinto, P. S., Lopes, L., Agostinho, and Vieira, A. (1999) "Seismic behaviour of solid waste Grândola landfill", *Proc. of 2nd International Conference on Earthquake Geotechnical Engineering Lisbon,* Ed. P. S. Sêco e Pinto, Balkema, Rotterdam, 2: 661-666

Sêco e Pinto, P. S., Lopes, L., Agostinho, and Vieira, A. (1998) "Seismic analysis of solid waste landfills", *Proc. of 3rd. International Congress on Environmental Geotechnics,* Lisbon, Ed. P. S. Sêco e Pinto, Balkema, Rotterdam

Sêco e Pinto, P. S. (1997) "Analysis of the behaviour of solid waste landfills", *Seminar on Geoenvironmental Aspects of Industrial Solid Waste Landfills,* Lisbon (in Portuguese)

Sêco e Pinto, P. S., Dakoulas, P. C., Harder, L., Watanabe, H., and Chugh, A. (1995) "Stability of slopes and earth dams under earthquakes", General Report of Session VI, *Proc. of 3rd International Conference on Recent Advances in Geotechnical Earthquake Engineering and Soil Dynamics,* St. Louis, 3: 323-332

Sêco e Pinto, P.S (1993) "Dynamic analysis of embankment dams", *Proc. of Seminar on Soil Dynamics and Geotechnical Earthquake Engineering,* Ed. P. S. Sêco e Pinto, Balkema, Rotterdam, 159-269

Sêco e Pinto, P.S. (1990) "Dynamic characterization of soils", *Natural Hazards and Engineering Geology - Prevention and Control of Landslides and other Mass Movements,* European School of Climatology and Natural Hazard Course, 1971, 1249-1273

Seed, H. B., Wong, R. T., Idriss, I. M. and Tokimatsu, K. (1986) "Moduli and Damping Factors for Dynamic Analysis of Cohesionless Soils", *Journal Geotechnical Eng. Division,* ASCE, 112(11): 1016-1032

Seed, H. B., Tokimatsu, K., Harder, L. F., and Chung, R. M. (1985) "Influence of SPT Procedures in soil liquefaction resistance evaluations." Journal of Geotechnical Engineering, ASCE, 111(12), 1425-1445

Seed, H. B., Ugas, C., and Lysmer, J. (1976) "Site-dependent spectra for earthquake-resistant design", Bulletin of the Seismological Society of America, 66(1): 221-243

Sekisui House (1996) Rokko Island city, Strong motion record during the 1995 Hyogoken-Nambu earthquake and its analysis, (in Japanese)

Shamoto, Y. and Hotta, H. (1996) "Measurement of lateral ground displacement in Rokko Island", *Proc., 31st Japan National Conf. on Geotechnical Engineering,* 1: 1251-1252, (in Japanese)

Shamoto, Y., Sato, M., Futaki, M., and Shimazu, S. (1996) "A site investigation of post-liquefaction lateral displacement of pile foundation in reclaimed land", *Tsuchi-to-Kiso, JSG,* 44(3): 25-27 (in Japanese)

Shamoto, Y., Zhang, J.-M., and Tokimatsu, K. (1997) "Methods for evaluating residual post-liquefaction ground settlement and horizontal displacement", *Soils and Foundations,* Special Issue, 69-83

Sharma, H. D. and Goyal, H. K. (1991) "Performance of a hazardous waste and sanitary landfill subjected to Loma Prieta earthquake", *Proc. of 2nd International Conference on Recent Advances in Geotechnical Earthquake Engineering and Soil Dynamics*, St. Louis, 2: 1717- 1725

Shewbridge, S. (1996) "Yield acceleration of lined landfills", *J. of Geotechnical Eng.*, 122(2): 156-158

Shibuya, S. and Mitachi, T. (1997) "Development of a fully digitized triaxial apparatus for testing soils and soft rocks", *Geotechnical Engineering*, 28(2): 183-207

Silva W., Darragh, R., Stark, C., Wong, I. Stepp, J. C., Schneider, J. F., and Chiou, S-J. (1990) "A Methodology to Estimate Design Response Spectra in the Near-source Region of Large Earthquakes Using the Band-limited-white-noise Ground Motion Model", *4th US Nat. Conf. Earth. Eng.*, 1: 487-494

Singh, S. and Sun, J. I. (1995) "Seismic evaluation of municipal solid waste landfills", *Proc. of Geoenvironment 2000*, ASCE Specialty Conference, New Orleans, Louisiana, 22-24 February

Singh, S. and Murphy, B. (1990) "Evaluation of the stability of sanitary landfils", *ASTM STP 1070 Geotechnics of Waste Landfils- Theory and Practice*, Eds. A. Landva, and G. D. Knowles, ASTM, Philadelfia, Pa., 240-258

Siro, L. (1982) "Emergency microzonations by Italian Geodynamics Project after November 23, 1980 earthquake: a short technical report", *Proceedings of the Third International Earthquake Microzonation Conference*, Univ. of Washington, Seattle, III: 1417-1427

Sipahioğlu, S. (1984) "Investigation of the earthquake activity of North Anatolian Fault and its surroundings", *Bulletin of Earthquake Research Institute*, 11(45) (in Turkish)

Somerville, P.G. (2000) "New Developments in Seismic Hazard Assessment", *Proceedings of the World Conference on Earthquake Engineering*, Auckland, New Zealand

Somerville, P.G., Smith, N. F., Graves, R. W. and Abrahamson, N. A. (1997) "Modification of Empirical Strong Motion Attenuation Relations to Include the Amplitude and Duration effects of Rupture Directivity", *Seismological Research Letters*, 68: 199-222

Somerville, P. G., Sen, M. K., and Cohee, B. (1991) "Simulation of Strong Ground Motions Recorded During the 1985 Michoacan, Mexico and Valparasio, Chile Earthquakes", *Bulletin of the Seismological Society of America*, 81: 1-27

Sousa, M. L., Campos-Costa, A., and Oliveira, C. S. (1997) "Modelos probabilísticos para avaliação de perdas causadas por sismos: Aplicação à cidade de Lisboa", *Proceedings, 3rd Portuguese Meeting on Seismology and Earthquake Engineering*, IST, Lisbon, 109-118

Sousa, M. L. and Oliveira, C. S. (1997) "Hazard mapping based on macroseismic data considering the influence of geological conditions", *Natural Hazards*, 14: 207-225

Sousa, L., Martins, A., and Oliveira, C. S. (1992) Compilação de catálogos sísmicos da região Ibérica, Relatório técnico 36/92 – NDA, LNEC, Lisbon, (in Portuguese)

Spudich, P., Fletcher, J. B., Hellweg, M., Boatwright, J., Sullivan, C., Joyner, W. B., Hanks, T. C., Boore, D. M., McGarr, A., Baker, L. M., and Lindh, A. G. (1997) "SEA96 - A New Predictive Relation for Earthquake Ground Motions in Extensional Tectonic Regimes", *Seis. Research Letters*, 68(1): 190-198

Seismological Society of America, SSA (1997) Special Issue, *Seismological Research Letters*, 68(1)

SSHAC (1997) *Recommendations for Probabilistic Seismic Hazard Analysis-Guidance on Uncertainty and Use of Experts*, Senior Seismic Hazard Analysis Committee, U.S. Nuclear Regulatory Commission NUREG/CR-6732

Standards New Zealand, NZS 4203 (1992) Code of practice for general design and design loadings for buildings - Loadings Standard, 2 Vol., Standards New Zealand, Wellington

Stepp, J.Carl, Wong, I. Whitney, J., Quittmeyer, R., Abrahamson, N., Toro, G., Youngs, R., Coppersmith, K., Savy, J., Sullivan, T. and Yucca Mountain PSHA Project Members, (2001) "Probabilistic seismic hazard analyses for ground motions and fault displacement at Yucca Mountain, Nevada", *Earthquake Spectra*, 17(1): 113-151

Stewart, J. P., Bray, J. D., Seed, R. B., and Sitar, N. (1994) Preliminary report on the principal geotechnical aspects of the January 17, 1994 Northridge Earthquake, Report No. UCB/EERC- 94/08, College of Engineering, University of California at Berkeley, Berkeley, California, 238 p.

Stiedl, J. H., Tumarkin, A. G., and Archuleta, R. J. (1996) "What is a reference site?", *Bulletin of the Seismological Society of America*, 86: 1733-1748

Stiedl, J. H. (1993) "Variation of site response estimates at the UCSB dense array of portable accelerometers", *Earthquake Spectra*, 9(2): 289-302

Stokoe, K. H. II, Darendeli, M. B., Andrus, R.D., and Brown, L.T. (1999) "Dynamic soil properties: Laboratory, field and correlation studies", Theme Lecture, *Proc. of 2nd International Conference on Earthquake Geotechnical Engineering*, Lisbon, Ed. P. Sêco e Pinto, Balkema, Rotterdam, 3: 811-845

Stokoe, K. H. II, Hwang, S. K., Lee, J. N. K., and Andrus, R. D. (1995) "Effects of Various Parameters on the Stiffness and Damping of Soils at Small to Medium Strains", Keynote Lecture 2, *IS Hokkaido*, Balkema, Rotterdam, 2: 785-816

Stokoe, K. H. II, Lee, J. N. K., and Lee, S. H. H. (1991) "Characterization of Soil in Calibration Chambers with Seismic Waves", *Proc. ISOCCT1*, Postdam, NY

Stokoe, K. H. II, Mok, Y. S., Lee, N., and Lopez, R. (1989) "In Situ Seismic Methods; Recent Advances in Testing, Understanding and Applications", *Proc. XIV CGT*, Politecnico di Torino, Department of Structural Engineering

Stokoe, K. H. II and Hoar, R. J. (1978) "Variable Affecting In Situ Seismic Measurements", *Proc. of Conference on Earthquake Engineering and Soil Dynamics*, ASCE, Pasadena, CA, 2: 919-939

Sykora, D. W. and Stokoe, K. H. II (1983) Correlations of in Situ Measurements in Sands with Shear Wave Velocity, Geotechnical Engineering Report GR83-33, University of Texas at Austin

Takahashi, T., Mimoto, K., and Hayakawa, T. (1997) "Present state of applications of geophysical methods to characterization of active faults", *Journal of Japan Society of Engineering Geology*, 38: 118-129

Tamura, C., Kanyo, S., Uesaka, T., Nagayama, I., and Wakizaka, Y. (2000) "Survey and evaluation of active faults on dam construction in Japan", *Proc. of 12th World Conference on Earthquake Engineering*, Auckland, New Zealand, Paper No. 2493

Tatsuoka, F., Jardine, R. J., Lo Presti, D. C. F., Di Benedetto, H., and Kodaka, T. (1997) "Characterising the Pre-Failure Deformation Properties of Geomaterials", Theme Lecture, Plenary Session 1, *XIV ICSMFE*, Hamburg, Balkema, Rotterdam

Tatsuoka, F. and Kohata, Y. (1995) "Stiffness of Hard Soils and Soft Rocks in Engineering Applications", Keynote Lecture 8, *IS Hokkaido 1994*, Balkema, Rotterdam, 2: 947-1066

Tatsuoka, F. and Kohata Y., Ochi, K. and Tsubouchi, T. (1995a) "Stiffness of Soft Rocks in Tokyo Metropolitan Area from Laboratory Tests to Full-Scale Behaviour", *International Workshop on Rock Foundation of Large Scaled Structures*, Tokyo

Tatsuoka, F., Lo Presti, D. C. F., and Kohata, Y. (1995b) "Deformation Characteristics of Soils and Soft Rocks Under Monotonic and Cyclic Loads and Their Relations", *3rd Int. Conference on Recent Advances in Geotechnical Earthquake Engineering and Soil Dynamics*, State of the Art, 2: 851-879

Tatsuoka, F., Sato, T., Park, C-S, Kim, Y-S, Mukabi, J. N., and Kohata, Y. (1994) "Measurements of Elastic Properties of Geomaterials in Laboratory Compresion Tests", *Geotechnical Testing J.*, 17(1): 80-94

Tatsuoka, F., Siddique, M. S. A., Park, C. S., Sakamoto, M., and Abe, F. (1993) "Modelling stress-strain relations of sand" *Soil and Foundations*, 33(2): 60-81

Tatsuoka, F. and Shibuya, S. (1992) "Deformation Characteristics of Soil and Rocks from Field and Laboratory Tests", Keynote Lecture, *IX Asian Conference on SMFE*, Bangkok, 2: 101-190

Tatsuoka, F. (1988) "Some recent developments in triaxial testing systems for cohesionless soils", ASTM STP 977, *Advanced Triaxial Testing of Soil and Rock*, 7-67

TC4-ISSMGE (1999) *Manual For Zonation on Seismic Geotechnical Hazard*, Revised edition, Technical Committee for Earthquake Geotechnical Engineering (TC4) of the International Society of Soil Mechanics and Geotechnical Engineering (ISSMGE), 209p

Teachavorasinskun, S. (1989) Deformation Characteristics of Sands at Small Strains, M.Sc. Thesis, University of Tokyo, Japan

Tento, A., de Franco, R., Franceschina, G., and Pagani, M. (2001) "Site effect zonation of the Fabriano municipality", *Italian Geotechnical Journal*, XXXV(2): 131-145

Tento, A. (1999) Computer codes for seismic input evaluation, Report CNR-IRRS, Milano, Italy

Teves-Costa, P., Costa Nunes, J. A., Senos, M. L., Oliveira, C. S., and Ramalhete, D. (1995). "Predominant frequencies of soil formations in the town of Lisbon using microtremor measurements", *Proceedings 5th International Conference on Seismic Zonation*, Nice, 1683-1690

Thomson, W.T. (1950) "Transmission of elastic waves through a stratified solid medium", *J. Applied Physics*, 21(1): 89-93

Tiedemann, H. (1992) *Earthquakes and Volcanic Eruptions: A Handbook on Risk Assessment,* Zurich, Swiss Re

Todorovska, M. I. and Trifunac, M. D. (1999) "Liquefaction opportunity mapping via seismic wave energy", *Journal of Geotechnical and Geoenvironmental Engineering*, 125(12): 1032-1042

Toki, S., Shibuya, S., and Yamashita, S. (1995) "Standardization of laboratory test methods to determine the cyclic deformation properties of geomaterials in Japan", Keynote Lecture, *IS Hokkaido*, Balkema, Rotterdam, 2: 741-784

Tokimatsu, K. (1999), "Performance of pile foundations in laterally spreading soils", *Earthquake Geotechnical Engineering*, 3: 957-964

Tokimatsu, K. and Asaka, Y. (1998), "Effects of liquefaction-induced ground displacements on pile performance in the 1995 Hyogoken-Nambu earthquake", *Soils and Foundations*, Special Issue, 163-177

Tokimatsu, K., Oh-oka, H., Shamoto, Y., Nakazawa, A., and Asaka, Y. (1997) "Failure and deformation modes of piles caused by liquefaction-induced lateral spreading in the 1995 Hyogoken-Nambu earthquake", *Proc., 3rd Kansai Int. Geotechnical Forum on Comparative Geotechnical Eng.*, 239-248

Tokimatsu, K., Mizuno, H., and Kakurai, M. (1996), "Building damage associated with geotechnical problems", *Soils and Foundations*, Special Issue, 219-234

Tokimatsu, K. (1995) "Geotechnical Site Characterisation using Surface Waves", *Proc. IS Tokyo 1995*, Balkema, Rotterdam, 1333-1368

Tokimatsu, K., Kuwayama, S., and Tamura, S. (1991) "Liquefaction potential evaluation based on Rayleigh wave investigation and its comparison with field behaviour", *Proc. of 2nd International Conference on Recent Advances in Geotechnical Earthquake Engineering and Soil Dynamics*, St. Louis, 1: 357-364

Tokimatsu, K. and Yoshimi, Y. (1983), "Empirical correlation of soil liquefaction based on N-value and fines content", *Soils and Foundations*, 23(4): 56-74

Trifunac, M .D. and Lee, V. W. (1989) "Empirical Models for Scaling Fourier Amplitude Spectra of Strong Ground Acceleration in Terms of Earthquake Magnitude, Source to Station Distance, Site Intensity and

Recording Site Conditions", *Journal of Soil Dynamics and Earthquake Eng.*, 8(3): 110-125

Trifunac, M. D. (1976) "Preliminary Empirical Model for Scaling Fourier Amplitude Spectra of Strong Ground Acceleration in Terms of Earthquake Magnitude, Source to Station Distance and Recording Site Conditions", *Bulletin of the Seismological Society of America*, 66: 1342

Trifunac M. D. and Brady, A. G. (1975) "A study on the Duration of Strong Earthquake Ground Motion", *Bulletin of the Seismological Society of America*, 65: 581-626

Trifunac, M. D. (1973) "Scattering of plane SH waves by a semi-cylindrical canyon", *International Journal of Earthquake Engineering Structural Dynamics*, 1: 267-281

Tselentis, G. A. and Delis, G. (1998) "Rapid assessment of S-wave profiles from the inversion of multichannel surface wave dispersion data", *Annali di Geofisica*, 41: 1-15

Tsukamoto, Y., Ishihara, K., Nakazawa, H., Yasuda, S. and Horie, Y. (2001) "Soil properties of the deposits in Adapazarı from laboratory tests", *Proc. of the Satellite Conference on Lessons Learned from Recent Strong Earthquakes, 15th ICSMGE*, 2001, 275-280

Turkish Earthquake Code (1997) *Specification for Structures to be Built in Disaster Areas*, Ministry of Public Works and Settlement, Ankara, Turkey, English Translation: Kandilli Observatory and Earthquake Research Institute http://www.koeri.boun.edu.tr/depremmuh/

Ukritchon, B., Whittle, A. J. and Sloan, S.W. (1998) "Undrained limit analysis for combined loading of strip footings on clay", *Journal of Geotechnical and Geoenvironmental Engineering*, 265-275

USEPA (1994) RCRA Subtitle D(258) Seismic design guidance for Municipal Solid Waste Landfill Facilities, Final Draft, United States Environmental Protection Agency

Vaid, Y. P., Sayao, A., Hou, E., and Negussey, D. (1990) "Generalized stress-path-dependent soil behaviour with a new hollow cylinder torsional apparatus", *Canadian Geotechnical Journal*, 27: 601-616

Vucetic, M. (1994) "Cyclic Threshold Shear Strains in Soils", *Journal of GED*, ASCE, 120(12): 2208-2228

Watanabe, K. (1981) Development of Kobe Port Island and foundation of structures in the island, *The Foundation Engineering and Equipment*, 9(1): 83-91 (in Japanese)

Wills, C. J., Peterson, M., Bryant, W. A., Reichle, M., Saucedo, G. J., Tan, S., Taylor, G. and Treiman, J. (2000) "A Site Conditions Map for California based on Geology and Shear Wave Velocity", *BSSA*, (90)6B: 187-S208

Wills, C. J. and Silva, W. (1998) "Shear Wave Velocity Characteristics of Geologic Units in California", *Earthquake Spectra*, (14)3:533-566

Wolf, J.P. (1994) *Foundation vibration analysis using simple physical models*, Prentice Hall Inc

Wong, H. L. and Trifunac, M. D. (1974) "Scattering of plane SH waves by a semi-elliptical canyon", *International Journal of Earthquake Engineering Structural Dynamics*, 3: 157-169

Woods, R.D. (1994) "Borehole Methods in Shallow Seismic Exploration". *Proc. of XIII ICSMFE* New Delhi, India, TC # 10, 91-100

Woods, R. D. (1991) "Field and Laboratory Determination of Soil Properties at Low and High Strains", *Proc. of 2nd International Conference on Recent Advances in Geotechnical Earthquake Engineering and Soil Dynamics*, St. Louis, Missouri, SOA1

Woods, R. D. and Stokoe, K. H. II (1985) "Shallow Seismic Exploration in Soil Dynamics", *Richart Commemorative Lectures*, ASCE, 120-156

Woods, R. D. (1978) "Measurement of dynamic soil properties", *Earthquake Engineering and Soil Dynamics*, Pasadena, CA, 1: 91-179

Wroth, C. P. and Hughes, J. H. O. (1973) "An Instrument for the In-Situ Measurement of the Properties of Soft Clays", *Proc. VIII ICSMFE*, 1.2: 487-494.

Wu, W. H. and Lee, W. H. (2002) "Systematic lumped-parameter models for foundations based on polynomial-fraction approximation", *Earthquake Engineering and Structural Dynamics*, 31: 1383-1412

Yagi, Y. and Kikuchi, M. (2000) "Source rupture process of the Kocaeli, Turkey earthquake of August 17, 1999, obtained from the joint inversion of near-field data and teleseimic data", online: http://wwweic.eri.u-tokyo.ac.jp/yuji/PDF/turkey.pdf

Yamashita, S. and Suzuki, T. (1999) "Young's and shear moduli under different principal stress directions of sand", *Proc. of IS Torino 99*, Balkema, Rotterdam, 1: 149-158

Yasuda, S., Irisawa, T. and Kazami, K. (2001a) "Liquefaction-induced settlements of buildings and damages in coastal areas during the Kocaeli and other earthquakes", *Proc. of the Satellite Conference on Lessons Learned from Recent Strong Earthquakes, 15th ICSMGE*, 33-42

Yasuda, S., Yoshida, N. and Irisawa, T. (2001b) Settlement of buildings due to liquefaction during the 1999 Kocaeli earthquake, *Proc. of the Satellite Conference on Lessons Learned from Recent Strong Earthquakes, 15th ICSMGE*, 77-82

Yasuda, S., Abo, H., Yoshida, N., Kiku, H. and Uda, M. (2001c): Analyses of liquefaction-induced deformation of grounds and structures by a simple method, *4th International Conference on Recent Advances in Geotechnical Earthquake Engineering and Soil Dynamics,* Paper No.4.34

Yasuda, S. and Berrill, J. (2000) Observation of the earthquake response of foundations in soil profiles containing saturated sands, *1st International Conference on Geotechnical and Geological Engineering –GeoEng2000*, Issue Lecture, pp.1441-1470.

Yasuda, S. (1999) "Seismic Design Codes for Liquefaction", *Proc. of the 2nd International Conference on Earthquake Geotechnical Engineering*, 1117-1122

Yasuda, S., Yoshida, N., Adachi, K., Kiku, H. and Gose, S. (1999a) "A simplified analysis of liquefaction-induced residual deformation", *2nd Int. Conf. on Earthquake Geotechnical Engineering*, 555-560.

Yasuda, S., Yoshida, N., Adachi, K., Kiku, H., and Gose, S. (1999b) "A simplified practical method for evaluating liqueaction-induced flow", *Journal of JSCE*, 638/3-49: 71-89, (in Japanese)

Yasuda, S., T. Terauchi, H. Morimoto, A. Erken and N. Yoshida (1998) "Post liquefaction behaviour of several sands", *Proc. of the 11th European Conference on Earthquake Engineering,* Balkema

Yasuda, S., Ishihara, K., Harada, K., and Nomura, H. (1997) "Area of ground flow occurred behind quay walls due to liquefaction", *Proc., 3rd Kansai Int. Geotechnical Forum on Comparative Geotechnical Engineering*, 85-93

Yasuda, S., Ishihara, K., Harada, K. and Shinkawa, N. (1996) "Effects of soil improvement on ground subsidence due to liquefaction", *Soils and Foundations,* Special Issue on Geotechnical Aspects of the January 17, 1995 Hyogoken-Nambu Earthquake, JGS, 99-107

Yasuda, N. and Matsumoto, N. (1993) "Dynamic deformation characteristics of sands and rockfill materials", *Canadian Geotechnical Journal*, 30: 747-757

Yegian, M. K. and Kadakai, H. V. (1998) "Seismic response of landfills with geosynthetic liners", *Proc. of the 6th International Conference on Geosynthetics*, 1: 227-230

Yegian, M. K., Yee, Z. Y. and Harb, J. N. (1995) "Response of geosynthetics under eartquake excitations", *Proceedings of Geosynthetics'95*, 2: 677-689

Yokota, K., Imai, T. and Konno, M. 1981. Dynamic Deformation Characteristics of Soils Determined by Laboratory Tests, OYO Tec.

Yokoyama, Tamura, and Matsuo (1997) "Design methods of bridge foundations against soil liquefaction and liquefaction-induced ground flow", *Second Italy-Japan Workshop on Seismic Design and Retrofit of Bridge*, Rome, Italy, 1-23

Yong, C., Qi-Fu, C., and Ling, C. (1997) "Worldwide Seismic Risk Analysis Based on Limited Data", Proceedings 1st International Earthquake and Megacities Workshop, Seeheim, Germany

Yoshimi, Y. and Tokimatsu, K. (1977) "Settlement of buildings on saturated sand during earthquakes", *Soils and Foundations,* 17(1): 23-38

Youd, T. L. and Gilstrap, S. D. (1999) "Liquefaction and deformation of silty and fine-grained soils", General Report, *Proc. of 2nd International Conference on Earthquake Geotechnical Engineering*, Lisbon, Ed. P. Sêco e Pinto, Balkema, Rotterdam, 3: 1013-1020

Youngs, R. R., Chiou, S.-J., Silva, W. J. and Humphrey, J. R. (1997) "Strong Ground Motion Attenuation Relationships for Subduction Zone Earthquakes", *Seismological Research Letters*, 68(1): 58-74

Youngs, R. R. and Coppersmith, K. (1985) "Implications of fault slip rates and earthquake recurrence models to probabilistic seismic hazard estimates", *Bulletin of the Seismological Society of America*, 75: 939-964

Zeng, X. and Steedman, R. S. (1998) "Bearing capcity failures of shallow foundations in earthquakes", *Geotechnique*, 48(2) 235-256

Zywicki, D. and Rix, G.J. (1999) "Frequency-wavenumber analysis of passive su\rface waves", *Proc. Symp. on the Appl. of Geophysics to Environm. and Eng. Problems*, Oakland, 75-84

INDEX